Physiology

PHYSIOLOGY

A Review with Questions and Explanations

Benjamin Hsu, M.D.

Resident, Department of Internal Medicine,
Stanford University Medical Center,
Stanford, California

Charles H. Tadlock, M.D.

Resident, Department of Anesthesia,
Stanford University Medical Center, Stanford California

Saied Assef, M.D.

Resident, Department of Orthopedic Surgery,
Mayo Graduate School of Medicine, Rochester, Minnesota

Jack M. Percelay, M.P.H.

Medical Student, University of California,
San Francisco, School of Medicine, San Francisco

Foreword by Roy H. Maffly, M.D.
Professor of Medicine and Chairman, Department of Physiology,
Stanford University School of Medicine, Stanford, California

Little, Brown and Company   Boston/Toronto

First Edition

Fourth Printing

Library of Congress Catalog Card No. 87-45378

ISBN 0-316-37835-6

Printed in the United States of America

SEM

Contents

Foreword

What can be more fulfilling for a teacher than working with students who are turned on by his subject? And how better for students to show their enthusiasm for the subject than to write about it? The authors have done just this, and it is with pleasure and pride that I congratulate them. As an internist and nephrologist, I have an abiding love for the discipline of physiology; its principles underlie what I teach my students and how I treat my patients. Ben, Charles, Saied, and Jack share my appreciation for physiology, and *Physiology: A Review with Questions and Explanations* shows it.

Their intention has been to provide a book that will be helpful to medical students, and they have written it from their first-hand perspective. They have tried to stress the important principles and relationships that are essential for mastery of the subject. How well they have succeeded can best be judged by other medical students, but from my special vantage point, they have succeeded admirably. They know what they are writing about and why they are writing. They are sensitive to the audience at which they are aiming. The result is an excellent blend of quality of content and practicality in utilization. I commend Ben, Charles, Saied, and Jack for a superb volume.

Roy H. Maffly, M.D.

Preface

This book was written by medical students from the Stanford and University of California, San Francisco medical schools. It reviews the essential points covered in a standard medical physiology course and in the physiology section of the National Medical Board Examination (Part I). The text is intended for medical students and other health professional students. When we were studying for the National Boards, we were unable to find any book suitable for reviewing physiology (i.e., one which weighs less than 20 lbs). We did find, however, Little, Brown's *Biochemistry: A Review with Questions and Explanations,* by Paul Friedman, to be an amazingly well-written book which could be read easily in a couple of days. It was an indispensable book for "cramming" in biochemistry during that last panic-stricken week before the Boards. Because Little, Brown was interested in producing a similar text for physiology, and because we were given excellent physiology courses in our medical schools, we got together and wrote *Physiology: A Review with Questions and Explanations.*

In writing this book, we used lecture notes as well as standard textbooks (the 20-lb ones) as sources. It is intended for students who have had some formal course work in medical physiology. Lengthy explanations are thus minimized. Experimental and trivial data are judiciously left out. Discussions that overlap with other National Board sections (such as anatomy and pathology) are included only when they elucidate the physiology. Each chapter contains a set of problems along with their answers and explanations. We have made much effort in making the book concise, comprehensive, and most importantly, "reader-friendly."

We are grateful to the following faculty members who have reviewed our chapters and provided invaluable suggestions: Roy H. Maffly, M.D., Mark G. Perlroth, M.D., E. William Hancock, M.D., Eugene D. Robin, M.D., Andrew R. Hoffman, M.D., Timothy Meyers, M.D., H. Barrie Fairley, M.B., B.S., and John Kerner, M.D. We are also indebted to students at Stanford and University of California, San Francisco for their constructive criticism of each chapter. Finally, we are especially thankful to Jim Krosschell of Little, Brown for his support and his patience during our long periods of procrastination.

We hope you enjoy reading Physiology: A Review with Questions and Explanations, and we welcome any comments or criticisms.

B. H.
C. H. T.
S. A.
J. M. P.

Physiology

NOTICE

The indications and dosages of all drugs in this book have been recommended in the medical literature and conform to the practices of the general medical community. The medications described do not necessarily have specific approval by the Food and Drug Administration for use in the diseases and dosages for which they are recommended. The package insert for each drug should be consulted for use and dosage as approved by the FDA. Because standards for usage change, it is advisable to keep abreast of revised recommendations, particularly those concerning new drugs.

1 General Physiology

Jack M. Percelay and Saied Assef

TRANSPORT

The primary forces producing movement of water and molecules between body compartments are diffusion, osmosis, and active transport. Diffusion, sometimes termed *passive transport,* is the net movement of particles from an area of high concentration to an area of low concentration due to the random molecular movement of the particles. Diffusion of charged particles is affected by the electrical gradient in addition to the concentration gradient.

Simple diffusion of a neutral particle through a boundary such as the cell membrane is described by Fick's law.

$$\text{flux} = \frac{pA\,(C_1 - C_2)}{d} \tag{1-1}$$

Flux is the amount of material moved, p is the permeability of the material across the boundary, A is the cross sectional area of the boundary, C_1 and C_2 are the concentrations of the material in the respective two regions, and d is the thickness of the boundary. The high lipid solubility of such small neutral molecules as O_2, N_2, and the alcohols enables them to pass directly through the interstices of the lipid bilayer (Fig. 1-1A). Ions, because of their charge, are much

Fig. 1-1. *A. Simple diffusion directly through the bilayer and via protein channels. B. Facilitated diffusion. C. Active transport — the sodium-potassium pump, the Na^+-K^+ ATPase D. Cotransport via a symport protein. (From A. C. Guyton.* Textbook of Medical Physiology *[7th ed.]. Philadelphia: Saunders, 1986. Pp. 89, 97, 99.)*

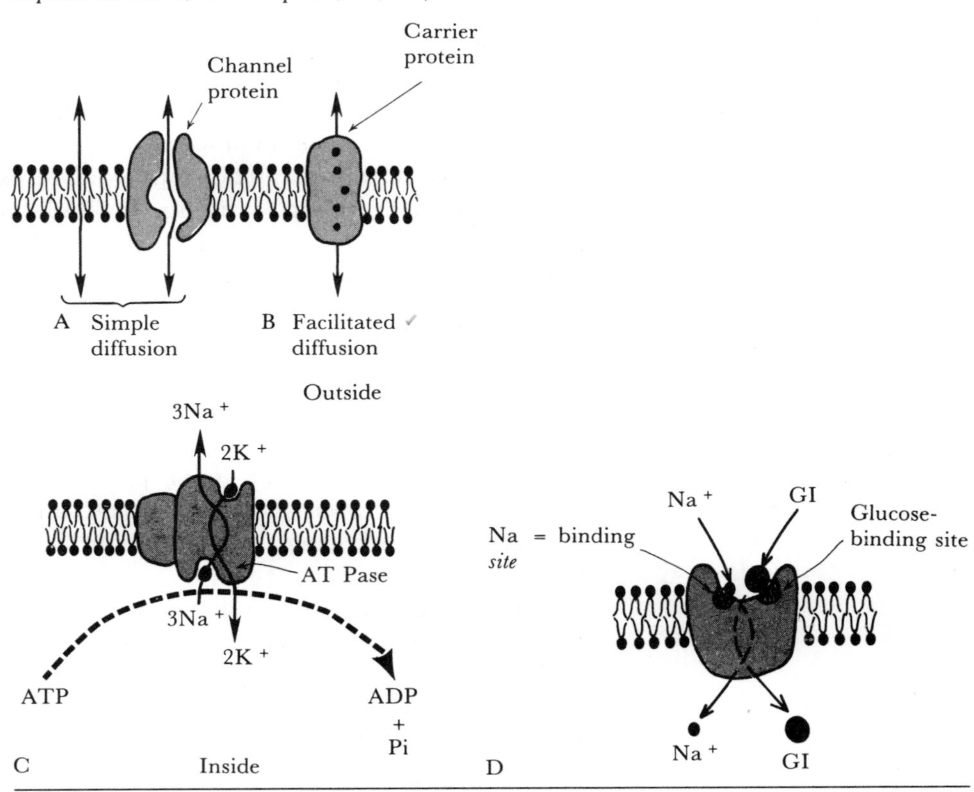

less able to pass through the small hydrophobic pores of the lipid bilayer but can pass through the membrane via channels created by protein molecules (Fig. 1-1B). These protein channels are selectively permeable to specific compounds based on channel characteristics such as shape, diameter, and charge. In addition, many of these channels can be opened or closed by "gates"—conformational changes in the protein molecule which either allow or prevent passage through the channel. Gates may be controlled by voltage or by binding of a specific ligand. A given channel may have more than one gate, an activation gate which opens the channel and an inactivation gate which closes the channel.

Facilitated or carrier-mediated diffusion differs from simple diffusion in that the transported molecule actually binds to the carrier. The carrier then undergoes a conformational change to allow the molecule to pass through to the other side of the cell membrane. This intermediate complex is analogous to the enzyme-substrate complex of Michaelis-Menten kinetics. Similarly, carrier-mediated diffusion can be saturated and reaches a V_{max} with increasing substrate concentration. Simple diffusion does not saturate and continues to increase in rate with increasing substrate concentration.

Osmosis can be viewed as the tendency of water, or more generally any solvent, to diffuse down its concentration gradient across a selectively permeable membrane. It is a passive process requiring no energy expenditure by the organism. The concentration of water is greater in a more dilute solution because addition of solute to solvent decreases solvent concentration. Thus, one can regard osmosis as the tendency of water to move from a dilute solution to a more concentrated solution. For example, a 1 mM salt solution is more pure water than a 10 mM salt solution, and water will flow from the 1 mM solution into the 10 mM solution when the two solutions are separated by a selectively permeable membrane. Osmotic pressure is determined by the number of osmotically active particles in solution and is effectively independent of particle type. A solution of 1 M glucose, therefore, has an osmolarity of 1, but a solution of 1 M NaCl has an osmolarity of approximately 2 because in solution NaCl exists as separate Na^+ and Cl^- ions.

In active transport, energy is used to move a substance against its electrochemical gradient from an area of low concentration to an area of high concentration. Such transport depends upon protein pumps which couple ATP hydrolysis to transport across the cell membrane. As an enzymatic process, active transport, like facilitated diffusion, can be saturated. One example of active transport is the sodium-potassium pump—the Na^+-K^+ ATPase. It pumps three Na^+ out of the cell and two K^+ into the cell for each ATP consumed (Fig. 1-1C). Note that both ions are pumped against their concentration gradients.

Cotransport, sometimes called secondary active transport, uses the energy of the Na^+ concentration gradient created by the Na^+-K^+ ATPase to drive the transport of other substances against their concentration gradients. The sodium moves into the cell and provides the energy to move another substance against its concentration gradient, either into the cell via a symport protein or out of the cell via an antiport protein. Symport proteins are used in the mucosa of the small intestine to couple Na^+ influx to glucose and amino acid uptake (Fig. 1-1D). The Na^+-Ca^{2+} exchange protein of cardiac muscle is an important antiport protein.

Transepithelial transport is the movement of substances across a sheet of cells, such as the renal or intestinal epithelium. The basic mechanism involves active transport through the cell membrane on one side of the sheet, and then diffusion (facilitated or simple) through the membrane on the opposite side as shown in Figure 1-2. For example, in the renal tubule, Na^+ is actively absorbed by ion pumps at the lumenal end, increasing the intracellular Na^+ concentration.

Fig. 1-2. *Transepithelial transport. Passive transport at the lumen combines with active transport at the base of the cell and along the cell borders to move substances through a sheet of cells. (From A. C. Guyton.* Textbook of Medial Physiology *[7th ed.]. Philadelphia: Saunders, 1986. P. 99.)*

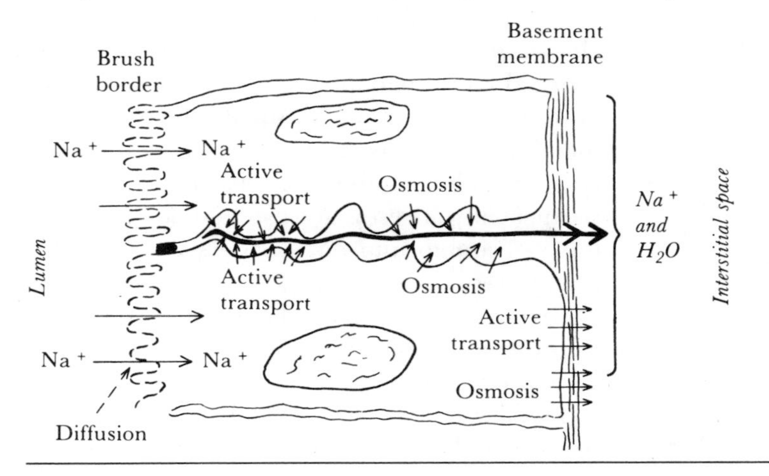

The Na^+ then diffuses down its concentration gradient into the interstitial space via passive diffusion along the base and sides of the cell. This movement of cation results in a corresponding movement of anions to balance charge and of water to maintain osmolarity. The sodium has thus been reabsorbed from the glomerular filtrate into the extracellular space, but it has passed through a layer of cells to do so. The role of transport between cells is as yet undetermined.

QUESTIONS

1. Name three factors which influence membrane permeability.
2. If at baseline $C_1 = 10$ and $C_2 = 5$, what effect will doubling C_1 have on flux?
3. Normal cell osmolarity is approximately 280 mosm. What will happen to a cell placed in 280 mM glucose? in 280 mM NaC1?
4. What transport process(es) would be disrupted with a disturbance of ATP synthesis?
5. Name three advantages of protein-mediated transport.

THE RESTING POTENTIAL

The selective permeability of cell membranes allows for the differences in the composition of extracellular and intracellular fluid which, in turn, are responsible for the generation of a membrane potential. The equilibrium potential is the point at which the forces of the concentration gradient and electrical gradient balance. This potential can be calculated by the Nernst equation,

$$E_X = -\frac{RT}{nF} \ln \frac{[X]_i}{[X]_o} \qquad (1\text{-}2)$$

in which E_x = equilibrium potential (in mv) of X, R = gas constant, T = absolute temperature, n = valence of X (including sign of charge), F = Faraday's constant, $[X]_i$ = concentration of X inside the cell, and $[X]_o$ = concentration of X outside the cell. For a T of 37°C and an n of +1,

$$E_X = -61.5 \log \frac{[X]_i}{[X]_o} \qquad (1\text{-}3)$$

Consider a membrane selectively permeable to potassium. Because the intracellular K^+ concentration is greater than the extracellular K^+ concentration, K^+

diffuses out of the cell down its concentration gradient. Continued diffusion results in a separation of charge. There is a net excess of positive charge just outside the membrane (from the additional K^+ ions) and a net excess of negative charge just inside the membrane (from the loss of K^+ ions). This separation of charge produces an electrical potential opposing further efflux of positively charged K^+. When the force of the electrical potential exactly balances the force of the concentration gradient, there is no net movement of K^+ and the membrane remains at the equilibrium potential for K^+. In reality, the membrane is permeable to other ions (such as Na^+) in addition to K^+. Because K^+ permeability at rest is at least 35 times greater than Na^+ permeability, however, the K^+ equilibrium potential (E_K) is a good approximation of the cell membrane's resting potential (E_m).

Membrane potential and ion concentrations vary among species and cell types. Representative ion concentrations for human neurons are shown in Table 1-1. Here the K^+ equilibrium potential is -90 mv (inside negative with respect to the outside). A potential of $+60$ mv (inside positive with respect to the outside) is required to balance the tendency of Na^+ to flow down its concentration gradient and enter the cell. The normal membrane resting potential is approximately -70 mv. This reflects the much greater contribution of K^+ because of the greater permeability of K^+. Because neither E_K nor E_{Na} are at the membrane potential, one would expect the cell to gradually gain Na^+ and to lose K^+ (see question 8.) In both cases, the electrical potential is insufficient to prevent the tendency of the ions to move down their respective concentration gradients; however, active transport by the Na^+-K^+ pump maintains the concentration gradients. The Na^+-K^+ ATPase also contributes to the separation of charge because of its electrogenic properties. Each cycle results in a net movement of one positive charge out of the cell (three Na^+ out, two K^+ in). There are no active chloride pumps; thus, chloride is subject only to the passive forces of electrical and chemical gradients, and the chloride equilibrium potential equals the membrane potential.

QUESTIONS

6. What will happen to E_K if the K^+ concentration gradient is
   A. Increased 10 times
   B. Increased 100 times
   C. Decreased 10 times
   D. Reversed
7. What would be E_{Ca} if the Ca^{2+} concentration gradient were equal to the K^+ concentration gradient?
8. Voltage clamping is a technique that allows membrane potential to be set at a given voltage by constantly infusing current into the axoplasm. Predict the direction of passive Na^+ and K^+ flow at the following membrane potentials:
   a. 100 mv
   b. -90 mv
   c. -70 mv

Table 1-1. *Ion Concentrations in Representative Mammalian Neurons*

Ion	Concentration (mM)			
	Inside	Outside	Relative Permeability	Equilibrium Potential (mv)
Na^+	15.0	150.0	1	+60
K^+	150.0	5.5	50–100	−90
Cl^-	9.0	125.0	25–50	−70
Ca^{2+}	0.0001	1.0	—	—

d. 0 mv
e. +60 mv
f. +70 mv

9. Provide two reasons why computing E_K from the Nernst equation does not yield the actual cell membrane potential, Em.
10. What effect would increasing Na^+ permeability have on the membrane potential?

ACTION POTENTIALS

If the resting membrane potential is altered by passing a current through the membrane, the electrical gradients affecting ions are altered, and there is a change in ion flow to restore the membrane to its resting potential. For example, if the potential is decreased (depolarized, made less negative), the electrical gradient that keeps K^+ inside the cell is decreased. The force of the concentration gradient now exceeds the electrical force, and there is a corresponding increase in K^+ diffusion out of the cell until the forces balance and the membrane is restored to its resting potential. When the membrane potential increases (is made more negative or is hyperpolarized), ions move in the opposite direction. At low levels of stimulation this response is proportional to stimulus intensity and decays exponentially with distance and time. These short lived, local, graded potentials are also called local or generator receptor potentials (Fig. 1-3). This ion flow reflects passive properties of the membrane not unique to excitable cells and can be predicted from the Nernst equation and the electrical characteristics of the membrane.

If, however, the membrane is depolarized above a threshold of approximately 15 mV (from −70 mv to −55 mv), an action potential results. This is a self-propagating spike response unique to excitable tissues. A typical action potential is depicted in Figure 1-4. It consists of a period of rapid depolarization, an overshoot period where polarity may be reversed (i.e., becomes positive), a period of rapid repolarization, a period of slower repolarization, and a relatively prolonged period of hyperpolarization. Action potentials travel along nerve axons and are the impulses that carry information in the nervous system. Action potentials are all or none. Depolarization above threshold may result in more frequently repeated action potentials but will not cause a greater action potential

Fig. 1-3. *Local potentials and threshold. Changes in membrane potential following application of subthreshold and then threshold stimuli. At the threshold (−55mV) an action potential is produced. Subthreshold depolarizing stimuli and hyperpolarizing stimuli produce potentials which decay over time (shown here) and distance (not shown). (From W.F. Ganong.* Review of Medical Physiology *[11th ed.]. Los Altos, CA: Lange, 1983. P. 37.)*

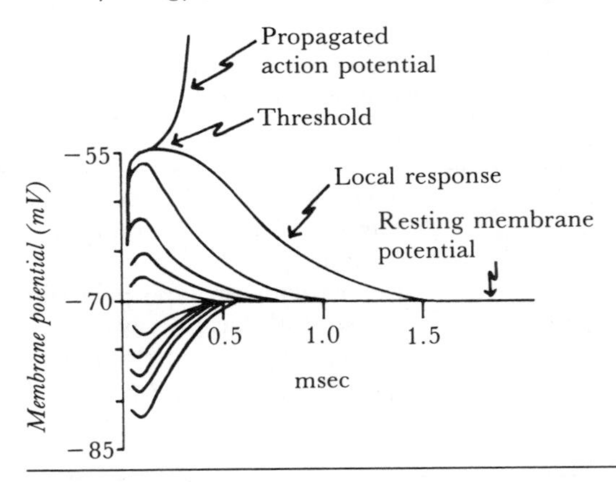

Fig. 1-4.

The action potential. The components of the action potential are displayed above. Na^+ and K^+ conductances are displayed below, illustrating the molecular events which underly the action potential. These curves are deprived from the experimental system of Hodgkin and Huxley which differ from that described in the text, thus the resting membrane potential of approximately −90 mV. (From A. C. Guyton. Textbook of Medical Physiology *[7th ed.]. Philadelphia: Saunders, 1986. P. 109.)*

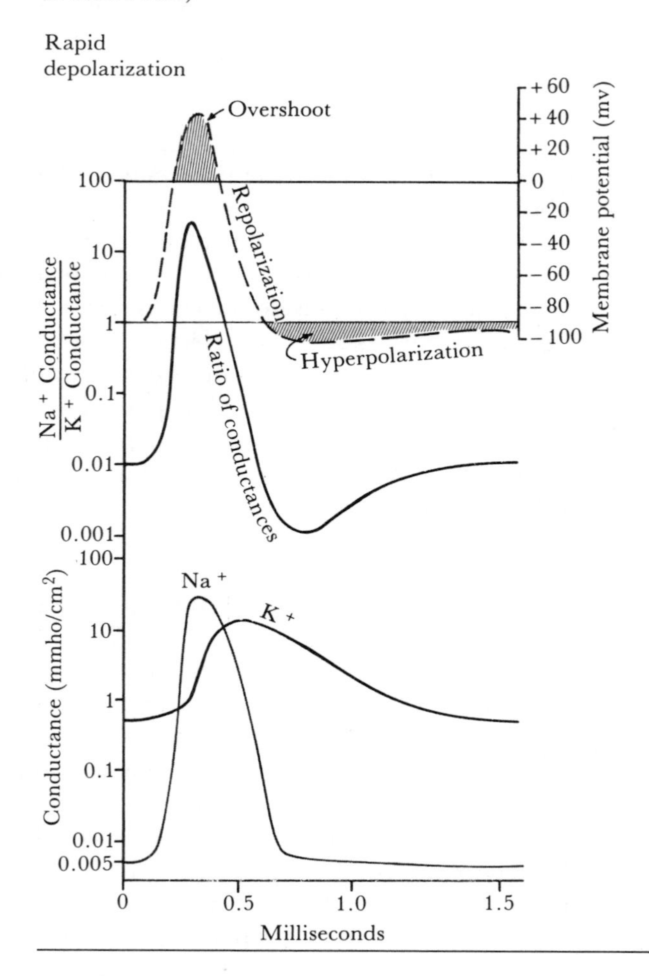

(Fig. 1-5). Once the nerve has been stimulated, the neuron is refractory to further stimulation. This refractory period consists of an initial absolute refractory period during which no stimulus, no matter how strong, will excite the nerve and a later relative refractory period during which stronger than normal stimuli are required to cause excitation. Table 1-2 summarizes the differences between local and action potentials.

Table 1-2

Comparison of Local and Action Potentials

Feature	Local Potential	Action Potential
Amplitude	Small (0.1–10 mv)	Large (70–110 mv)
Duration	Variable (5 msec to minutes)	Brief (1–10 msec)
Form	Graded	Spike, all or none
Polarity	Depolarizing or hyperpolarizing	Depolarizing
Propagation	Passive with local decay	Active, self-propagating
Channels	Na^+-K^+ leak channels, other special receptors (e.g., neutrotransmitters, sensory receptors)	Voltage-gated Na^+, K^+ channels

Fig. 1-5. *Threshold and action potentials. Once a threshold stimulus is reached (here −65 mV) an action potential results. Increasing receptor potential increases the frequency of the action potential, but it does not change shape. (From A. C. Guyton.* Textbook of Medical Physiology *[7th ed.]. Philadelphia: Saunders, 1986. P. 574.)*

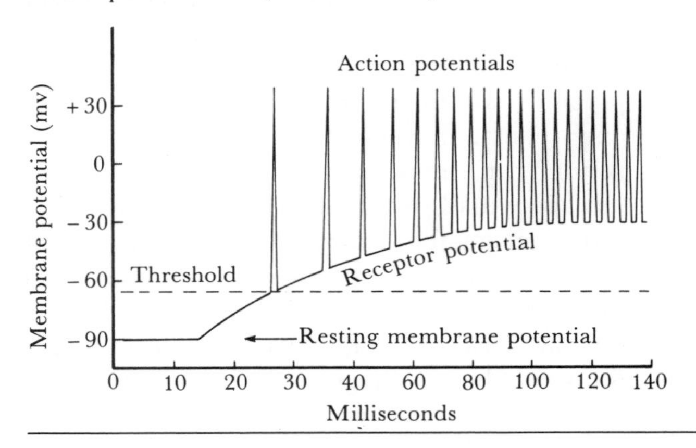

Voltage sensitive Na^+ and K^+ channels distinct from the K^+-Na^+ leak channels underly the molecular basis of the action potential. The K^+-Na^+ leak channel is the channel discussed in the previous section and is crucial to the maintenance of the resting potential and the generation of local potentials. The key feature of this channel is that K^+ permeability (also called conductance) is much greater than Na^+ permeability. When the membrane depolarizes by more than 7–15 mV, a voltage-sensitive Na^+ channel opens (see Fig. 1-4). This dramatically increases Na^+ permeability 500–5000 fold. At the resting potential of −70 mv, both the Na^+ concentration and electrical gradients favor an influx of Na^+, which tends to further depolarize the membrane towards the Na^+ equilibrium potential. The potassium efflux through the leak channels and the Na^+-K^+ pump oppose the depolarizing effect of Na^+ influx. The threshold value is the point at which Na^+ influx overcomes K^+ efflux and produces an action potential.

The same depolarization which opened the activation gate acts more slowly to close the inactivation gate of the Na^+ channel. This restores Na^+ conductance to its low normal level. The reversal of membrane potential during the overshoot period also reverses the Na^+ electric gradient and further limits Na^+ influx; thus, depolarization is short lived because of the brief period of increased Na^+ permeability. Opening the voltage-sensitive K^+ channels also contributes to repolarizing the membrane. These voltage-sensitive K^+ channels respond slowly to depolarization and reach peak permeability at approximately the same time as the Na^+ channel inactivation gates close. The increase in K^+ permeability facilitates K^+ efflux, as does the less negative membrane potential during the repolarization phase. The K^+ channels remain open for a few milliseconds after repolarization is complete. This hyperpolarizes the cell (brings the cell closer to E_K) because K^+ permeability is greater than in the resting state. This hyperpolarization is the cause of the relative refractory period. The absolute refractory period is due to the voltage-sensitive Na^+ channel, which, once inactivated, cannot be activated again until the membrane is repolarized.

The number of ions involved in the production of the action potential is minute relative to the total ion concentration. There is no measureable change in ion concentrations after a single action potential. Nonetheless, repeated stimulation will lead to significant changes in concentration. The Na^+-K^+ pump maintains the concentration gradients which maintain membrane polarization. This is the site at which the cell expends energy to provide for signaling.

Propagation of the action potential is a result of local currents produced by the action potential at one point of the membrane depolarizing adjacent parts above threshold to induce an action potential. This is demonstrated schematically in Figure 1-6. In A the nerve fiber is at rest with the inside negatively charged. During the action potential this polarity is reversed. This produces a local circuit of current flow. Positive charges move towards the area of negativity. In B the potential is reduced (made less negative) by this electrical current. Note that extracellular movement is in a direction opposite to intracellular movement, resulting in circular current flow. These local currents are the basis by which local potentials spread along the membrane. When the threshold is reached, an action potential is generated, and the cycle continues.

Myelination accelerates the rate of axonal transmission and conserves energy by limiting the depolarization process to the nodes of Ranvier. Myelin is a very effective insulator and thus limits transmembrane ion flow to the nodes of Ranvier where voltage-sensitive channels are found. Myelin also decreases capacitance so that fewer ions are required to charge the membrane. The net result is that more ions can participate in the circular ion flow. Action potentials are thus transmitted from node to node in a process termed *saltatory conduction* (Fig. 1-7). By causing the depolarization process to jump long intervals, conduction velocity increases. Conduction velocity also increases with axonal diameter due to decreased resistance to ion flow within the axon. This is true for both unmyelinated and myelinated neurons. Myelination also decreases the metabolic energy required to restore the resting potential because total ion loss is reduced with the limited ion currents.

QUESTIONS

11. What is the molecular basis of the rapid depolarization phase of the action potential?
12. What is the molecular basis of the repolarization phase of the action potential?
13. If extracellular Na^+ is increased, how would the action potential change?
14. Tetrodotoxin (TTX) is a substance that specifically blocks the Na^+ channels. Predict the result of a suprathreshold stimulus applied to a TTX-treated neuron.
15. Tetraethylammonium (TEA) is a substance that specifically blocks the K^+

Fig. 1-6. *Propagation of action potentials. See text for explanation. Although the figure illustrates propagation of action potentials with reversal of polarity during the overshoot period, local potentials also have a similar circular current flow. (From A. C. Guyton.* Textbook of Medical Physiology *[7th ed.]. Philadelphia: Saunders, 1986. P. 111.)*

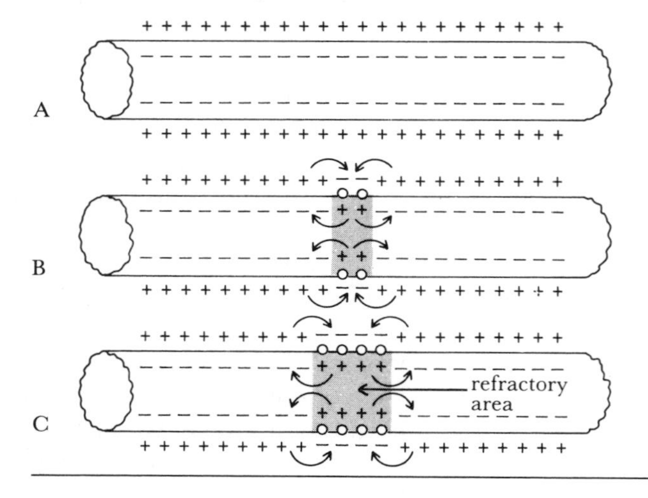

Fig. 1-7 *Saltatory conduction in myelinated axons. Note that conduction is unidirectional and that transmembrane ion flow occurs only at the node of Ranvier. (From A. C. Guyton.* Textbook of Medical Physiology *[7th ed.]. Philadelphia: Saunders, 1986. P. 115.)*

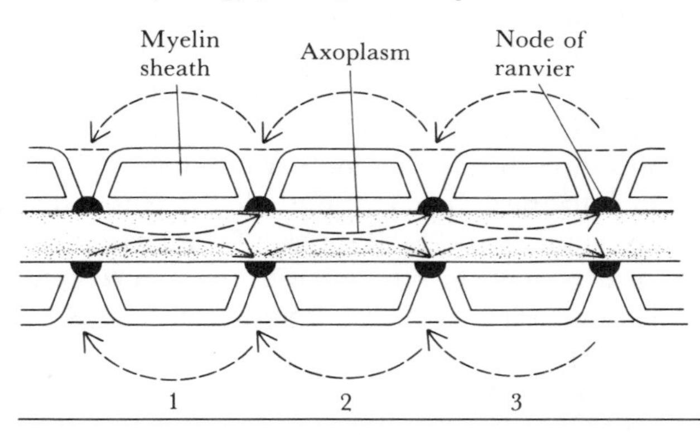

channel. Predict the result of a suprathreshold stimulus applied to a TEA-treated neuron.

16. Predict the result of a suprathreshold stimulus applied to a neuron treated with both TEA and TTX.
17. What would be the effect of changing E_m to
    A. −60 mv
    B. −80 mv
18. Figure 1-6 displays bidirectional impulse conduction. Why is in vivo conduction unidirectional?
19. How do Na^+ and K^+ flow compare at
    A. Subthreshold
    B. Threshold
    C. Suprathreshold levels
20. Current flow in local potentials resembles that depicted in Figure 1-6, yet polarity is not reversed. Explain this apparent contradiction.

SYNAPTIC TRANSMISSION IN THE NERVOUS SYSTEM

Synapses are specialized junctions for the transmission of impulses from one nerve cell to another. The extensive, intricate and varied connections among nerve cells and the plasticity of the chemical synapses are essential for information transmission, processing, and storage.

Chemical transmission involves the release by the presynaptic cell (PrSC) of a chemical neurotransmitter (NT) which diffuses across the synaptic cleft to bind to receptors on the postsynaptic cell (PoSC) and leads to a change in membrane potential. These intermediate chemical processes result in a time delay between the arrival of the impulse in the PrSC and the generation of a postsynaptic potential. Because of the synaptic delay, conduction along a chain of neurons slows as the number of synapses in the circuit increases. The minimum synaptic delay is 0.5 msec. Known NTs include acetylcholine (Ach), several different amines (e.g., the catecholamines and serotonin), a number of amino acids, and a variety of peptides.

The molecular events of neurotransmitter release are well studied and are essentially independent of the specific NT or morphologic type of synapse. When the action potential depolarizes the PrSC nerve terminal, voltage gated Ca^{2+} channels open and allow Ca^{2+} to enter the cell down its electrochemical gradient. The Ca^{2+} influx stimulates the fusion of neurotransmitter-containing vesicles with the PrSC cell membrane. This process of exocytosis results in the release of NT. The number of vesicles released is directly proportional to the Ca^{2+} influx. The NT diffuses across the synaptic cleft and binds to receptors on the PoSC.

Postsynaptic cell response does vary with NT. Moreover, the same NT can produce different responses if the PoSC receptor is different. PoSC responses are of two general types: ion permeability may be altered or internal metabolic processes of the PoSC may be activated. The postsynaptic potential (PSP) is produced by altering ion permeability in the PoSC through the inactivation of ion channels. It is a local graded potential that decays with distance and time. Excitatory PSPs (EPSPs) result from increasing membrane permeability to Na^+, such as that which occurs with the Ach receptor. Inhibitory PSPs are produced by increasing membrane permeability to K^+ and Cl^-. Even if this does not hyperpolarize the cell, this can "short-circuit" excitatory stimuli, because an increase in Na^+ permeability has less of an effect on membrane potential if K^+ and Cl^- permeability are also increased. These changes in permeability are short-lived and decay in approximately 15 msec after the NT is removed from the receptor. Slower, longer lasting PSPs have been described in autonomic ganglia, cardiac and smooth muscle, and cortical neurons. These potentials depend on decreasing ion permeability.

Besides affecting ion channels, NTs may also activate internal metabolic processes of the PoSC such as protein kinases or receptor protein synthesis. Prolonged stimulation or a lack thereof may lead to down regulation or up regulation of receptors. Such metabolic changes produce long-lasting changes in the reactivity of the synapse. These modulations may be important in memory and learning. The action of the NT is terminated by removing the NT from the synaptic cleft by either diffusion, active reuptake by the PrSC, or enzymatic inactivation of the NT. New transmitter is synthesized in the PrSC nerve terminal or cell body and transported from the cell body to the nerve terminal by axonal transport.

The response of the PoSC is the result of the integration of all the cells' input. A single neuron such as a spinal motor neuron receives convergent input from hundreds of sources on its soma and dendrites. No single stimulus is sufficient to induce the cell to fire or to prevent firing. Input is summed algebraically both spatially and temporally. Temporal summation occurs when a nerve terminal fires rapidly before the previous PSP has had a chance to decay. Spatial summation occurs when activity is simultaneously present at more than one of the cell's synapses. The action potential is initiated at the initial segment, the axon hillock. This segment of the axon has a threshold for firing which is approximately 15 mv lower that that of the rest of the cell membrane because of the high concentration of voltage-gated Na^+ channels at the initial segment. Because PSPs are local potentials which decay with distance, synapses closer to the initial segment have a greater influence on the cell's firing status than do more distant synapses. When an action potential occurs, it travels anterograde down the axon but also retrograde to the cell soma and dendrites. This retrograde firing into the soma serves to "wipe the slate clean" so that the summation process starts fresh.

Transmission at the neuromuscular junction (NMJ) follows the same basic mechanism as described above. Acetylcholine is the NT released by the PrSC (the spinal motorneuron) when an action potential produces a Ca^{2+} influx. The Ach receptor on the muscle cell binds Ach, Na^+ permeability increases, and an end-plate potential is produced which spreads along the muscle leading to excitation contraction coupling. A unique feature of transmission at the NMJ is that an action potential always produces a PSP sufficient to cause the muscle membrane to fire. At rest, miniature end-plate potentials of approximately 0.5

mV can be detected at the muscle end plate. These represent single vesicles (quanta) of Ach which are randomly released at rest. The action of ACh is terminated by the enzyme acetylcholinesterase, which is located in the synaptic cleft.

In addition to excitatory and inhibitory effects on the PoSC, the PrSC can also be affected. Presynaptic inhibition or facilitation occurs by an alteration of Ca^{2+} influx and thus the amount of NT released by the PrSC. Axoaxonal synapses are involved as shown in Figure 1-8. The value of a mechanism such as presynaptic inhibition is that it permits selective modification of input from a specific source. In contrast, postsynaptic inhibition results in generalized depression of the PoSC. The calcium ion influx can be altered by either direct modification of the voltage-sensitive Ca^{2+} channels or by altering the voltage of the action potential at the nerve terminal.

QUESTIONS

21. Given an experimental system in which extracellular calcium is removed, what would be the effect on the PoSC of
    A. Depolarizing the PrSC
    B. Depolarizing the PoSC
    C. Injecting NT into the synaptic cleft
    D. Injecting Ca^{2+} into the synaptic cleft
22. What factors contribute to integration of information by the PoSC?
23. Name the three ways by which NT action may be terminated.
24. What effect does repetitive stimulation of the PrSC have on the PoSC?
25. Which of the following are known effects of neurotransmitters: altered membrane permeability, activation of internal metabolic processes, alteration of receptor numbers?
26. In cardiac muscle, direct electrical synapses are used via gap junctions which form between adjacent cells. Electrical excitation can thus spread from cell to cell without synaptic delay. What are some advantages of chemical synapses over the more rapid electrical synapses?

Fig. 1-8. *Pre- and postsynaptic inhibiton. Postsynaptic inhibition affects all inputs to the cell while presynaptic inhibition selectively inhibits a single input. (Reprinted by permission of the publisher from* Principles of Neural Science, *by E. K. Kandel and G. H. Schwartz. P. 90, 1981, by Elsevier Science Publishing Company, Inc.)*

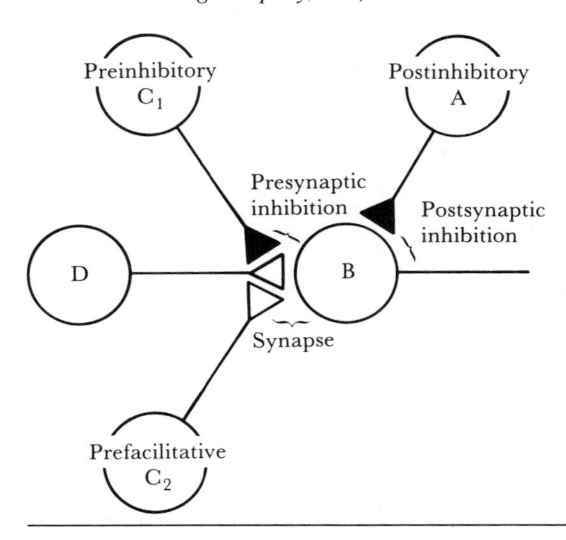

SKELETAL MUSCLE

STRUCTURE OF SKELETAL MUSCLE. The individual muscle cell or muscle fiber is a large, multinucleate, highly organized cell, specialized for contraction. The cytoplasm contains many myofibrils which consist of two types of myofilaments, thick and thin. These filaments are organized in a regular repeating pattern forming the basic functional unit of skeletal muscle, the sarcomere (Fig. 1-9).

Thick filaments are composed of myosin molecules. Myosin is a protein that can be proteolytically digested into two fragments, light meromyosin (LMM) and heavy meromyosin (HMM). The LMM is rod-shaped and assembles into filaments, while HMM contains the myosin globular head (myosin cross bridges) with ATPase activity and a binding site for actin. Myosin light chain, a smaller peptide that may modulate myosin ATPase activity, is also found in association with the globular portion of HMM. As seen in Figure 1-9, the myosin cross bridges are arranged in a helical array extending outward from the myosin

Fig. 1-9. *Organization of skeletal muscle from gross to molecular level. (From D. W. Fawcett.* A Textbook of Histology *[11th ed.]. Philadelphia: Saunders, 1986. P. 282.)*

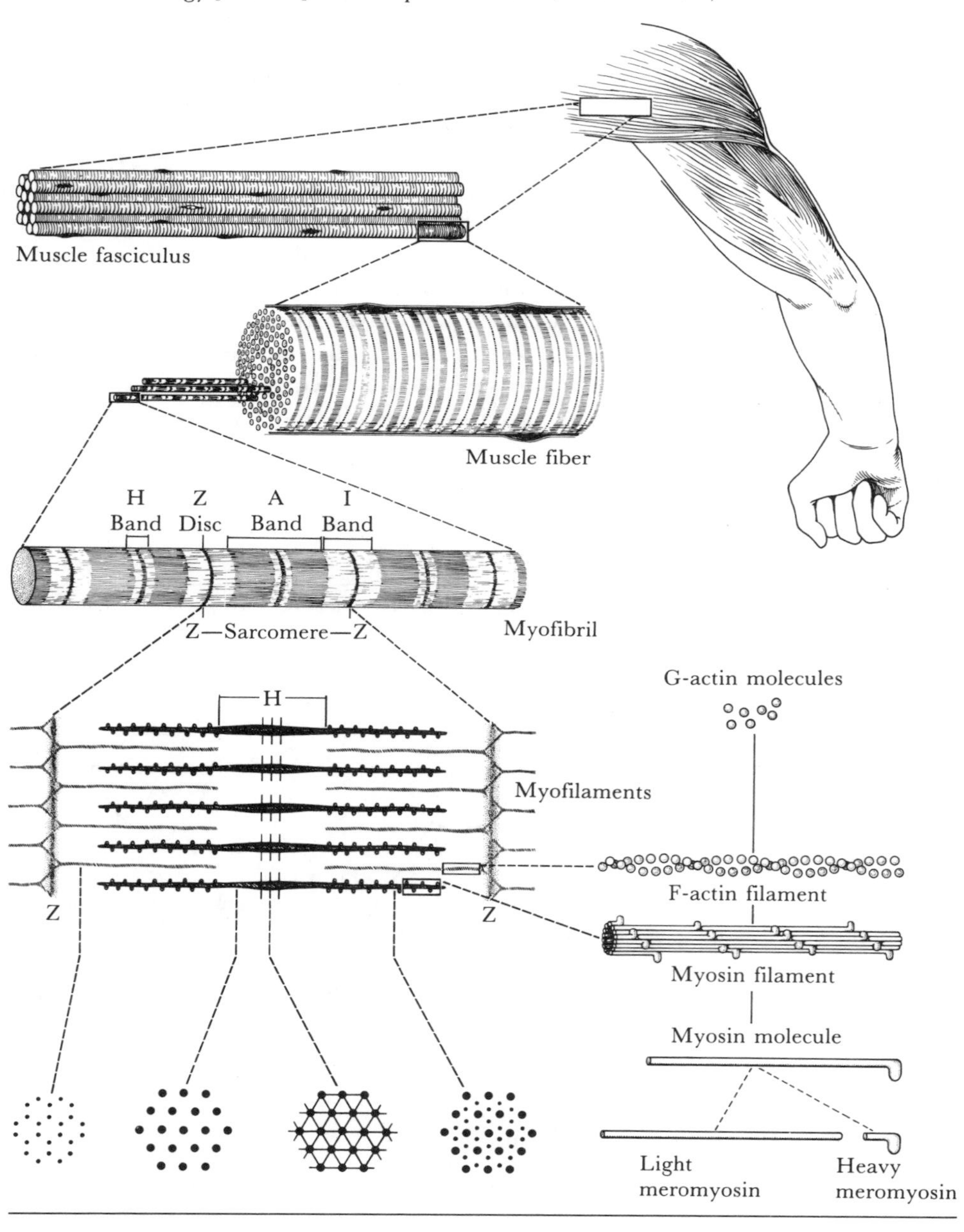

filament. Myosin molecules self-assemble into filaments with definite polarity. In the sarcomere the myosin thick filament reverses its polarity in the middle of the sarcomere, forming a central "bare zone" which is devoid of myosin cross bridges (Fig. 1-10).

The principal constituent of thin filaments is actin. Actin monomer (G actin) is a 42,000-dalton molecular weight protein which has the capacity to self-assemble into a double-stranded α-helix of actin monomers (F actin). In addition to actin, thin filaments also contain troponin and tropomyosin. Tropomyosin is also a double-stranded α-helix. It is situated in the groove between the actin strands. A molecule of tropomyosin spans approximately seven actin monomers. Troponin is a complex of three separate proteins: (1) TN-T, which binds to tropomyosin, (2) TN-C, which binds Ca^{2+}, and (3) TN-I, which binds to actin. There is one troponin complex for each molecule of tropomyosin (Fig. 1-11).

THE SLIDING FILAMENT THEORY OF MUSCLE CONTRACTION. During contraction, the overall length of the sarcomeres decrease, while individual thick and thin filaments remain the same length. According to the sliding filament theory of muscle contraction, during contraction thick and thin filaments slide past one another, resulting in increased overlap.

Excitation-Contraction (E-C) Coupling. The critical link between the electrical events of the muscle cell action potential and the organized interaction between actin and myosin, which causes contraction of skeletal muscle, is an increase in intracellular Ca^{2+} concentration from approximately 10^{-7} M in the relaxed state to 2×10^{-5}M or greater during contraction. Of special importance in E-C coupling are the "triads" composed of two terminal cisterns located in close proximity to a T tubule. Terminal cisterns are specialized evaginations of the muscle cell sarcoplasmic reticulum. The T tubules are extensive, deep invaginations of the muscle cell membrane. In the triads, extensions of the muscle cell membrane are in close proximity to extensions of the sarcoplasmic reticulum. The muscle cell action potential is propagated down the T tubules and, by an unknown mechanism, activates the terminal cisterns and sarcoplasmic reticulum to release their stored Ca^{2+} into the cytoplasm. The sarcoplasmic reticulum has a high-capacity Ca^{2+} ATPase which can actively pump Ca^{2+} back into the sarcoplasmic reticulum and return the cytoplasmic concentration of Ca^{2+} back to resting levels in order for muscle relaxation to take place.

Fig. 1-10. *A. Diagrammatic representation of the myosin molecule.*
B. Arrangement of myosin molecules into filaments. Note the reversal of polarity in the center which gives rise to the central bare zone. (Reprinted by permission of the publisher from Histology: Cell and Tissue Biology. *[5th ed.], by L. Weiss, P. 267, 1983, by Elsevier Science Publishing Company, Inc.)*

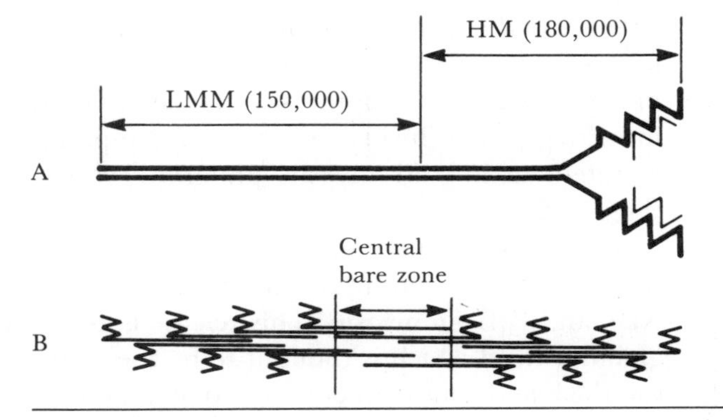

Fig. 1-11. *The proposed model of a thin filament in the resting state. (After C. Cohen. Protein switch of muscle contraction. Copyright © 1975 by Scientific American, Inc. All rights reserved.)*

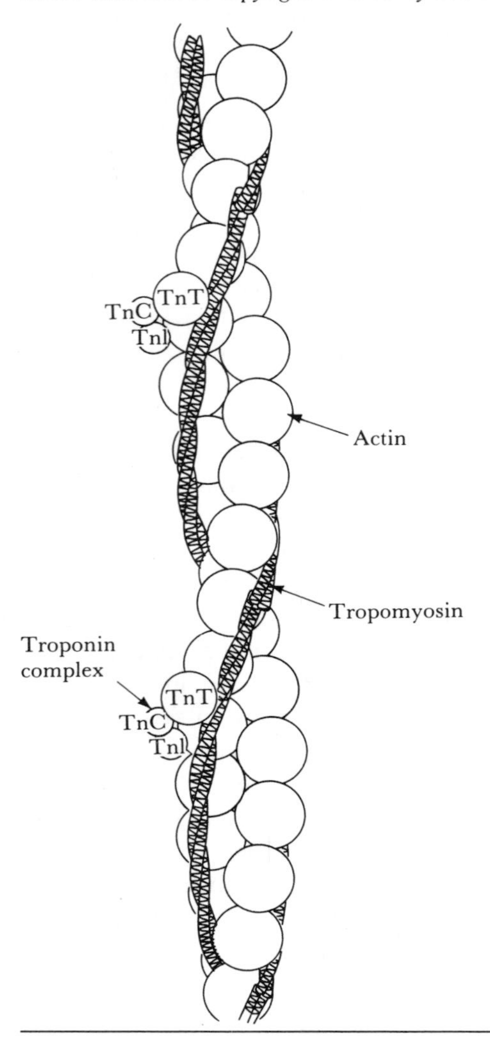

Molecular Events in Muscle Contraction. In the resting state, prior to the rise in cytoplasmic Ca^{2+} concentration, the tropomyosin molecules are located in the groove between the strands of actin filaments in such a way that they sterically block the myosin binding sites on actin. They inhibit therefore, the formation of actin-myosin cross bridges.

During activation, Ca^{2+} is released from the sarcoplasmic reticulum. The TN-C component of troponin binds Ca^{2+}. This binding causes conformational changes in the troponin complex which in turn result in conformational changes in tropomyosin. The tropomyosin then alters its position with respect to actin in such a way that binding sites on actin become available for interaction with myosin globular heads. As long as cytoplasmic Ca^{2+} concentrations remain elevated, actin and myosin undergo cycles of association and dissociation (Fig. 1-12).

In the resting state, myosin globular heads have bound ADP and inorganic phosphate (P_i). Binding of actin to myosin globular heads causes the release of ADP and P_i from myosin. This release is associated with a tilting of the myosin globular head which is thought to be the "power stroke" causing the sliding of the thick and thin filaments past one another.

Fig. 1-12. *Proposed mechanism (A–D) by which the ATP-driven cycles of actin and myosin association and dissociation generate the mechanical force needed for the sliding of thick and thin filaments during contraction. (From L. Stryer.* Biochemistry *[2nd ed.]. San Francisco: Freeman, 1981. P. 825.)*

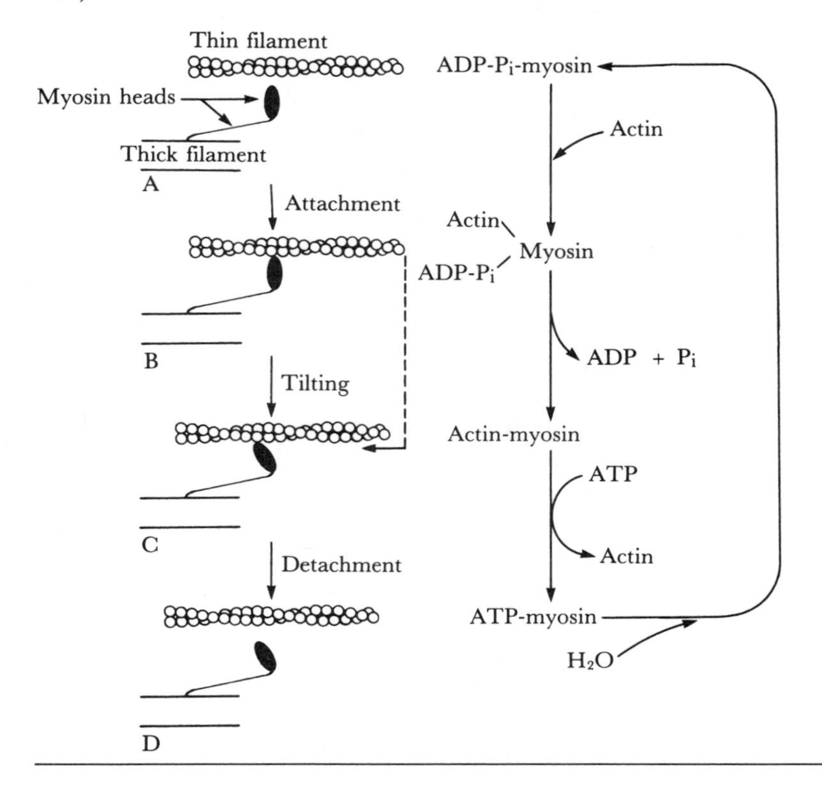

For the actin-myosin complex to dissociate, ATP is needed. Actin has a low affinity for myosin-ATP complex, and the binding of ATP by myosin causes the detachment of actin. Myosin globular head has ATPase activity, resulting in the hydrolysis of the bound ATP to ADP and P_i. The cycle is now complete and can be repeated as long as cytoplasmic Ca^{2+} concentrations remain greater than 2×10^5 M.

FAST AND SLOW MUSCLE FIBERS. Two different types of muscle fibers can be recognized, slow and fast. Slow fibers are well endowed with cellular machinery for aerobic metabolism. These include abundant mitochondria, a high concentration of myoglobin, and a rich blood supply. The high myoglobin concentration and rich blood supply impart a red color to slow fibers. Slow fibers are adapted for sustained contraction over long periods of time. Fast fibers, on the other hand, are adapted for intense but sporadic contractions. They have a high concentration of glycolytic enzymes as well as large stores of glycogen which enable them to carry out anaerobic metabolism (Table 1-3). Although both slow and fast fibers are present in most muscles of the body, one or the other usually predominates. For example, in the calf muscles, which must sustain contraction for long periods of time, slow fibers predominate, while the extraocular muscles, which are adapted for rapid contraction, have a predominance of fast fibers.

MUSCLE TWITCH. A muscle twitch is a cycle of contraction and relaxation that occurs in response to a neural stimulus in vivo or an electrical impulse passed directly into the muscle in an experimental setting. The duration of the twitch

Table 1-3

Distinguishing Features of Slow and Fast Muscle Fibers

Feature	Slow Fiber	Fast Fiber
Color	red	white
Mitochondria content	+ + +*	+
Myoglobin content	+ + +	+
Glycolytic enzymes	+	+ + +
Glycogen content	+	+ + +
Myofibrillan ATPase	+	+ + +

* Number of plus signs indicate relatıve amount.

varies with different muscles. The time lag between the stimulus and the initiation of contraction is called the latent period (Fig. 1-13).

REFRACTORY PERIOD. When a muscle has been stimulated to contract, it remains totally refractory to further stimulation for a period of time called the absolute refractory period. During this time, no stimulus, regardless of its amplitude, can produce a contraction. Following the absolute refractory period, there is a period, called the relatively refractory period, during which a muscle does not respond to the usual threshold stimulus, but can be made to contract by a large suprathreshold stimulus.

SUMMATION AND TETANUS. Let us imagine a muscle fiber that is undergoing a twitch response. What would happen if it were stimulated again, prior to completion of the first twitch response and before complete relaxation had taken place? Assuming that the stimulus came after the refractory period was over, tension would rise to a new peak which would be higher than the peak tension for the first twitch (Fig. 1-14A). This phenomenon is called summation (wave summation or temporal summation) and may be explained by the fact that after the second stimulus, the Ca^{2+} concentration in the sarcoplasm is higher than it was after the first stimulus, because some of the Ca^{2+} released from the sarcoplasmic reticulum by the first stimulus remained in the cytoplasm (i.e., reuptake was not completed when the second stimulus arrived).

As the frequency of stimulation is increased, higher and higher peak tensions are developed in the muscle, until a certain stimulus frequency is reached where

Fig. 1-13. *Muscle twitch. Note the variation in the duration of muscle twitch with different muscles. Also note the latent period between action potential and muscle contraction. (From A. C. Guyton.* Textbook of Medical Physiology *[7th ed.]. Philadelphia: Saunders, 1986. P. 130.)*

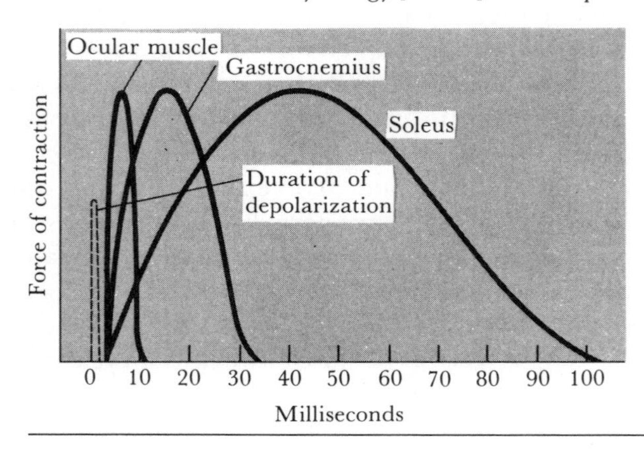

there is no relaxation of muscle between successive stimuli and a peak plateau of tension is reached. The muscle fiber is said to be in tetany (Fig. 1-14B).

MOTOR UNIT. The functional unit in skeletal muscle is the motor unit which consists of an alpha motor neuron in the anterior horn of the spinal cord and all the muscle fibers that it innervates. A motor unit may contain as few as two or three muscle fibers when precise control of muscular action is needed, e.g., laryngeal muscles, or it may consist of several hundred muscle fibers when fine control of muscle activation is not needed, e.g., gluteal muscles.

SPATIAL VERSUS TEMPORAL SUMMATION. Each individual motor unit contracts in an "all or none" fashion. In other words, a suprathreshold stimulus produces a twitch, the amplitude of which is independent of the stimulus intensity. Whole skeletal muscle, however, exhibits a graded response. Increasing intensity of stimulus produces higher amplitudes of twitch. The reason for this graded response is the fact that different motor units in a muscle have different thresholds of stimulation, and increasing stimulus intensity activates more motor units leading to increased amplitude of response. This phenomenon is called spatial summation. Temporal summation refers to the fact that increasing frequency of stimulation can cause increased tension developed by each motor unit (Fig. 1-14). Thus, skeletal muscle can be made to develop increasing tension in a graded fashion in response to increases in the frequency (temporal summation) and intensity (spatial summation) of the stimulus.

ISOMETRIC AND ISOTONIC CONTRACTIONS. The molecular events of actin and myosin interaction result in transduction of chemical energy (ATP) to mechanical energy. This mechanical energy can be in the form of tension developed in the muscle or in the form of work done by the muscle as it shortens. If two ends of a muscle are fixed such that it cannot shorten, and the muscle is stimulated to contract, tension develops in the muscle, but the length of the muscle does not change during the contraction. This is called an isometric contraction. On the other hand, if a muscle contracts such that tension remains constant while the muscle shortens, the contraction is called an isotonic contraction.

LENGTH-TENSION RELATIONSHIP. The tension developed in a muscle fiber upon neural stimulation depends on the extent to which the muscle was stretched prior to stimulation. If one were to passively stretch muscle to various lengths and measure the tension at each length, a passive length-tension curve could be plotted as in Figure 1-15. If the muscle were stimulated to contract isometrically

Fig. 1-14. *Twitch, summation, and tetanus. S = stimulus. (From J. B. West.* Best and Taylor's Physiological Basis of Medical Practice *[11th ed.] P. 71. © 1985 The Williams & Wilkins Co., Baltimore.)*

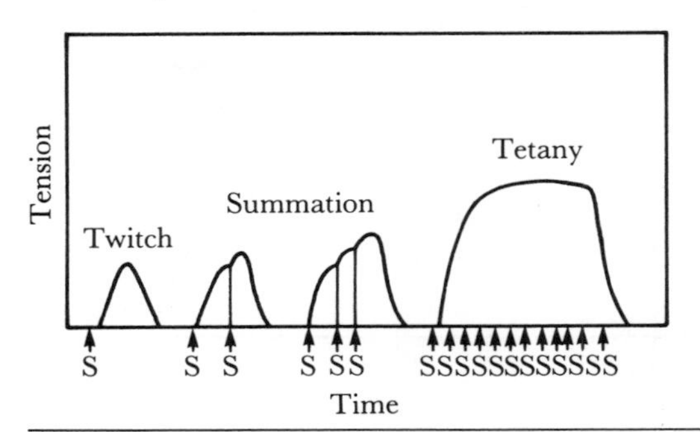

Fig. 1-15. *Resting, active, and total tension in a skeletal muscle undergoing isometric contraction at various muscle lengths. Resting tension = tension in muscle prior to contraction = preload. Total tension = peak total tension developed in muscle after contraction. Active tension = total tension = resting tension = tension developed in a muscle as a result of contraction.*

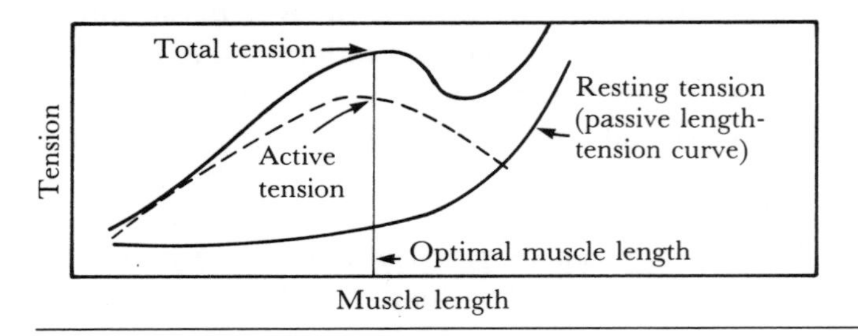

at each length and the peak tension recorded for each length, one could calculate the active tension (tension developed as a result of contraction) at each length (Fig. 1-15).

It is apparent from Figure 1-15 that there is an optimal length of muscle in which the greatest tension is developed upon contraction. This phenomenon can be explained based on the degree of overlap between thick and thin myofilaments at each muscle length (Fig. 1-16).

Based on this model, increasing overlap between thin and thick filaments produces increasing tension with stimulation. Note that between points B and C there is no change in the degree of thin and thick filament overlap because the central bare zone of the thick filament is devoid of myosin cross bridges and does not contribute to the overall number of actin and myosin cross links. The strength of the contraction, therefore, is constant at sarcomere lengths between B and C. If sarcomere length is decreased from B or increased from C, the degree of thick and thin filament overlap is decreased, resulting in decreased strength of contraction.

LOAD-VELOCITY RELATIONSHIP. Afterload is defined as the load against which the muscle exerts its contractile force. When a muscle is undergoing isotonic contraction, the force available to cause muscle shortening is the difference between the force generated by muscle contraction and the force against which

Fig. 1-16. *Length-tension diagram for a single sarcomere, illustrating maximum strength of contraction when the sarcomere is 2.0 to 2.2 microns in length. At the upper right are the relative positions of thin and thick filaments at different sarcomere lengths. (From A. C. Guyton.* Textbook of Medical Physiology *[7th ed.]. Philadelphia: Saunders, 1986. P. 125.)*

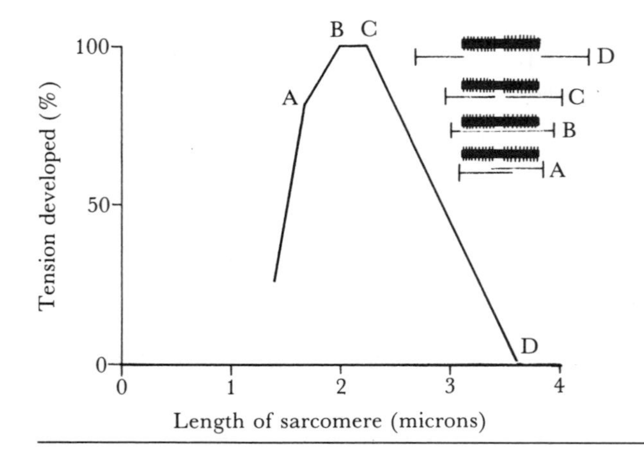

the muscle has to work (afterload). As afterload is decreased, the force available to cause muscle shortening is increased, and therefore, velocity of contraction is increased. The maximum velocity of contraction is reached when afterload approaches zero.

Conversely, as afterload is increased, the velocity of contraction is decreased. Contraction velocity approaches zero when the afterload is equal to the maximum force that the muscle can generate (Fig. 1-17).

SMOOTH MUSCLE

The structural, biochemical, and mechanical properties of smooth muscle are not as well understood as those of skeletal muscle. Smooth muscle from different sites of the body vary considerably in their properties and physiologic function. There are two broad categories of smooth muscle, multiunit and single unit smooth muscle.

In multiunit smooth muscle the individual muscle cells are "insulated" from one another and operate independently. Their contractile activity is controlled by neural input from the autonomic nervous system. Nonneural stimuli do not greatly influence contractile activity of multiunit smooth muscle. Examples of this type of smooth muscle include erector pilli muscles in the skin and ciliary muscles in the eye.

By far the most abundant type of smooth muscle in the body is single unit smooth muscle. This type is found in walls of vessels and hollow organs, such as the bladder and organs of the gastrointestinal system. In single unit smooth muscle the individual muscle cells are connected by means of gap junctions which allow the passage of ions and small molecules from one cell to the next. This essentially couples the cells electrically so that electrical stimulation of any one cell is passed on to other cells, and the muscle functions as a single unit. Contractile activity of single unit smooth muscle is influenced not only by the autonomic nervous system, but also by nonneural stimuli, such as hormones, and by local tissue factors, such as temperature and pH.

Smooth muscle contains thin filaments of actin and thick filaments of myosin. These interact in a similar fashion as they do in skeletal muscle. The relative ratio of these filaments is different in smooth muscle, however, as is their organization. Smooth muscle contains relatively fewer thick filaments and relatively more thin filaments. The organization of these filaments in smooth muscle is not well understood. "Dense bodies," the smooth muscle counterparts of Z lines in skeletal muscle, are electron-dense structures in the cytoplasm and cell membrane which serve to anchor the thin filaments.

The mechanism of E-C coupling is significantly different in smooth muscle. Changes in the intracellular concentration of Ca^{2+} controls the contractile state of smooth muscle just as in skeletal muscle. Smooth muscle, however, does not

Fig. 1-17. *Force-velocity relationship for a muscle undergoing isotonic contraction with different afterloads.*

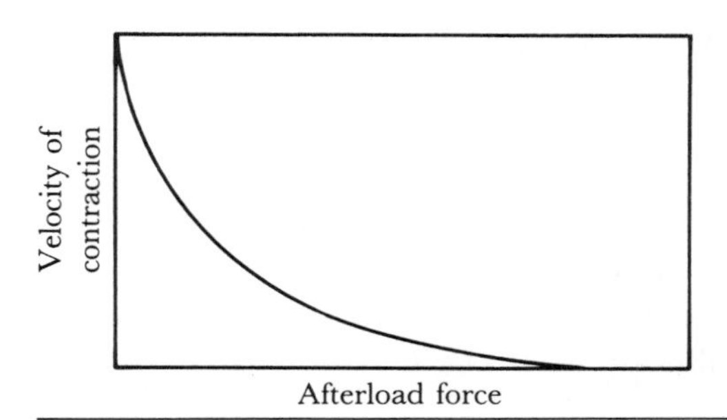

contain troponin (there is some tropomyosin in smooth muscle, but its function is unknown). When Ca^{2+} concentration increases in the cytoplasm, it binds to a protein called calmodulin. The Ca^{2+}-calmodulin complex activates a protein kinase which phosphorylates the myosin light chain. In the resting state, the ATPase activity of the myosin head is very low. The phosphorylated myosin light chain greatly enhances the ATPase activity of the myosin head, allowing it to interact with actin in a similar fashion to the process described for skeletal muscle.

The source of Ca^{2+} involved in E-C coupling is different in skeletal and smooth muscle. The Ca^{2+} released into the cytoplasm in smooth-muscle E-C coupling comes from both intracellular sources (sarcoplasmic reticulum) and extracellular sources through Ca^{2+} channels in the cell membrane.

QUESTIONS

27. Based on the sliding filament theory, which of the following changes would you expect in the sarcomere during contraction?
    1. The length of A zone should increase.
    2. The length of H zone should decrease.
    3. The length of A zone should decrease.
    4. The length of I zone should decrease.

    Answer: A. 1, 2, and 3
    B. 1 and 3
    C. 2 and 4
    D. 4 only
    E. 1, 2, 3, and 4
28. Ca^{2+} channel blockers would have a significant effect on muscle function.
    A. Skeletal muscle
    B. Smooth muscle
    C. Both
    D. Neither
29. Ca^{2+} binding proteins are involved in excitation contraction coupling.
    A. Skeletal muscle
    B. Smooth muscle
    C. Both
    D. Neither
30. The total tension developed by a skeletal muscle can be increased by
    1. Increasing frequency of neural stimulation because this increases the number of motor units that will contract (spatial summation).
    2. Increasing frequency of neural stimulation because this results in increased tension developed by individual contracting motor units (temporal summation).
    3. Increasing intensity of electrical stimulation because this increases the tension developed by each individual contracting motor unit (temporal summation).
    4. Increasing intensity of electrical stimulation because this increases the number of motor units that will contract (spatial summation).

    Answer: A. 1, 2, and 3 are correct
    B. 1 and 3 are correct
    C. 2 and 4 are correct
    D. 4 only is correct
    E. 1, 2, 3, and 4 are correct
31. If you were developing a drug to treat the muscle spasticity of several neurologic diseases such as cerebral palsy or multiple sclerosis, which of the following would be more useful?

A. A drug that inhibited the Ca^{2+} ATPase in the sarcoplasmic reticulum
B. A drug that inhibited the Ca^{2+} channels in the cell membrane
C. A drug that inhibited protein kinases
D. A drug that inhibited Ca^{2+} release from the sarcoplasmic reticulum

ANSWERS

1. Particle size, temperature, interactions of particle with the membrane, and membrane pore size.
2. Flux will increase 3 times.
3. a. There would be no net movement of water in isotonic 280 mM glucose
   b. The cell would shrink when placed in hypertonic 280 mM NaCl (approximately 560 mosm).
4. Simple diffusion and carrier-mediated diffusion would remain intact, though the flux would change with the concentration gradients. There would be no active transport and no cotransport once the Na^+ gradient dissipated. Osmosis would not be affected.
5. Large and/or charged particles may be transported. Transport can be controlled through voltage or ligand gating. Substances can be transported against their electrochemical gradients.
6. A. E_K will increase by -61 mV (-61 log10)
   B. E_K will increase by -122 mV (-61 log100)
   C. E_K will decrease by 61 mV (-61 log 1/10)
   D. E_K will reverse polarity
7. -47 mv. Recall in the Nernst equation n for Ca^{2+} = $+2$, so that a smaller membrane potential can balance a larger concentration gradient.
8. A. Na^+ influx, K^+ influx
   B. Na^+ influx, K^+ no net flow
   C. Na^+ influx, K^+ efflux
   D. Na^+ influx, K^+ efflux
   E. Na^+ no net flow, K^+ efflux
   F. Na^+efflux, K^+ efflux
9. In reality, the membrane is permeable to other ions in addition to K^+ and a more complex equation, the Goldman equation, is necessary to describe passive membrane behavior. Additionally, the electrogenic properties of the Na^+-K^+ pump contribute a few millivolts to E_m.
10. E_m would increase (become less negative). This is the basis of the action potential.
11. Opening the voltage sensitive Na^+ channel to allow Na^+ influx.
12. Closing the inactivation gates of the Na^+ channel to prevent further Na^+ influx and opening the K^+ channel to increase K^+ efflux.
13. The amplitude would increase, but the overall time sequence would remain the same because gate activity depends on the initial depolarization to threshold. The overshoot does not activate any voltage gates.
14. See Figure 1-4. There would be no depolarization beyond threshold because of the blocked Na^+ channels. The threshold stimulus would, however, open the K^+ channels which would tend to hyperpolarize the neuron following the time course of increased K^+ conductance.
15. There may be an increase in amplitude due to the now unopposed Na^+ influx. The period of depolarization would also last much longer. The relatively short period during which the Na^+ gate is open and the activity of the pump Na^+-K^+ would bring the resting potential back to normal. There would be no period of hyperpolarization.
16. There would be a local potential that decayed with time and distance. Neither TTX nor TEA affects the leak channels.

17. A. Excitability would increase, a lower stimulus would be required to reach threshold
    B. Excitability would decrease
18. Action potentials are initiated at the axon hillock or analogous structure. Once initiated, a moving impulse does not depolarize the area behind it because the Na^+ channels are refractory.
19. A. K^+ efflux is greater than Na^+ influx
    B. K^+ efflux equals Na^+ influx
    C. Na^+ influx is greater than K^+ efflux
20. Although there is no reversal of polarity, there are areas of relative cation and anion excess, and there is flow of charge along these pathways.
21. A. None, depolarization of the PoSC in the absence of Ca^{2+} influx is insufficient to cause NT release
    B. A local graded EPSP
    C. A PSP corresponding to the amount and type of NT released
    D. None, only intracellular Ca^{2+} stimulates vesicle exocytosis
22. Strength of the PSP, algebraic, spatial and temporal summation, duration of the PSP, and proximity of the input to the axon hillock.
23. Diffusion, active reuptake, and enzymatic inactivation.
24. Repetitive stimulation eventually results in a decrease of NT stores in the PrSC. Thus, for the same stimulus in the PrSC, there would be a decreased amount of NT released and a correspondingly decreased PSP.
25. All.
26. Transmission is unidirectional. A greater flexibility of response (the capability of integrating and storing information) is possible with the more plastic chemical synapses.
27. C. Based on the sliding filament theory and your knowledge of the structure of the sarcomers (see Fig. 1-9), you can predict that the length of the A zone depends only on the length of thick filaments which remain constant during contraction. The length of the H and I zones, on the other hand, depend on the degree of thick and thin filament overlap and would therefore decrease during contraction (see Fig. 1-18).
28. B. Only smooth muscle depends on influx of extracellular Ca^{2+} through Ca^{2+} channels in the cell membrane for excitation contraction coupling.
29. C. Both smooth and skeletal muscle contain Ca^{2+}-binding proteins that play an important role in E-C coupling. In skeletal muscle, the binding of Ca^{2+} by troponin brings about conformational changes that ultimately result in availability of binding sites on actin for interaction with myosin. In smooth muscle, binding of Ca^{2+} by calmodulin ultimately results in enhanced ATPase activity of myosin globular heads.
30. C. Spatial summation caused by increasing stimulus intensity involves activation of more motor units. Temporal summation caused by increasing stimulus frequency involves development of more tension by the individual contracting motor unit.

Fig. 1-18. *Diagrammatic representation of a sarcomere in relaxed and contracted states.*

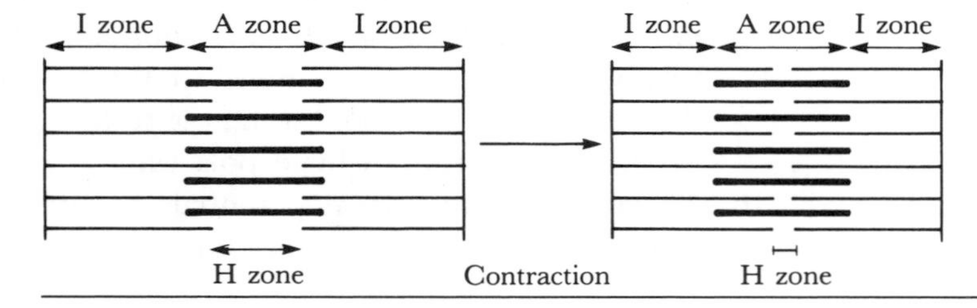

31. D. Inhibition of Ca^{2+} release from sarcoplasmic reticulum would be expected to decrease spasticity in skeletal muscle. Smooth muscle function would be much less influenced by this drug because the endoplasmic reticulum Ca^{2+} stores play less of a role in smooth muscle than they do in skeletal muscle.

BIBLIOGRAPHY

1. Fawcett, D. W. *A Textbook of Histology* (11th ed.). Philadelphia: Saunders, 1986.
2. Ganong, W. F. *Review of Medical Physiology* (12th ed.). Los Altos, CA: Lange, 1985.
3. Guyton, A. C. *Textbook of Medical Physiology* (7th ed.). Philadelphia: Saunders, 1986.
4. Kendel, E. K. and Schwartz, G. H. *Principle of Neuro Science* (2nd ed.). New York: Elsevier, 1985.
5. Stryer, L. *Biochemistry* (2nd ed.). San Francisco: Freeman, 1981.
6. Vander A. J., Sherman J. H., and Luciano D. S. *Human Physiology — The Mechanisms of Body Function* (4th ed.). New York: McGraw-Hill, 1985.
7. Weiss, L. *Histology: Cell and Tissue Biology* (5th ed.). New York: Elsevier, 1983.
8. West, J. B. *Best and Taylor's Physiological Basis of Medical Practice* (11th ed.) Baltimore: Williams & Wilikins, 1985.

2 Endocrine Physiology

Benjamin Hsu

PEPTIDE HORMONES AND STEROID HORMONES

PEPTIDE HORMONES. The mechanism of action for many peptide hormones involves the 3′,5′-cyclic adenosine monophosphate (cAMP) system (Fig. 2-1). Adenylate cyclase is a membrane-associated holoenzyme made up of three subunits; receptor (R), nucleotide regulatory protein (N), and catalytic subunit (C). Hormone binding to R causes N to take up guanosine triphosphate (GTP). The N-GTP complex activates C to catalyze the conversion of adenosine triphosphate (ATP) to cAMP (Mg^{2+} is also required). The GTP is degraded by GTPase activity associated with N to guanosine diphosphate (GDP), resulting in deactivation of N and dissociation of N from C. Cholera toxin inhibits the GTPase activity of N by ADP ribosylation. The result is persistent activation of N and persistent catalytic activity of C.

Cyclic AMP, the "second messenger," acts on the protein kinase complex, which consists of a regulatory subunit and a catalytic subunit. The regulatory subunit inhibits the enzyme activity of the catalytic subunit. Cyclic AMP binds to the regulatory subunit causing the two subunits to dissociate. The catalytic subunit is then free to phosphorylate potentially any protein and the resulting phosphoprotein mediates the physiologic effect of the hormone. To prevent prolonged activity of cAMP, cyclic nucleotide phosphodiesterase hydrolyzes cAMP to 5′-AMP. This enzyme is inhibited by several pharmacologic agents such as methylxanthines and nonsteroidal anti-inflammatory agents.

There is increasing evidence for the role of calcium as a mediator of peptide hormone action. In the absence of extracellular calcium, the action of some peptide hormones is inhibited even though their ability to increase or decrease

Fig. 2-1. *The cAMP system. See text.*

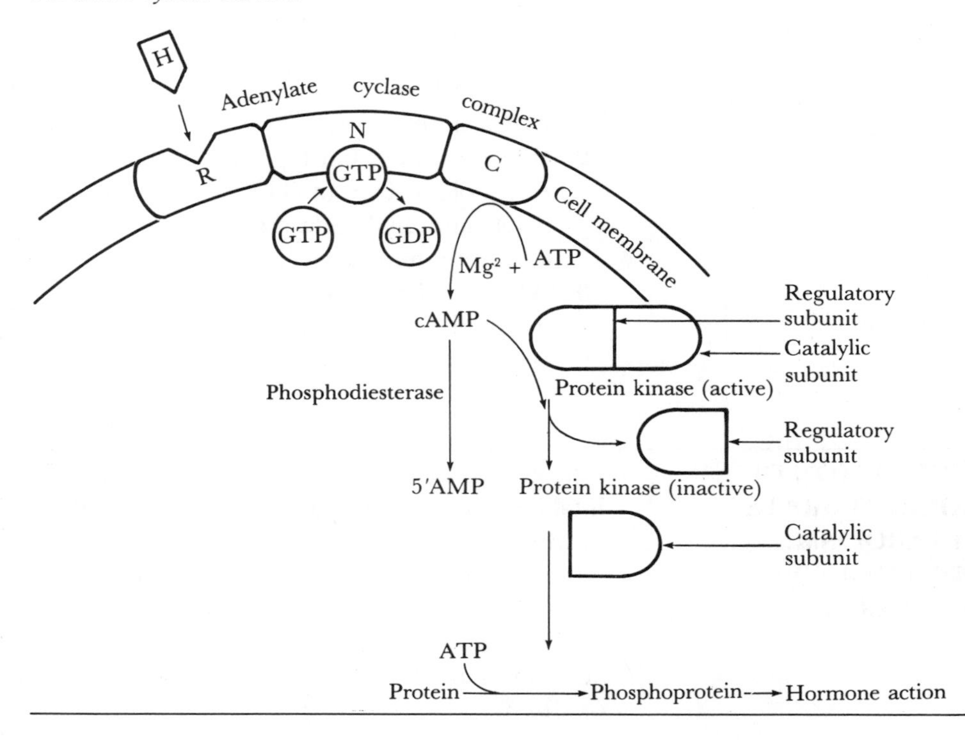

cAMP is unimpaired. Peptide hormones increase cytosolic calcium by either opening calcium channels in the membrane or releasing calcium from intracellular stores, or by both mechanisms. Calcium binds to calmodulin, a protein similar in structure to troponin C. Calcium-bound calmodulin activates various enzymes and has been shown to mediate several hormonal effects. It seems that the second messenger may involve both cAMP and calcium, or an interaction between them.

STEROID HORMONES. The steroid hormones include the mineralocorticoids, glucocorticoids, androgens, estrogens, and progestins. Vitamin D, derived from cholesterol, shares many similarities with the steroid hormones. Thyroid hormone also shares several features with steroid hormones. Steroid hormones are hydrophobic and bound to proteins for transport in plasma. Hormone action begins by binding of the free hormone to specific cytosol receptor proteins in target cells. This changes the receptor conformation and the resulting complex is translocated into the nucleus of the cell where it interacts with chromatin. This interaction produces an increase in specific messenger RNA (mRNA) molecules which are translated into the proteins that carry out the hormone's effects.

Steroid hormones differ from peptide hormones in several ways: (1) steroids are small hydrophobic molecules which are protein bound in plasma while peptides are larger hydrophilic molecules that are usually free in plasma; (2) steroids have longer plasma half-lives than peptides; (3) steroids have cytosol receptors while peptides have membrane receptors; (4) steroids act directly through a nuclear mechanism, peptides through a second messenger; (5) a longer time is required before the effects of steroids are initiated or terminated because proteins have to be synthesized and destroyed; and (6) steroid hormones are synthesized in the smooth endoplasmic reticulum (ER) and are released without being stored in granules; peptide hormones are synthesized in the rough ER and are stored in granules.

QUESTIONS

1. Which of the following play roles as second messengers for peptide hormones?
   A. Adenylate cyclase
   B. GTP
   C. AMP
   D. cAMP
   E. Calcium
2. Which of the following may inhibit the action of peptide hormones?
   A. Cholera toxin (a GTPase inhibitor)
   B. Aminophylline (a methylxanthine)
   C. Lanthanum (a calcium-channel blocker)
3. Where is the steroid hormone receptor located in the cell and what is the mechanism of action of steroid hormones?
4. What are some differences between steroid and peptide hormones?

REGULATION OF CARBOHYDRATE METABOLISM: INSULIN AND GLUCAGON

INTERMEDIARY METABOLISM. In the anabolic phase, as after a meal, energy intake exceeds the energy requirements of the body, and under the direction of insulin, energy fuel is stored in the form of glycogen, structural protein, and fat. After a meal, the major product of carbohydrate digestion is glucose. Most of the glucose is taken up by the liver where it is stored as glycogen or used to form triglycerides, which are subsequently transported to adipose tissue. Glucose is also taken up by skeletal muscle, where it is incorporated into glycogen or oxidized.

The catabolic phase occurs, as during a fast, when the body's energy needs are met only by endogenous sources. Mediated largely by glucagon, catabolism involves the breakdown of glycogen, structural protein, and triglyceride stores to maintain a constant energy supply to the body. The brain can function on two fuel sources, glucose and ketone bodies. During the first 12 to 24 hours of fasting glycogen breakdown in the liver provides sufficient glucose for the brain. When glycogen stores become depleted, gluconeogenesis takes place, in which noncarbohydrate molecules are converted to glucose. Gluconeogenic substrates include amino acids, glycerol, and lactate. Fatty acids cannot be converted to glucose because the oxidation of pyruvate to acetyl-CoA is irreversible. Thus the breakdown of triglycerides cannot be a quantitatively significant source of new glucose, because one triglyceride molecule gives only one glycerol, which gives only half a glucose. Amino acids derived from the breakdown of structural protein of skeletal muscle are the major gluconeogenic substrates. As a consequence, muscle wasting is a price paid to maintain brain function during the first few days of fasting. Amino acids are sent from muscle to the liver, where the bulk of gluconeogenesis takes place. The other site is the kidney.

The body's fat stores are critical to survival during prolonged starvation. Although lipolysis of triglyceride from adipose tissue cannot supply much glucose to the brain, it does supply the rest of the body with free fatty acids. Free fatty acids are also converted to ketone bodies in the liver. After 2 to 4 days of fasting, most tissues of the body use free fatty acids and ketone bodies for energy, allowing only the central nervous system (CNS) the privilege of consuming glucose. Protein loss from muscle is thus minimized. When the fasting period extends into weeks, the brain eventually makes use of ketones as its major fuel source.

The pancreatic hormones, insulin and glucagon, regulate not only carbohydrate metabolism but also protein and fat metabolism.

INSULIN. Insulin is produced by the beta cells of the pancreatic islets. It is a polypeptide hormone made up of the A chain and the B chain which are connected by disulfide bridges. The action of this hormone is anabolic, increasing the storage of glucose, fatty acid, and amino acids. Insulin facilitates the entry of glucose into cells of all tissues except brain, kidney tubules, intestinal mucosa, and red blood cells. In the liver, insulin increases glycogenesis by stimulating glucokinase and glycogen synthetase and inhibits gluconeogenesis as well as glycogenolysis by inhibiting phosphorylase. The incorporation of glucose into glycogen and the decrease in gluconeogenesis give a net result of glucose entry into hepatocytes. Hence, insulin facilitates glucose entry into liver cells by an indirect mechanism, while it does so in other tissues by a direct action on cell membranes. Insulin also increases lipid and protein synthesis in the liver.

In adipose tissue, insulin increases formation of glycerol and long-chain fatty acids from glucose and their esterification to form triglycerides. Insulin also inhibits the hormone-sensitive lipase in adipocytes and, thereby, decreases lipolysis. In skeletal muscle, insulin increases amino acid uptake as well as glucose entry, glycogen synthesis, and protein synthesis.

The mechanism of insulin action begins with binding to highly specific receptors in the cell membrane. The resulting hormone-receptor complex is taken into the cell by endocytosis (internalization). The insulin receptor is a protein kinase which phosphorylates tyrosine residues on the receptor itself and on other proteins. From this point, the exact mechanism leading to insulin's ultimate biologic effect remains unclear. Insulin receptors show a phenomenon called "down regulation," in which high levels of insulin cause a decrease in receptor number. Down regulation is associated with the insulin resistance in obesity and noninsulin-dependent diabetes mellitus.

The most important stimulator of insulin secretion is a rise in blood glucose. Amino acids (which enter the bloodstream after a protein meal), beta-keto acids, glucagon, acetylcholine, and intestinal hormones such as gastric inhibitory peptide (GIP) play minor roles in the stimulation of insulin secretion. Inhibitors of insulin secretion include catecholamines, sympathetic nerve stimulation, and somatostatin.

Insulin insufficiency leads to several metabolic derangements that constitute diabetes mellitus. Hyperglycemia results from decreased glucose entry into tissues and increased gluconeogenesis and glucose release by the liver. The plasma glucose concentration can exceed the ability of the kidney's proximal tubules to reabsorb all the filtered glucose and results in an osmotic diuresis. This may ultimately lead to polyuria with urinary loss of sodium and potassium, hypovolemic hypotension, dehydration, and polydipsia. In muscle tissue, decreased insulin causes increased protein catabolism, leading to muscle wasting. Amino acids are released into the bloodstream from the breakdown of protein and become substrates for hepatic gluconeogenesis. The lack of glucose entry into cells produces an intracellular glucose deficiency. Cells in the hypothalamic ventromedial nuclei interpret this as if it were a fasting state and respond by an increased appetite (polyphagia). Adipose cells respond to the lack of circulating insulin with increased fat catabolism, releasing free fatty acids into the bloodstream. In the liver, the free fatty acids are converted to acetyl-CoA which enters the tricarboxylic acid cycle. Excess acetyl-CoA is converted to ketone bodies (acetone, acetoacetate, β-hydroxybutyrate) resulting in ketoacidosis.

Insulin excess leads to hypoglycemia, the effects of which are most pronounced in the CNS (confusion, convulsions, coma). The body compensates by releasing glucagon and epinephrine, both of which raise the blood glucose concentration. Glucocorticoids and growth hormone also elevate the blood glucose concentration, but their role is of less importance.

GLUCAGON. Glucagon is a polypeptide hormone secreted by the alpha cells of the pancreatic islets. Its actions are catabolic, generally opposing insulin's actions. It activates glycogen phosphorylase via cAMP to increase glycogenolysis and it increases gluconeogenesis. Glucagon increases proteolysis and the flow of amino acids from muscle to liver for gluconeogenesis. It increases lipolysis and ketogenesis. Stimulators of glucagon secretion include amino acids (following a protein meal), beta-adrenergic stimulation, fasting hypoglycemia, exercise, cholecystokinin, gastrin, and cortisol. Inhibitors include glucose, somatostatin, free fatty acids, ketones, and insulin.

SOMATOSTATIN AND PANCREATIC POLYPEPTIDE. Pancreatic islet delta cells secrete somatostatin. Somatostatin inhibits secretion of the other three pancreatic islet hormones and its role may only be one of local regulation. It is the same somatostatin as that released by the hypothalamus to inhibit growth hormone secretion.

QUESTIONS

5. What substrates can the brain use for fuel?
6. Which of the following substrates can be used to make new glucose?
   A. Amino acids
   B. Free fatty acids
   C. Glycerol
   D. Lactate
7. The following are increased, decreased, or unchanged by insulin:
   A. Glucose entry into brain
   B. Liver gluconeogenesis
   C. Liver glycolysis

D. Liver lipogenesis
E. Glycerol synthesis in adipose tissue
F. Amino acid uptake in skeletal muscle

8. In which of the following tissues does insulin act directly on cell membranes to facilitate glucose entry into the cell?
   A. Kidney tubule
   B. Red blood cell
   C. Intestinal mucosa
   D. Liver
9. The following stimulate, inhibit, or have no effect upon insulin secretion:
   A. Catecholamines
   B. Acetylcholine
   C. Increased blood glucose
   D. Somatostatin
   E. ACTH
   F. Glucagon
   G. GIP
10. Which hormones increase the blood glucose?
11. List two stimulators and two inhibitors of glucagon secretion?
12. Why is amino acid stimulation of glucagon secretion useful?
13. Compare the metabolic state of a patient in diabetic ketoacidosis with that of a person who has not eaten for several days.

REGULATION OF CALCIUM METABOLISM: PTH, CALCITONIN AND VITAMIN D

CALCIUM. The adult human body contains about 1 to 2 kg of calcium, of which over 98 percent is contained in the skeleton and 0.03 percent in plasma. The total plasma calcium concentration is normally about 10 mg per deciliter, half of which is protein bound (mainly with albumin), and half exists as free ions. Plasma calcium is in equilibrium with bone calcium, but only 0.5 percent of bone calcium is readily exchangeable, and the remainder is only slowly exchangeable. The plasma concentration of free calcium is critical in maintaining neuormuscular and other cellular functions. It is under tight hormonal control and the body chooses to sacrifice mineral calcium stores in order to maintain serum calcium levels. Calcium enters the plasma via intestinal absorption and bone resorption and is removed from extracellular fluid by urinary excretion, secretion into the gastrointestinal tract, and deposition in bone. The intestinal absorption of calcium is regulated by a vitamin D metabolite, its urinary excretion by parathyroid hormone (PTH) and calcitonin, and its bone resorption by PTH and calcitonin. Short-term control of serum calcium is achieved by hormonal effects on bone, while long-term control by hormonal effects on the intestinal tract and kidney.

PHOSPHORUS. The average adult body contains about 1 kg of phosphorus, 85 percent of which is in the skeleton and 15 percent in muscle and other tissues. Fasting plasma phosphorus concentrations range from 3 to 4 mg per deciliter, about 12 percent of which is protein bound. Unlike calcium, serum phosphate levels can vary widely, because phosphate not only moves between extracellular fluid (ECF) and bone, but also between ECF and intracellular fluid (ICF) (because it is involved in almost all intracellular metabolic processes). Phosphate enters the ECF via the intestinal tract, ICF and bone; it leaves the ECF via urine and movement to ICF or bone. Phosphorus absorption from the intestine is very efficient, from 70 to 90 percent. The major site of control of the body's phosphorus is the kidney, in which urinary excretion parallels dietary intake. Parathyroid hormone modifies urinary excretion of phosphate.

PARATHYROID HORMONE. Parathyroid hormone (PTH) is produced by the chief cells of the parathyroids. This peptide hormone is first synthesized as prepro-PTH, processed by cleavage to pro-PTH, and further cleaved to PTH, the active form. PTH has the following actions: (1) increases bone resorption to mobilize calcium, (2) increases reabsorption of calcium in the kidney's distal tubules, (3) decreases reabsorption of phosphate in the renal tubules, and (4) increases the production of 1,25-dihydroxycholecalciferol, the vitamin D metabolite which enhances intestinal calcium absorption. The net result is increased serum calcium, decreased serum phosphate, and increased urine phosphate. The major regulator of PTH secretion is serum calcium. Increased serum calcium depresses PTH secretion, while decreased serum calcium elevates PTH secretion. Primary hyperparathyroidism may be defined as the inappropriate hypersecretion of PTH resulting in hypercalcemia, and secondary hyperparathyroidism as the hypersecretion of PTH in response to a hypocalcemic stress. Hyperparathyroidism gives rise to hypercalcemia, hypercalciuria (the kidney cannot reabsorb the entire increased calcium load even in the presence of elevated PTH), hypophosphatemia, hyperphosphaturia, and bone demineralization. If all parathyroid glands are inadvertently removed during thyroid surgery, serum calcium declines and neuromuscular hyperexcitability develops. Unless the patient is given supplemental calcium, a fatal hypocalcemic tetany may occur.

CALCITONIN. Calcitonin is produced in the parafollicular (also called clear cells or C cells) of the thyroid gland. It is a peptide hormone whose principal pharmacologic actions include inhibition of bone resorption and increased urinary excretion of calcium and phosphate. The result is a lowered serum calcium and phosphate. Hormonal secretion is controlled by serum calcium, i.e., increased serum calcium increases secretion and vice versa. The exact physiologic role of calcitonin is not entirely clear, because neither an excess of hormone (e.g., medullary carcinoma of the thyroid) nor a deficiency (e.g., thyroidectomy) leads to any defects in bone or calcium metabolism.

VITAMIN D. Vitamin D_3, or cholecalciferol, is produced in the skin from 7-dehydrocholesterol after exposure to ultraviolet light (Fig. 2-2). Vitamin D_2, or ergocalciferol (Fig. 2-3), is derived from plants and used to supplement dairy products. This dietary supplement is not required for humans adequately exposed to sunlight. Vitamin D_2, as with other fat-soluble vitamins, does require the fat-absorbing mechanisms of the intestine to be intact. Vitamins D_2 and D_3 are biologically equally potent. Vitamins D_2 and D_3 are transported to the liver, where they are hydroxylated to produce 25-hydroxy-D_2 and 25-hydroxy-D_3, respectively. The 25-hydroxy D_3 is the most abundant form of vitamin D in the circulation and the major storage form of vitamin D. It is further hydroxylated in the kidneys to 1,25-dihydroxy-D_3 (Fig. 2-2), the active form of vitamin D. The major action of 1,25-dihydroxy-D_3 is to increase the absorption of dietary calcium and phosphate from the intestinal tract. It initially binds to a specific cytoplasmic receptor molecule. The resulting complex is translocated to the nucleus and causes an increased synthesis of an intestinal calcium-binding protein.

The production of 1,25-dihydroxy-D_3 in the kidney is regulated by PTH, serum phosphate, and its own level. An increase in PTH or a decrease in serum phosphate stimulates 1,25-dihydroxy-D_3 production. The hormone is also regulated by its own feedback inhibition. When PTH is low, serum phosphate is high, or 1,25-dihydroxy-D_3 is high, the kidney hydroxylates another position and produces the inactive metabolite 24,25-dihydroxy-D_3 instead.

Vitamin D deficiency may be caused by either inadequate sunlight exposure or fat malabsorption. Decreased vitamin D causes decreased absorption of calcium

Fig. 2-2. *Vitamin D synthesis.*

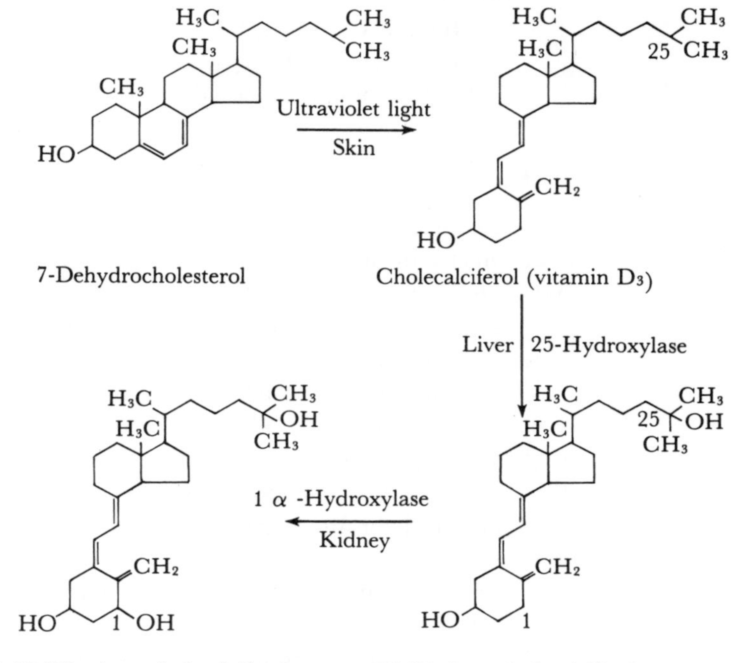

7-Dehydrocholesterol

Ultraviolet light
Skin

Cholecalciferol (vitamin D_3)

Liver
25-Hydroxylase

25-Hydroxycholecalciferol

1 α -Hydroxylase
Kidney

1,25-Dihydroxycholecalciferol

Fig. 2-3. *Vitamin D_2 (ergocalciferol).*

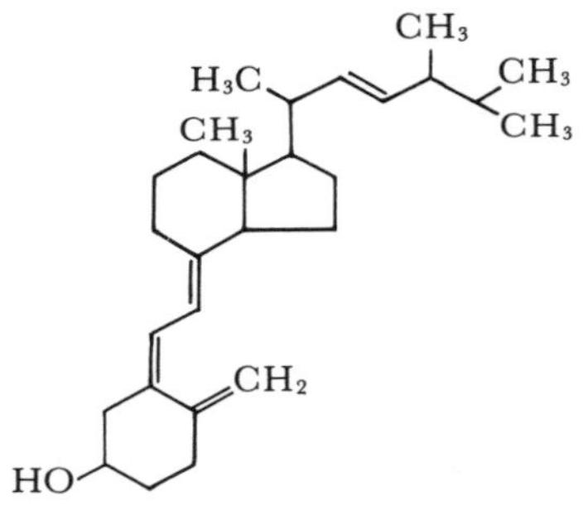

Vitamin D_2 (ergocalciferol)

and phosphate, decreased serum calcium and phosphate, increased PTH, and increased bone resorption. Bones are not adequately calcified and in adults osteomalacia results. In children, vitamin D deficiency causes rickets, in which bone fails to mineralize, epiphyses fail to fuse, epiphyseal plates widen, and fractures occur.

QUESTIONS

14. Which of the following is not an action of PTH?
    A. Increases production of 25-hydroxycholecalciferol
    B. Increases the reabsorption of phosphate in the renal tubules
    C. Increases bone resorption to mobilize calcium
    D. Increases the reabsorption of calcium in the kidneys
15. Which of the following directly regulates PTH secretion?
    A. Phosphate
    B. Calcium
    C. Calcitonin
    D. 1,25-dihydroxycholecalciferol

16. Which of the following decreases 1,25-dihydroxy-D_3 production in the kidney?
    A. High serum phosphate
    B. Low serum PTH
    C. High serum 1,25-dihydroxy-D_3

HYPOTHALAMIC PITUITARY RELATIONSHIPS

Neurons in the supraoptic and paraventricular nuclei of the hypothalamus send axons to the posterior pituitary (neurohypophysis). The portal hypophyseal vessels form a vascular connection between the median eminence of the ventral hypothalamus and the anterior pituitary.

The hormones oxytocin and vasopressin (antidiuretic hormone [ADH]) are synthesized in the cell bodies of neurons in the supraoptic and paraventricular nuclei, transported down their axons to the posterior pituitary, and released into the bloodstream. The ADH increases water reabsorption in the kidney to conserve fluid volume and oxytocin stimulates milk secretion from the mammary glands.

The median eminence of the hypothalamus secretes the hypothalamic releasing and inhibiting hormones: growth hormone–releasing hormone (GRH), growth hormone–inhibiting hormone (GIH, also called somatostatin), corticotropin-releasing hormone (CRH), thyrotropin-releasing hormone (TRH), gonadotropin-releasing hormone (GnRH), prolactin-releasing factor (PRF), and prolactin-inhibiting hormone (PIH). These hypophysiotropic hormones are polypeptides which regulate the secretion of six anterior pituitary hormones: growth hormone, adrenocorticotropic hormone (ACTH), thyroid-stimulating hormone (TSH), follicle-stimulating hormone (FSH), luteinizing hormone (LH), and prolactin. The GnRH stimulates the secretion of both FSH and LH.

Growth hormone promotes body growth via stimulation of somatomedin secretion by the liver. The ACTH stimulates corticosteroid secretion and TSH stimulates thyroid secretion. In the female, FSH stimulates early growth of the ovarian follicle and in the male spermatogenesis. The LH stimulates ovulation and luteinization of the ovarian follicle in the female and testosterone secretion in the male. Prolactin stimulates lactation. These hormonal relationships are summarized in Figure 2-4 and each hormone will be discussed in further detail later in this chapter.

The ACTH comes from a larger precursor protein, called pro-opiomelanocortin (POMC), which is synthesized in the hypothalamus and anterior pituitary where it is cleaved into ACTH, beta lipotropin (beta-LPH) and gamma melanocyte-stimulating hormone (gamma-MSH). A small amount of beta-LPH is cleaved into gamma-LPH and beta endorphin. All these substances are secreted from the anterior pituitary. The physiologic function of only ACTH is known. Beta endorphin is an endogenous opiate that binds to opioid receptors. In pharmacologic amounts, gamma-MSH stimulates melanin synthesis in melanocytes. The function of gamma-LPH is not known.

The secretion of hypothalamic and pituitary hormones is controlled by feedback mechanisms to preserve homeostasis. For example, high plasma levels of cortisol exert a negative feedback on CRH and ACTH secretion. Superimposed on the feedback control are chronobiologic and emotional factors. The ACTH displays a circadian rhythm in that pulses of secretion are more frequent in the early morning hours and less frequent in the evening. Release of growth hormone is also strongly associated with sleep. The circadian clock responsible

Fig. 2-4. *Hypothalamic-pituitary relationships.*

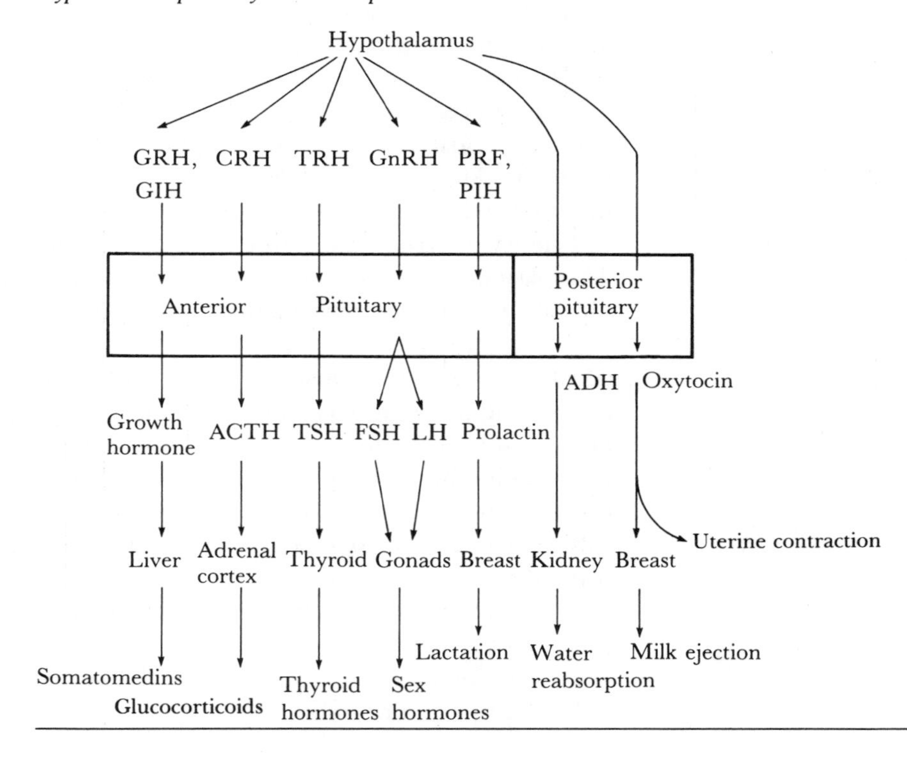

for these diurnal rhythms appears to be located in the suprachiasmatic nuclei of the hypothalamus. With long-term stress, GnRH activity is depressed (e.g., "boarding-school amenorrhea" in college students who move away from home). Emotional stress, fear, or anxiety cause an increase in CRH and ACTH. These emotional signals are probably relayed to the median eminence of the hypothalamus via the limbic system.

POSTERIOR PITUITARY HORMONES: VASOPRESSIN AND OXYTOCIN

VASOPRESSIN (ANTIDIURETIC HORMONE [ADH]). A vasopressin polypeptide precursor, called prepropressophysin, is synthesized in the cell bodies of the hypothalamus. It is cleaved into vasopressin, neurophysin, and glycopeptide during transport down the axon to the posterior pituitary. The actions of the latter two peptide fragments are not known.

The ADH increases the permeability of the collecting ducts in the kidneys and, thereby, increases reabsorption of solute-free water. This results in an increased urine osmolality, decreased urine volume, decreased plasma osmolality, and increased extracellular fluid volume. In pharmacologic doses, ADH raises arterial blood pressure by vasoconstriction, hence its other name vasopressin.

The ADH is regulated by plasma osmolality. Osmoreceptor cells located in the anterior hypothalamus near the supraoptic and paraventricular nuclei are activated when plasma osmolality rises above 280 mosm per kilogram. A signal is relayed to the adjacent nuclei to synthesize and release ADH. A hyperosmolar plasma stimulates separate osmoreceptors in the hypothalamus resulting in a sensation of thirst. Decreases in ECF volume stimulates ADH secretion via stretch receptors in the great veins, atria, and pulmonary vessels. Other stimulators of ADH secretion include angiotensin II, nicotine, nausea, pain, and some emotions. Secretion of ADH is depressed with decreased plasma osmolality, increased ECF volume, and alcohol.

In the syndrome of "inappropriate" hypersecretion of ADH (SIADH), ADH is autonomously released from the posterior pituitary (CNS disorders, drug-induced) or from tumor tissue (e.g., oat-cell carcinoma of lung). The excess ADH causes water retention which results in hyponatremia. Urine is inappropriately concentrated and the plasma is hypoosmolar.

Diabetes insipidus is characterized by ADH deficiency. It can be caused by either a lesion anywhere along that part of the hypothalamic-pituitary axis involved with ADH production, or an inability of the kidney to respond to ADH. Deficiency of ADH results in excretion of large amounts of hypoosmotic urine. The plasma becomes hyperosmotic and stimulates the thirst mechanism (polydipsia). It is the water intake that keeps these patients from severe dehydration.

OXYTOCIN. Oxytocin causes contraction of the uterus during labor and contraction of the myoepithelial cells of the lactating breast. In late pregnancy, both uterine sensitivity to oxytocin and oxytocin secretion is increased. Complex neuroendocrine reflexes take place during labor, involving activation of afferent fibers to the oxytocin-containing neurons in the hypothalamus so that there is sufficient release of oxytocin to maintain uterine contractions until delivery is completed.

The milk ejection reflex includes the stimulation of touch receptors in the breast by infant sucking, activation of afferent fibers to the supraoptic (SO) and paraventricular (PV) nuclei, release of oxytocin, contraction of myoepithelial cells, and ejection of milk.

QUESTIONS

17. The following are increased, decreased, or not changed with ADH administration in pharmacologic amounts
    A. Urine osmolality
    B. Extracellular fluid volume
    C. Urine volume
    D. Plasma osmolality
    E. Serum sodium concentration
    F. Thirst
    G. Blood pressure
18. In which of the following situations would ADH secretion be increased?
    A. Continuous intravenous infusion of a hypertonic (3%) NaCl solution
    B. A patient who acutely lost four units of blood from a bleeding duodenal ulcer
    C. A plasma osmolality of 300 mosm/kg
    D. A psychotic who drinks 12 liters of water a day

GROWTH HORMONE

Growth hormone (GH, somatotropin) is a peptide hormone synthesized and secreted by the anterior pituitary. Its major action is to promote linear growth by its effect on the skeletal system. This action, however, is mediated by the induction of somatomedins, a group of small peptides produced in the liver which have insulin-like and mitogenic activity. Somatomedins stimulate the proliferation of chondrocytes by increasing the synthesis of DNA, RNA, protein, and hydroxyproline. As a result, the proliferating cartilage widens the epiphyseal plates, allowing new bone to be laid down, and thereby causes linear growth. The somatomedins are regulated by GH and transported in the blood by carrier proteins.

Growth hormone also has many metabolic actions, the most important of which include increased protein synthesis, a positive nitrogen balance, increased lipolysis, insulin-resistance (diabetogenic), increased intestinal absorption of calcium, and increased renal reabsorption of phosphate. Growth and development require a net anabolic effect on protein metabolism. Lipolysis releases free fatty acids which can be used by the body so that protein is spared from catabolism. The anti-insulin action of GH on carbohydrate metabolism includes increased hepatic glucose output and decreased glucose uptake in some tissues. The effects on calcium and phosphate increase their availability for skeletal growth. It is unclear whether these metabolic actions are exerted through GH directly or through somatomedins.

The secretion of GH is regulated by several factors. Secretion is increased by the hypothalamic GH-releasing hormone (GRH), sleep, physical or emotional stress, exercise, hypoglycemia, increase in blood amino acids following a protein meal, and dopaminergic agonists. Factors which decrease GH secretion include the hypothalamic GH-inhibiting hormone (GIH, somatostatin), somatomedins, obesity, hyperglycemia, elevated free fatty acids, GH, and cortisol.

Growth hormone deficiency in a child causes retardation of growth, of epiphyseal development, and of bone age. Excess growth hormone (e.g., in the case of anterior pituitary tumor) in children causes gigantism with massive skeletal growth. In adults, whose epiphyseal plates have fused, excess GH causes acromegaly, characterized by enlarged acral parts and enlarged visceral organs.

QUESTIONS

19. Which of the following statements about GH is true?
    A. Its effect on blood sugar is opposite that of insulin
    B. It acts directly on bone to promote linear growth
    C. The intestinal absorption of calcium is inhibited by GH
    D. A short (5′ 4″), 25-year-old man who takes exogenous GH has a good chance of becoming tall enough to play for the Boston Celtics
20. The following stimulate or inhibit GH secretion:
    A. Sleep
    B. Somatostatin
    C. Hyperglycemia
    D. Stress
    E. GRH
    F. Hypoglycemia
    G. Exercise

THE ADRENAL GLAND

ADRENAL MEDULLA. The adrenal medulla is the inner part of the adrenal gland and secretes the catecholamines epinephrine, norepinephrine, and dopamine. Because they respond to stimulation by preganglionic sympathetic nerve fibers (splanchnic nerve), endocrine cells of the medulla can be considered as postganglionic sympathetic neurons with missing axons. Unlike other postganglionic neurons, the medulla cells can convert norepinephrine to epinephrine. The biosynthesis of catecholamines is shown in Figure 2-5. The conversion of norepinephrine to epinephrine is catalyzed by phenylethanolamine-N-methyltransferase (PNMT). This enzyme is induced by glucocorticoids from the adrenal cortex carried in the blood by a direct vascular connection between the cortex and medulla. The medulla secretes about 80 percent epinephrine and 20 percent norepinephrine.

The catecholamines are stored in chromaffin granules bound to ATP and protein. Release of the granules is mediated by the release of acetylcholine from preganglionic fibers. The acetylcholine increases the cell membrane's permeability to calcium, and calcium moves in to trigger exocytosis of these storage

Fig. 2-5. *Catecholamine synthesis.*

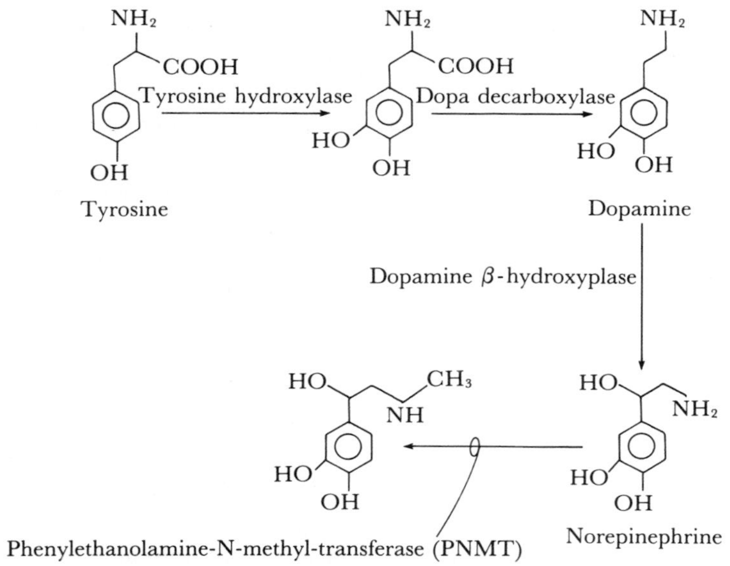

Tyrosine → (Tyrosine hydroxylase) → Dopa → (Dopa decarboxylase) → Dopamine → (Dopamine β-hydroxyplase) → Norepinephrine → (Phenylethanolamine-N-methyl-transferase (PNMT)) → Epinephrine

vesicles. Also released are dopamine, ATP, dopamine beta hydroxylase, and other proteins. The half-life of catecholamines in the circulation is about 2 minutes before they are metabolized and excreted.

The actions of catecholamines are on the visceral organs and mediated by two classes of receptors, alpha- and beta-adrenergic receptors, which are further divided into alpha-1, alpha-2, beta-1, and beta-2 receptors. The effects of circulating catecholamines on these receptors are identical to the effects of catecholamines released from noradrenergic nerve terminals of sympathetic postganglionic axons (see Chap. 8) and are shown in Table 2-1. Epinephrine and norepinephrine act on beta-1 receptors to increase the force and rate of myocardial contraction. They also increase myocardial excitability. Alpha-1-

Table 2-1. *Adrenergic Responses of Selected Tissues*

Organ	Receptor	Effect
Heart	Beta-1	Increased inotropy Increased chronotropy
Blood vessels	Alpha	Vasoconstriction
	Beta-2	Vasodilatation
Kidney	Beta	Increased renin release
Gut	Alpha, beta	Decreased motility Increased sphincter tone
Pancreas	Alpha	Decreased insulin release Increased glucagon release
	Beta	Increased insulin and glucagon release
Liver	Alpha, beta	Increased glycogenolysis
Adipose	Beta	Increased lipolysis
Skin	Alpha	Increased sweating
Bronchioles	Beta-2	Bronchodilation
Uterus	Alpha, beta	Contraction, relaxation

Modified from N. Weiner, and P. Taylor. Neurohumoral Transmission: The Autonomic and Somatic Motor Nervous Systems, In A. G. Gilman, L. S. Goodman, T. W. Rall and F. Murad (eds.). *Goodman and Gilman's The Pharmacological Basis of Therapeutics* (7th ed.). New York: Macmillan, 1985. (Copyright © 1985 by Macmillan Publishing Company.)

receptor-mediated vasoconstriction is produced by both catecholamines. Epinephrine causes a beta-2-receptor-mediated vasodilatation in skeletal muscle and liver and a net decrease in total peripheral resistance (TPR) results. Norepinephrine produces only alpha-1 vasoconstriction which increases TPR and elevates both systolic and diastolic blood pressures. The rise in blood pressure produces a reflex bradycardia which overrides the chronotropic effect of norepinephrine. In summary, epinephrine decreases TPR, increases heart rate, increases cardiac output, and does not change the mean blood pressure. Norepinephrine increases TPR and increases the mean blood pressure, resulting in a reflex decrease in heart rate and cardiac output.

The metabolic effects of the catecholamines include increased glycogenolysis in liver and muscle producing a rise in blood glucose, increased lipolysis in adipose tissue and release of free fatty acids, and an increased metabolic rate (i.e., increased oxygen consumption and heat production). The adrenal medullary hormones also play a role in stimulating the CNS. They enhance alertness and arousal by lowering the threshold of neurons in the reticular formation in the brainstem. The physiological function of dopamine is not known.

Adrenal medullary secretion is regulated by the sympathetic nervous system. Situations which evoke a "flight or fight" response include fear, anxiety, pain, trauma, hemorrhage, exercise, extreme cold or heat, hypoglycemia, and hypotension. The increase in medullary secretion is part of the diffuse sympathetic discharge which prepares the individual for emergency situations. The body responds with an increase in heart rate and cardiac output, increased blood flow to skeletal muscles, decreased blood flow to the viscera, mobilization of glucose and free fatty acids to provide a quick charge of energy, and enhanced CNS arousal. The role of circulating catecholamines in mediating vascular changes is less important than that of sympathetic innervation of the vessels. The metabolic actions of circulating catecholamines are more significant during such emergency situations. The adrenal medulla, however, is not essential for life.

Pheochromocytomas are rare tumors, arising from chromaffin cells in the sympathetic nervous system, which secrete catecholamines.

ADRENAL CORTEX. The adrenal cortex is the outer part of the adrenal gland. It secretes glucocorticoids, which have widespread effects on carbohydrate and protein metabolism, and mineralocorticoids, which regulate the body's sodium, potassium, and fluid volume. It also secretes small amounts of androgens, which normally play a less important role than the androgens secreted by the gonads. The adrenal cortex comprises three zones: the zona glomerulosa (outermost), the zona fasciculata, and the zona reticularis (innermost). The zona glomerulosa produces aldosterone and the other two zones produce cortisol and androgens.

The C19 steroids contain 19 carbon atoms with a keto or hydroxyl group at position 17 and possess androgenic activity. The C21 steroids contain 21 carbons and a two-carbon side chain at position 17. Steroids that have glucocorticoid or mineralocorticoid activity contain 21 carbons. The structure of a steroid is shown in Figure 2-6. The synthesis of adrenal cortical steroids (Fig. 2-7) begins with cholesterol, which is converted to pregnenolone. This rate-limiting step is mediated by a cholesterol hydroxylase and desmolase. The ACTH via the adenylate cyclase–cAMP system stimulates this conversion. The ACTH also increases the uptake of lipoprotein (the major source of adrenal cholesterol) by the adrenal cortex and the hydrolysis of stored cholesterol esters to free cholesterol. Many of the intermediates of steroid synthesis are secreted to some extent, but the steroids that are secreted in physiologically significant amounts include aldosterone, cortisol, corticosterone, dehydroepiandrosterone (DHEA), and androstenedione.

Fig. 2-6. *Steroid structure.*

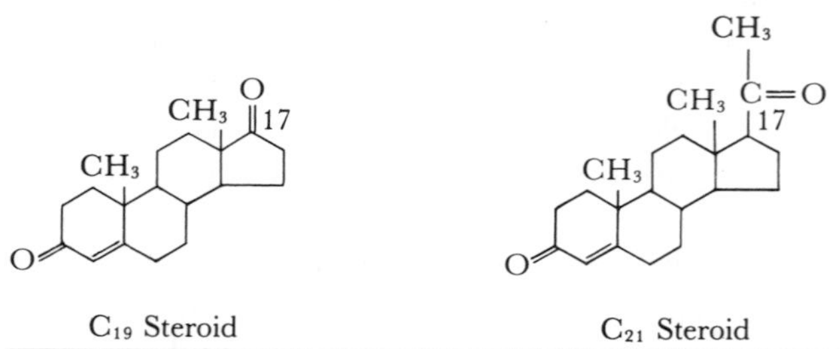

CH_3 O 17 CH_3 O

CH_3 C=O 17 CH_3 CH_3 O

C_{19} Steroid

C_{21} Steroid

Fig. 2-7. *Steroid synthesis.*

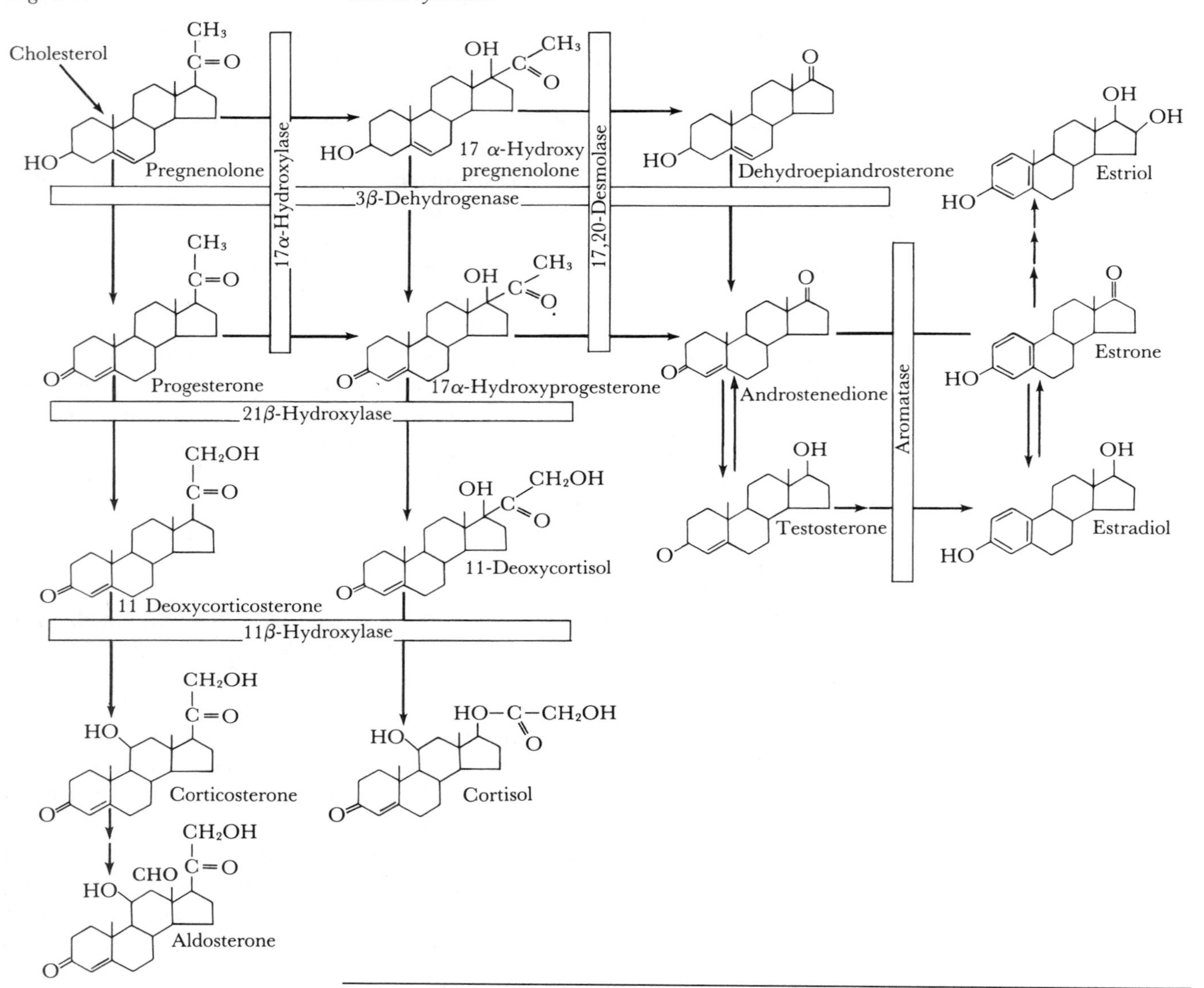

Glucocorticoids. About 10 percent of circulating cortisol is free, 75 percent is bound to corticosteroid-binding globulin (CBG, or transcortin), and 15 percent is bound to albumin. Only the free form is biologically active. The CBG has high-affinity binding and low-capacity binding. The opposite is true of albumin. Estrogen increases CBG synthesis in the liver while protein deficiency states such as cirrhosis or nephrosis decrease CBG production. The hypothalamic-pituitary-adrenal axis is regulated by the free cortisol level, not the total cortisol level.

Cortisol is largely metabolized and conjugated in the liver before excretion in the urine.

The physiological effects of glucocorticoids are as follows:

1. Intermediary metabolism
   Increased hepatic glycogenesis
   Increased hepatic gluconeogenesis
   Increased blood glucose
   Increased protein catabolism
   Increased plasma amino acids
   Increased lipolysis
   Increased plasma lipids and ketone bodies
2. Feedback inhibition of ACTH secretion
3. Cardiovascular
   Maintenance of vascular sensitivity to catecholamines' vasoconstricting effects
4. Growth and development
   Increased pulmonary surfactant production
   Development of hepatic and gastrointestinal enzymes
5. Hematologic
   Increased circulating neutrophils
   Decreased circulating lymphocytes, eosinophils, and monocytes
   Inhibition of migration of white blood cells
   Lympholysis
   Decreased size of lymph nodes and thymus
6. Renal
   Increased ability to excrete free water
7. CNS
   Personality changes
8. Resistance to stress
   Adrenalectomized animals die when exposed to noxious stimuli

The effects of high (nonphysiologic) levels of glucocorticoids can be seen with chronic administration of exogenous steroids, glucocorticoid-producing adrenocortical tumors, and hypersecretion of ACTH. The result is Cushing's syndrome, which consists of the following clinical manifestations:

1. Connective tissue — thinning of skin, easy bruising, stria formation, and poor wound healing due to decreased collagen and inhibition of fibroblasts
2. Bone — inhibition of intestinal absorption of calcium, decreased serum calcium, increased PTH secretion, decreased bone formation, and increased bone resorption leading to osteoporosis
3. Muscle — muscle wasting, weakness, and fatigability
4. Fat — redistributed to trunk (truncal obesity), face (moon facies), and upper back (buffalo hump)
5. Skin — hirsutism and acne secondary to increased adrenal androgen secretion
6. Endocrine — impaired glucose tolerance, amenorrhea
7. Renal — salt and water retention and hypokalemia due to the mineralocorticoid activity of the excess glucocorticoids
8. Cardiovascular — hypertension
9. CNS — euphoria, irritability, emotional lability, and depression

The anti-inflammatory and anti-allergic effects of pharmacologic doses of steroids make them therapeutically useful. Glucocorticoids inhibit the conversion of membrane phospholipids to arachidonic acid. This reduces the formation of mediators of inflammation such as leukotrienes, prostaglandins, and prostacyclin. Steroids also prevent the release of histamine from mast cells.

The ACTH stimulates both the basal and the stress-induced secretion of glucocorticoids. As mentioned earlier, ACTH comes from a larger precursor protein, pro-opiomelanocortin (POMC), which is cleaved after secretion into ACTH, beta lipotropin, and gamma-melanocyte-stimulating hormone. The adrenals respond to an increase in ACTH with a rapid synthesis and secretion of steroids. Chronic stimulation with ACTH leads to adrenocortical hyperplasia and hypertrophy, while chronic depletion of ACTH results in adrenal atrophy and decreased adrenal responsiveness. Other actions of ACTH include stimulation of acute aldosterone secretion and melanocyte stimulation. The ACTH is released from the pituitary in response to hypothalamic corticotropin-releasing hormone (CRH). Three factors control ACTH secretion. The first is severe stress. Afferent signals relayed to the hypothalamus include emotions via the limbic system or traumatic injury via the spinothalamic pathways. Secondly, ACTH is secreted episodically as well as with a diurnal rhythm. It is released in irregular bursts throughout the day, but most frequently in the early morning and least frequently in the evening. The increased morning secretion occurs before waking up. (Perhaps the thought of waking up is traumatic.) The biologic clock driving the circadian rhythm is located in the suprachiasmatic nuclei of the hypothalamus. Lastly, ACTH secretion is controlled negatively by feedback inhibition by cortisol. The inhibition takes place mostly at the pituitary level.

The most common cause of Cushing's syndrome is iatrogenic. The most common noniatrogenic cause is Cushing's disease, which is the hypersecretion of ACTH from the pituitary. Nonpituitary (ectopic) ACTH hypersecretion and glucocorticoid-producing adrenal tumors are less common causes. Adrenal insufficiency is called Addison's disease. Primary Addisons's disease refers to adrenal disorders and secondary Addison's disease refers to hypothalamic or pituitary disorders. Clinical manifestations include weakness, fatigue, anorexia, weight loss, hyperpigmentation (only in primary Addison's caused by increased ACTH), hypotension, and hypoglycemia. In patients receiving long-term exogenous steroid therapy, not only does the adrenal gland become atrophied and less responsive to ACTH, but the pituitary also becomes suppressed and is unable to secrete normal amounts of ACTH. Thus the termination of steroid therapy should involve a careful tapering schedule to prevent a fatal Addisonian crisis.

Mineralocorticoids. Aldosterone is the major mineralocorticoid hormone. In the plasma, it is bound to protein only to a slight extent. Aldosterone is metabolized before excretion. The main action of aldosterone is increasing the renal reabsorption of sodium in the distal tubule and collecting duct in exchange for potassium and hydrogen ions. This leads to sodium retention, increased ECF volume, hypertension, and hypokalemic alkalosis. As with other steroids, aldosterone acts at the DNA level, stimulating synthesis of specific mRNAs, but the exact function of the aldosterone-induced protein is not known.

The renin-angiotensin system is the most important regulator of aldosterone secretion. Renin is a peptide hormone secreted from the juxtaglomerular (JG) cells of the kidney, specialized cells located in the renal afferent arterioles as they enter the glomerulus (see Chap. 5). It is an acid-protease whose action is to form angiotensin I by cleaving angiotensinogen. The secretion of renin is stimulated by the following factors: (1) decreased ECF volume (e.g., dehydration, diuretics, sodium depletion), (2) decreased systemic blood pressure (standing, hemorrhage) or, particularly, renal arterial pressure (constriction of renal artery), (3) increased sympathetic renal nerve discharge or circulating catecholamines, and (4) decreased delivery of chloride and sodium to the distal tubule sensed by the macula densa.

Angiotensinogen is a glycoprotein produced by the liver. Renin causes the liver to release angiotensin I, a decapeptide. Angiotensin I is then converted to the octapeptide angiotensin II (AII). Most of this conversion takes place as blood passes through the lungs and is mediated by converting enzyme located in endothelial cells. The AII acts directly on the adrenal cortex to stimulate aldosterone secretion. It is also the most potent vasoconstrictor known and causes arteriolar constriction in all vascular beds. Angiotensin II acts on the CNS to increase thirst and vasopressin secretion. It is rapidly metabolized by a group of enzymes called angiotensinases. One such enzyme, aminopeptidase, cleaves AII to form the heptapeptide angiotensin III (AIII), which is equally potent as AII in stimulating aldosterone secretion but only partially as potent in pressor activity.

Other stimulators of aldosterone secretion include ACTH acutely, increased plasma potassium, and decreased plasma sodium. The increase in aldosterone release from ACTH lasts only 1 to 2 days. The mechanism by which aldosterone secretion is regulated is at the level of biosynthesis. The ACTH, AII, and potassium stimulate the conversion of cholesterol to pregnenolone. Potassium and AII also stimulate the conversion of corticosterone to aldosterone.

Primary hyperaldosteronism (Conn's syndrome) is caused by aldosterone-secreting adrenal tumors or hyperplasias. The effects of chronic mineralocorticoid excess include increased ECF volume, hypertension, potassium depletion, and metabolic alkalosis. When the ECF expansion reaches a certain point, sodium retention is decreased in spite of continued aldosterone stimulation. This "aldosterone-escape" phenomenon can be seen after 2 or 3 days of continued mineralocorticoid stimulation. Urinary excretion of sodium may return to normal levels, so that patients with Conn's are not volume expanded to the point of edema. No "escape" is seen with potassium excretion. Secondary hyperaldosteronism may be caused by congestive heart failure, cirrhosis with ascites, and nephrosis. In these edematous states, ECF volume is distributed to extravascular spaces. The kidneys interpret this as a decrease in "effective" intravascular volume and respond by releasing renin. This increases aldosterone which causes even further retention of sodium and fluid. This only makes matters worse because the extra fluid accumulates as edema fluid.

In hypoaldosteronism (e.g., adrenal insufficiency), sodium is lost in the urine and potassium is retained. Plasma volume decreases and the resulting hypotension may develop to the degree of circulatory collapse.

Sex Hormones. Normally, the sex hormones produced in the adrenal cortex have minor physiological significance. The gonadal production of sex steroids plays the major role and the physiological function of these hormones will be discussed in the next section. As mentioned earlier, dehydroepiandrosterone (DHEA) and androstenedione are the only sex steroids that are normally secreted in physiologically significant amounts from the adrenals. Both are weakly androgenic and exert their effects by their peripheral conversion to more potent androgens. Estrogen is secreted in physiologically insignificant amounts.

Adrenal sex steroids become important in patients with congenital enzyme deficiencies. For example, congenital deficiencies in either 21-beta-hydroxylase, or 11-beta-hydroxylase lead to a deficiency in cortisol secretion (see Fig. 2-7). Because feedback inhibition is removed, ACTH secretion is increased. The excess ACTH causes adrenal hyperplasia as well as continued steroid synthesis. Due to the block in a biosynthetic pathway, the steroid intermediates pile up behind the block and are diverted to the remaining open pathways — in this case androgen synthesis. In genetic females, the excess of androgens may cause varying degrees of virilization. Virilization, or masculinization, consists of hirsutism, deepened voice, acne, clitoral enlargement, and male pattern baldness. These females may also have ambiguous external genitalia. From Figure 2-7, one can predict that a deficiency in 17,20-desmolase or 17-alpha-hydroxylase produces a deficiency in androgen synthesis. As a consequence, genetic males develop female or ambiguous genitalia and genetic females develop female genitalia. In addition, these enzyme deficiencies result in sexual infantilism, in which prepubertal sexual characteristics persist into adult life. The glucocorticoid deficiency in some of the above enzyme deficiencies may be severe enough to cause a fatal Addisonian crisis. Consequences of either mineralocorticoid excess or deficiency may also occur depending on where the biosynthetic block is and if the intermediates which build up have mineralocorticoid activity.

QUESTIONS

21. The following are increased or decreased by catecholamine action:
    A. Myocardial contractility
    B. Blood glucose
    C. Bronchodilation
    D. Anal sphincter tone
    E. Sweating
    F. CNS alertness
22. The regulation of catecholamine secretion involves which of the following?
    A. ACTH
    B. The sympathetic nervous system
    C. The parasympathetic nervous system
    D. The somatic motor nervous system
    E. None of the above
23. Which adrenal cortical zone(s) produce
    A. Aldosterone
    B. Cortisol
24. How many carbons are in
    A. Aldosterone
    B. Cortisol
    C. Testosterone
25. Which of the following are true of cortisol action?
    A. Increased plasma glucose
    B. Increased plasma amino acids
    C. Increased plasma lipids
    D. Increased plasma ACTH
26. Which of the following statements is not true of glucocorticoids?
    A. Glucocorticoids inhibit arachidonic acid formation
    B. Circulating cortisol is bound mostly to CBG
    C. Glucocorticoids suppress the immune system
    D. Glucocorticoids are effective vasodilators
    E. Cortisol secretion is stimulated by ACTH
27. Which of the following is not a characteristic of Cushing's syndrome?
    A. Osteomalacia
    B. Hirsutism
    C. Moon facies
    D. Hypertension
    E. Stria formation
28. ACTH secretion is
    A. Episodic
    B. Increased while taking the National Medical Board Examination
    C. More frequent in the early morning than in the evening
    D. Inhibited by cortisol
29. Renin release from the juxtaglomerular (JG) cells is stimulated by

A. Severe dehydration
B. Increased delivery of chloride and sodium to the distal tubule
C. Stimulation of the sympathetic renal nerve
D. Constriction of the renal artery
E. Angiotensin II

30. Angiotensin II
    A. Is a decapeptide
    B. Stimulates aldosterone secretion
    C. Stimulates ADH secretion
    D. Is a vasoconstrictor
    E. Is metabolized to AIII by converting enzyme
31. Aldosterone
    A. Increases the permeability of the collecting ducts to enhance free water reabsorption
    B. Is a peptide hormone
    C. Has no glucocorticoid activity
    D. Is produced in the zona reticularis
    E. Can cause a hypokalemic alkalosis in hyperaldosteronism

---

REGULATION OF THE REPRODUCTIVE SYSTEM: TESTOSTERONE, ESTROGEN, PROGESTERONE, AND PROLACTIN

MALE REPRODUCTIVE SYSTEM. The testis is made up of two components, seminiferous tubules for spermatogenesis and Leydig cells for androgen production. The seminiferous tubules are composed of Sertoli cells and germinal cells. The Sertoli cells line the epithelial basement membrane and form tight junctions with other Sertoli cells to create a blood-testis barrier through which testosterone penetrates easily but not protein. The Sertoli cells surround developing germ cells with their extensive cytoplasmic processes and provide nourishment essential for germ-cell differentiation. Sertoli cells also phagocytose damaged germ cells and unused portions of germ cell cytoplasm during differentiation. Finally, Sertoli cells secrete two substances: (1) androgen-binding protein (ABP), which enters the tubular lumen and ensures a high concentration of testosterone to the germ cells; and (2) inhibin, which inhibits FSH secretion from the anterior pituitary. The secretion of both substances is stimulated by FSH.

In males, gametogenesis (the production of haploid spermatozoa) begins at puberty and persists until death. Under the influence of FSH and testosterone, the spermatogonia, or primitive germ cells, mature into primary spermatocytes. Through meiotic division, primary spermatocytes divide into secondary spermatocytes, and then into spermatids, which contain the haploid number of 23 chromosomes. The spermatids mature into spermatozoa (sperm), which are transported into the head of the epididymis; however, they are still infertile and immobile. The spermatozoa undergo further morphologic and functional changes during their transit through the epididymis so that they are both fertile and mobile at the tail of the epididymis. They next enter the vas deferens, a muscular duct which peristaltically moves the spermatozoa into the ejaculatory duct. The ejaculatory duct also receives fluid from the seminal vesicles, which contain fructose for nourishment to the spermatozoa and other substances such as ascorbic acid and prostaglandins. The sperm then enters the prostatic urethra, into which prostatic secretions are also emptied. The prostate secretes citric acid, acid phosphatase, spermine, and other substances. The major components of fluid that is ejaculated at orgasm, the semen, include sperm, seminal vesical secretions, and prostatic secretions. An average ejaculate volume is 2.5 to 3.5 ml and an average sperm concentration is 100 million sperm per milliliter of semen. A concentration of less than 20 million sperm per milliliter usually results in sterility.

The male sexual response can be divided into five events: libido, erection, ejaculation, orgasm, and resolution. Libido, or sexual desire is produced by psychic factors and testosterone. Penile erection occurs as a result of dilatation of arteriolar vessels that supply the corpora cavernosa and spongiosa, engorgement of these erectile tissues with blood, and passive venous congestion. This vascular phenomenon is an involuntary reflex that can be mediated by either of the following mechanisms. Psychogenic stimuli are transmitted to the limbic system, which, in turn, stimulates the parasympathetic innervation to the penile vessels. The second mechanism involves a spinal reflex arc where direct genital stimuli are transmitted by the pudendal nerves (afferent limb) and activate sacral parasympathetic nerves (efferent limb). Ejaculation is mediated by the sympathetic nervous system and consists of two processes: the emission of semen into the urethra, and the true ejaculation of semen out of the urethra. Orgasm is a psychic phenomenon in which the rhythmic contraction of the bulbocavernosus and ischiocavernosus muscles is perceived as pleasurable. During the resolution phase, blood rapidly leaves the erectile tissue and the penis becomes flaccid.

Testosterone is the major hormone secreted by the testis, and it is synthesized from cholesterol in the Leydig cells (Fig. 2-7). In plasma, only 2 percent of testosterone exists in the free form and the remainder is protein bound, mainly to sex hormone-binding globulin (SHBG) and albumin. The actions of testosterone include: (1) the differentiation of internal and external genitalia in males during fetal development; (2) the enlargement of the penis, the scrotum, and the testis during puberty (primary sex characteristics); (3) the development and maintenance of male secondary sex characteristics (hair growth on face, chest, and pubis, male-pattern baldness, deeper voice, increased sebaceous gland secretion, increased musculature, broad shoulders, narrow pelvic outlet); (4) anabolic effects on protein metabolism, i.e., increased synthesis and decreased breakdown of protein causing an increase in the rate of growth (growth spurt at puberty); (5) fusion of epiphyses to long bones to stop growth; (6) maintenance of gametogenesis (along with FSH); (7) inhibitory feedback effect on pituitary LH secretion; and (8) libido. In most target cells, testosterone is converted to the more potent dihydrotesterone, which binds to a cytoplasmic protein and the resulting complex is translocated into the nucleus.

Testosterone secretion is stimulated by LH, which is also trophic to the Leydig cells. Testosterone feedback inhibits the secretion of LH both at the pituitary level and at the hypothalamic level. The FSH stimulates Sertoli cells to secrete ABP and inhibin. The latter feedback inhibits FSH secretion at the pituitary level.

FEMALE REPRODUCTIVE SYSTEM. The female reproductive system functions to accommodate both fertilization and pregnancy. It undergoes periodic changes known as the menstrual cycle (Fig. 2-8). Menarche, the first menstrual cycle, occurs at a mean age of 13 years. The cessation of menstrual cycles is called menopause, which occurs between the ages of 45 and 55 years. The length of each cycle is 28 days but can be quite variable even in the same woman. By convention, the first day of vaginal bleeding, or menstruation, is counted as the first day of the cycle. Each cycle results from the interaction of the hypothalamus, pituitary, and ovaries.

Unlike male gametogenesis, no new ova are formed in the female gonads (ovaries) after birth. At birth, the ova are arrested at prophase of the first meiotic division and are called primary oocytes. During each menstrual cycle, only one of these ova is stimulated to mature. Just before ovulation, the first meiotic division is completed to form a secondary oocyte and the first polar body. The secondary oocyte then begins the second meiotic division up to the metaphase

Fig. 2-8. *Hormone concentrations during the menstrual cycle. (From I. H. Thorneycroft, et al. The relation of serum 17-hydroxyprogesterone and estradiol-7β levels during the human menstrual cycle.* Am. J. Obstet. Gynecol. *111:950, 1921.)*

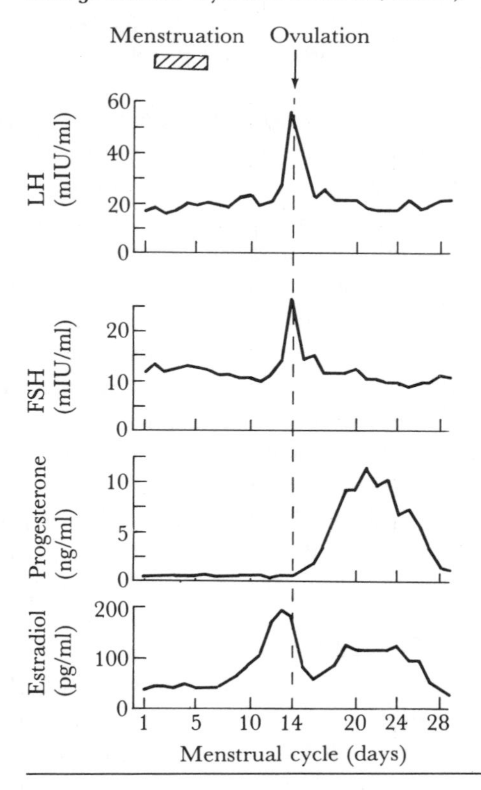

stage. The division is completed at fertilization and the second polar body extruded.

The menstrual cycle is divided into two phases, characterized by changes that take place in the ovaries. The follicular phase begins with the first day of menses and ends with ovulation. The rise in levels of FSH late in the preceeding menstrual cycle promotes the growth and recruitment of a group of primary follicles (each consisting of a primary oocyte and surrounding granulosa cells) in the ovary. By about day seven, one follicle becomes dominant, while the other recruited follicles begin to degenerate. In the dominant follicle there is an accelerated proliferation of granulosa cells and the development of a fluid-filled antrum. Follicle-stimulating hormone also stimulates the activity of aromatase enzymes in the granulosa cells converting androstenedione to estrogen. With further proliferation of granulosa cells, estrogen production increases, and FSH secretion is inhibited by the rising estrogen levels. Estrogen levels peak just prior to ovulation. Although low estrogen levels inhibit LH secretion (negative feedback), high estrogen levels stimulate LH secretion from the anterior pituitary (positive feedback). Thus an LH peak results (Fig. 2-8) which mediates the final maturation of the follicle and ovulation itself. The follicle is ruptured and the ovum extruded. The ovum is picked up by the fimbriated ends of the fallopian tubes (oviducts) and transported to the uterus. There is also a midcycle FSH peak, the significance of which is not clear.

The ruptured follicle becomes the corpus luteum. The luteal phase spans the interval from ovulation to the beginning of menses. The corpus luteum secretes both estrogen and progesterone, giving rise to an estrogen peak and a progesterone peak during the luteal phase. The LH stimulates progesterone and estrogen secretion from the corpus luteum. The FSH and LH levels then begin to decline, because estrogen suppresses the release of these gonadotropins. Because LH is required to maintain the corpus luteum, the decline in LH results in the regression of the corpus luteum unless conception occurs. The corpus luteum regresses into the corpus albicans. Estrogen and progesterone secretion falls and the loss of inhibition permits the pituitary to secrete FSH again to begin a new cycle.

During the menstrual cycle, changes take place in the reproductive tract as directed by estrogen and progesterone. After menses, estrogen stimulates glandular growth in the endometrium; therefore, this part of the cycle, from the menses to ovulation, is also called the proliferative phase. Estrogen also stimulates the endocervical glands to secrete a thin and alkaline mucous which promotes survival and transport of sperm. After ovulation, progesterone and estrogen secreted by the corpus luteum act on the endometrium to stimulate glandular secretion and curling of the spiral arterioles (blood supply to the endometrium). This secretory phase begins at ovulation and ends at the onset of menses. The endometrium is prepared for implantation during this phase. Progesterone causes the cervical mucous to be thick, tenacious, and cellular to prevent sperm transport. Progesterone also mediates a rise in basal body temperature of 0.3 to 0.5°C during the luteal phase. The fall in estrogen and progesterone levels late in the luteal phase causes vasospasm of the spiral arterioles. This results in tissue ischemia, desquamation and sloughing of the more superficial endometrial layer, and hemorrhage. Menses occurs on days one through five of the cycle.

The sexual response in the female begins with sexual arousal, which causes secretion of a lubricating fluid from the vestibular glands through the vaginal mucosa. Pelvic musculature under autonomic control begin to contract and results in a pleasurable sensation known as orgasm. Visual, auditory, olfactory, and tactile stimuli (especially of the clitoris) also play a role in the production of orgasm.

By their own motility, as well as by uterine and oviduct contractions, sperm reach the extruded ovum at about the midpoint of the oviduct. The fertilized egg (blastocyst) is transported to the uterus where it becomes implanted about 6 days after fertilization. A placenta eventually develops, which secretes human chorionic gonadotropin (HCG). This hormone structurally and functionally resembles LH and takes its place in supporting the corpus luteum during the luteal phase. The HCG is detectable 8 days after ovulation and reaches a maximum in 7 weeks. Its detection constitutes a positive pregnancy test. The HCG maintains the corpus luteum, which in turn secretes progesterone and estrogen to support the uterus and prevent endometrial sloughing. Thus a "missed period" is another indicator of pregnancy. The HCG supports the corpus luteum during the first trimester. During the subsequent two trimesters, the placenta is capable of producing its own estrogen and progesterone and the corpus luteum is no longer needed. While the ovaries produce mainly estradiol, the placental estrogen is largely estriol. Both fetus and placenta are required for the production of estriol. The fetal adrenal glands secrete dehydroepiandrosterone (DHAS), which is ultimately converted to estriol in the placenta.

Estrogen is secreted by the ovary, placenta, adrenal cortex, and testis. Under nonpregnant conditions, the ovary is the major source of estrogens. As described above, there are two peaks of estrogen secretion, during the late-follicular phase and during the midluteal phase. There is a dramatic decline in estrogen secretion during menopause and it remains low thereafter. The biosynthesis of estrogen is shown in Figure 2-7 and involves the aromatization of androgens. Estradiol is the major estrogen produced by the ovary. It is in equilibrium with estrone, which can be converted to estriol. Of the three, estradiol is the most potent, and estriol is the least potent. In the circulation, 2 percent of estradiol is free, 37 percent is bound to steroid hormone–binding globulin (SHBG), and 61 percent is bound albumin.

Estrogen's major actions include

1. Growth of the ovarian follicle
2. Endometrial proliferation and (with progesterone) endometrial glandular secretion
3. Enhanced excitability of uterine myometrium
4. Development of vagina, uterus, oviducts, and enlargement of external genitalia (mons pubis, labia majora, and labia minora)
5. Stromal development and ductal growth of breasts
6. Fat deposition in breasts, buttocks, and thighs
7. Libido
8. Hepatic synthesis of several hormone transport proteins
9. Feedback inhibition of FSH secretion

Progesterone is secreted by the corpus luteum, placenta, adrenal cortex, and testis. Its biosynthesis is shown in Figure 2-7. Its actions include

1. Stimulation of endometrial glandular secretions
2. Maintenance of pregnancy
3. Decreased excitability of the myometrium
4. Production of thick endocervical mucous
5. Development of breast lobules
6. Increased basal body temperature
7. Inhibition of gonadotropin secretion

The regulation of estrogen and progesterone secretion has already been described to some extent. To summarize, FSH and LH are required for ovarian follicle maturation, during which estrogen is secreted. In the luteal phase, LH maintains the corpus luteum, which secretes estrogen as well as progesterone. The secretion of FSH and LH by the pituitary is stimulated by gonadotropin-releasing hormone (GnRH), also called luteinizing hormone-releasing hormone (LHRH), which is released in a pulsatile fashion from the hypothalamus. A negative feedback control of the gonadotropin secretion occurs at lower estrogen levels, while a positive feedback control occurs at higher estrogen levels. The secretion of both gonadotropins is inhibited by estrogen and progesterone during the luteal phase. It is not known whether the steroid hormones exert their feedback control at the hypothalamic or pituitary level.

Prolactin is a polypeptide hormone produced in and released from the anterior pituitary. Its primary action is to stimulate lactation (milk secretion) in the postpartum period. An excess of prolactin disturbs the hypothalamic-pituitary control of gonadotropin secretion and ultimately leads to hypogonadism in both males and females. The secretion of prolactin is tonically inhibited by dopamine (the prolactin-inhibiting hormone) from the hypothalamus. Thyrotropin-releasing hormone (TRH) stimulates prolactin secretion. A yet unidentified prolactin-releasing factor (PRF) is also thought to be released by the hypothalamus. Physiologic factors which stimulate prolactin secretion include pregnancy, nursing, nipple stimulation, stress, and sleep.

QUESTIONS

32. Sertoli cells
    A. Secrete inhibin
    B. Form a blood-testis barrier

C. Secrete testosterone
D. Differentiate into primary spermatocytes
E. Secrete androgen-binding protein

33. Which of the following is the correct sequence in which sperm travel?
 A. Vas deferens, epididymis, ejaculatory duct, prostatic urethra
 B. Tail of epididymis, head of epididymis, ejaculatory duct, vas deferens
 C. Head of epididymis, tail of epididymis, vas deferens, ejaculatory duct
 D. Seminiferous tubules, tail of epididymis, ejaculatory duct, vas deferens, prostatic urethra
 E. None of the above
34. Semen contains all of the following except
 A. Sperm
 B. Fructose
 C. Ascorbic acid
 D. Citric acid
 E. Testosterone
 F. Acid phosphatase
35. Actions of testosterone include
 A. Maintenance of gametogenesis
 B. Differentiation of male genitalia
 C. Inhibitory feedback on LH secretion
 D. Male secondary sex characteristics
 E. All of the above
36. The menstrual cycle
 A. Begins after the first pregnancy and ends at menopause
 B. Has an average length of 9 months but can be variable
 C. Results from the interaction of hypothalamus, pituitary, and adrenal gland
 D. Continues through pregnancy
 E. None of the above
37. Which of the following is not true of female gametogenesis?
 A. The first meiotic division is completed just before ovulation
 B. New ova are continually formed after birth
 C. Only one mature ovum is produced during each menstrual cycle
 D. The second meiotic division is completed at fertilization
 E. At birth, all ova are arrested at prophase of meiosis I
38. Which hormone(s) is (are) most responsible for each of the following events of the menstrual cycle?
 A. Ovulation
 B. Maintenance of corpus luteum
 C. LH peak
 D. Glandular growth of endometrium
 E. Endometrial gland secretion
 F. Recruitment of a group of primary follicles
 G. Basal body temperature rise
 H. Secretion of thin endocervical mucous
 I. Inhibition of gonadotropin secretion
39. Human chorionic gonadotropin
 A. Is secreted from the placenta
 B. Is used to detect pregnancy
 C. Structurally and functionally resembles FSH
 D. Is required during the first two trimesters of pregnancy
 E. Is detectable 8 days after ovulation
40. Estrogen

A. Is bound mostly to SHBG and albumin
B. Is responsible for the development of female secondary sex characteristics
C. Decreases the excitability of the myometrium
D. Stimulates FSH secretion

Thyroid Hormones

The thyroid hormones are produced in the thyroid gland, which is made up of several follicles. Each follicle consists of a single layer of thyroid cells surrounding a viscous gel called colloid. The synthesis of thyroid hormones requires iodide, which is obtained from the diet and absorbed by the gastrointestinal tract. Free iodide in the extracellular space is removed either through uptake by the thyroid gland or by renal excretion. Iodide is actively taken up (trapped) by thyroid cells using a Na^+-K^+ ATPase pump. Thyroglobulin is a glycoprotein synthesized in the thyroid cells and secreted into the colloid by exocytosis of granules. Intracellular iodide is oxidized by thyroid peroxidase to neutral iodine, which then is bound to tyrosine residues of the thyroglobulin molecule, resulting in monoiodotyrosine (MIT). A second iodination forms diiodotyrosine (DIT). An oxidative condensation reaction (coupling) takes place between two DIT residues to form thyroxine (T_4). Triiodothyronine (T_3) comes from coupling of an MIT residue with a DIT residue. Both T_4 and T_3 are still linked to thyroglobulin by peptide bonds. By endocytosis, thyroid cells take up the thyroglobulin stored in the colloid and the resulting vesicle is joined by lysosomes. Proteases from the lysosomes hydrolyze the peptide bonds of thyroglobulin and thereby release T_4, T_3, MIT, and DIT into the intracellular space. Iodide is removed from MIT and DIT by deiodinase and returned to the intracellular iodide pool. Both T_4 and T_3 are released into the bloodstream. The thyroid gland secretes about 15 T_4 molecules for every T_3 molecule.

Only free (unbound) thyroid hormones in the circulation are biologically active. Only about 0.04 percent of the T_4 and 0.4 percent of the T_3 are free in the bloodstream. Most of T_4 is bound to thyroxine-binding globulin (TBG). The remainder is bound to thyroxine-binding prealbumin (TBPA) and albumin. The T_3 is bound only to TBG (mostly) and albumin. Although the serum level of free T_4 is about 7 times that of free T_3, T_3 is about 4 times as potent as T_4. The half-life of T_4 is 8 times that of T_3. Thyroxine is metabolized by deiodination, and it is deiodinated to form either T_3 or reverse T_3 (rT_3). While only 20 percent of the T_3 in circulation comes from thyroid secretion, 80 percent is derived from T_4 deiodination. Reverse T_3 does not have any known biologic activity. Both T_3 and rT_3 are further deiodinated to diiodothyronines.

Thyroid hormone stimulates calorigenesis. It increases oxygen consumption in all tissues except brain, spleen, and testis. Thyroid hormone exerts a wide range of effects on intermediary metabolism. It stimulates protein synthesis, glycogenolysis, gluconeogenesis, intestinal absorption of glucose, and synthesis, mobilization, and degradation of fat. Thyroid hormone increases heart rate and force of contraction. It enhances the sensitivity of tissues to catecholamines. Finally, thyroid hormone is required for fetal development, erythropoiesis, and vitamin A synthesis.

Thyroid hormone has its primary action at several sites within its target cell, including RNA transcription, protein translation, membrane-bound Na^+-K^+ ATPase, and mitochondria.

The secretion of thyroid hormone is stimulated by thyroid-stimulating hormone (TSH) from the anterior pituitary. The TSH stimulates the organification of iodide, exocytosis of thyroglobulin, endocytosis of colloid, proteolysis of thyroglobulin, and release of thyroid hormone. Thyroid hormone exerts a negative feedback control of TSH secretion. In the pituitary, T_4 is converted to T_3, which inhibits TSH secretion. The TSH-releasing hormone (TRH), a tripeptide secreted from the hypothalamus, stimulates TSH secretion. Thyroid hormone may also exert feedback inhibition at the level of the hypothalamus.

The thyroid gland itself exhibits autoregulation. In a state of intrathyroid organic iodine deficiency, the iodide transport mechanism becomes more active. An excess of intrathyroid organic iodine inhibits iodide transport. When the intracellular iodide concentration rises above a critical level, the iodide organification step is blocked (Wolff-Chaikoff block). Pharmacological doses of iodide result in a rapid and transient inhibition of thyroid hormone secretion by an unknown mechanism (different from the Wolff-Chaikoff block).

Hypothyroidism may be primary, secondary, or tertiary, depending on whether it is due to thyroid disease, pituitary disease, or hypothalamic disease, respectively. In adults, hypothyroidism results in cold intolerance, slow mentation, a husky voice, weight gain, decreased reflexes, and skin changes referred to as myxedema. Neonatal hypothyroidism results in cretinism, characterized by growth retardation as well as mental retardation.

Symptoms of hyperthyroidism (thyrotoxicosis) include heat intolerance, fatigue, palpitations, increased appetite, sweating, weight loss, nervousness, and hand tremors. The most common cause of hyperthyroidism is Graves' disease, in which thyroid-stimulating immunoglobulins (TSI) stimulate the TSH receptors on thyroid cells and produce the same hormone action as TSH.

Goiter simply means any enlargement of the thyroid gland and results from chronic stimulation of the gland by excessive amounts of TSH (or TSI). In iodine-deficiency goiter, the dietary iodine is inadequate to maintain synthesis of thyroid hormone. Thyroid hormone levels decline and feedback inhibition of TSH is decreased; thus, increased TSH secretion leads to hypertrophy of the thyroid gland.

QUESTIONS

41. Which is not true of thyroid hormone synthesis?
    A. Iodination of tyrosine residues of thyroglobulin gives rise to MIT and DIT
    B. Iodide passively diffuses into thyroid cells
    C. Thyroglobulin is hydrolyzed to release T_3 and T_4
    D. Coupling of MIT and DIT takes place in the thyroid cell cytoplasm
    E. Intracellular iodide is oxidized to neutral iodine before binding to tyrosine
42. Thyroid hormone
    A. Stimulates oxygen consumption in the brain
    B. Stimulates protein catabolism
    C. Stimulates TSH secretion
    D. Uses cAMP as a "second messenger"
    E. None of the above
43. T_3
    A. Is more potent than T_4
    B. Comes mostly from deiodination of T_4
    C. Normally has a lower serum concentration than T_4
    D. Is only slightly more active than rT_3
    E. Is not as active in inhibiting TSH secretion as T_4
44. Which of the following would decrease thyroid hormone secretion?
    A. TSH
    B. TSI
    C. Iodine deficiency
    D. Large pharmacological doses of iodine

ANSWERS

1. E, D
2. C
3. In the cytosol. The steroid binds to the receptor and activates the receptor. The steroid-receptor complex moves into the nucleus and interacts with the chromatin to increase the transcription of specific mRNAs coding for the proteins that carry out the physiologic effect.
4. Steroid hormones are protein bound; peptide hormones are free in plasma. Steroids have longer plasma half-lives than peptides. Steroids use a direct nuclear mechanism; peptides utilize a second messenger.
5. Glucose and ketone bodies
6. A, C, D
7. A. Unchanged
   B. Decreased
   C. Decreased
   D. Increased
   E. Increased
   F. Increased
8. None of the answers is correct.
9. A. Inhibit
   B. Stimulate
   C. Stimulate
   D. Inhibit
   E. No effect
   F. Stimulate
   G. Stimulate
10. Glucagon, epinephrine, glucocorticoids, growth hormone
11. Stimulators: hypoglycemia, beta-adrenergic stimulation. Inhibitors: glucose, insulin.
12. A meal high in protein and low in carbohydrate stimulates both insulin and glucagon secretion. If glucagon secretion were not stimulated, the relative insulin excess in the face of a minimal glucose load would lead to a hypoglycemic state.
13. Insulin levels in the diabetic patient are extremely low, while in the fasting person, insulin levels are only moderately low. As a result, the diabetic experiences an accelerated gluconeogenesis and lipolysis. Severe hyperglycemia and ketoacidosis are then expected. In the starved individual, gluconeogenesis takes place but at a lower rate and serum glucose is depressed. Lipolysis and ketogenesis occur at much lower rates, producing only a mild ketoacidosis.
14. A, B
15. B
16. A, B, C
17. A. Increased
    B. Increased
    C. Decreased
    D. Decreased
    E. Decreased
    F. No change
    G. Increased
18. A, B, C. Choice D is incorrect because drinking water lowers serum osmolality and decreases ADH secretion.
19. A
20. A. Stimulates
    B. Inhibits

    C. Inhibits
    D. Stimulates
    E. Stimulates
    F. Stimulates
    G. Stimulates
21. All are increased.
22. B
23. A. Zona glomerulosa
    B. Zona fasciculata and zona reticularis
24. A. 21
    B. 21
    C. 19
25. A, B, C
26. D
27. A
28. All are true.
29. A, C, D
30. B, C, D
31. E
32. A, B, E
33. C
34. D, E
35. E
36. E
37. B
38. A. LH
    B. LH
    C. Estrogen
    D. Estrogen
    E. Estrogen and progesterone
    F. FSH
    G. Progesterone
    H. Estrogen
    I. Estrogen and progesterone
39. A, B, E
40. A, B, D
41. B, D
42. E
43. A, B, C
44. C, D

Bibliography

Ganong, W. F. *Review of Medical Physiology* (12th ed.). Los Altos, CA: Lange, 1985. Pp. 254–369.

Greenspan, F. S., and Forsham, P. H. (eds.). *Basic and Clinical Endocrinology*. Los Altos, CA: Lange, 1983.

Guyton, A. C. *Textbook of Medical Physiology* (7th ed.). Philadelphia: Saunders, 1986. Pp. 875–982.

Martin, J. B., and Wilson, J. D. (eds.). *Harrison's Principles of Internal Medicine* (10th ed.). New York: McGraw-Hill, 1983. Pp. 580–750.

School of Medicine Faculty. *Endocrine Physiology Syllabus*. Lecture handout, Stanford University, 1984.

Wilson, J. D., and Foster, D. W. (eds.). *Williams Textbook of Endocrinology* (7th ed.). Philadelphia: Saunders, 1985.

3 Gastrointestinal Physiology

Saied Assef

The gastrointestinal (GI) system is composed of the alimentary canal (mouth, pharynx, esophagus, stomach, small intestine, large intestine, anus) and associated glands that empty their secretions into the alimentary canal (salivary glands, pancreas, hepatobiliary system).

The overall function of this system is the breakdown of nutrients into small absorbable molecules (digestion) and the subsequent uptake of these molecules across the GI mucosa into the blood or lymphatic circulation (absorption). In order to carry out digestive processes, a variety of ions, organic molecules, and enzymes must be present in the alimentary canal. These are provided in the 6 to 10 liters of fluids that are secreted by various GI exocrine glands and mucosal lining cells into the gut lumen each day. Gastrointestinal motility, brought about by contractions of the smooth muscle cells in the gut wall, causes mixing of substances in the gut lumen as well as their propulsion down the GI tract.

MOTILITY

ANATOMIC CONSIDERATIONS. In a cross section of the GI tract (Fig. 3-1) there are four anatomic subdivisions: mucosa, submucosa, muscularis, and adventitia.

The muscularis is composed of smooth muscle cells arranged in inner circular and outer longitudinal layers. The spindle-shaped smooth muscle cells in each layer are electrically coupled by means of gap junctions, thus forming a functional syncytium. The exception to this general organization of smooth muscle can be found in

1. Esophagus. The pharynx and the proximal one-half of the esophageal muscularis are composed of striated muscle. The cricopharyngeus muscle, a thickened band at the pharyngoesophageal junction, is the upper esophageal sphinctor.

   At the distal end of the esophagus, near the gastroesophageal junction, the circular layer of smooth muscle is functionally specialized to form the lower esophageal sphinctor. The lower esophageal sphinctor is anatomically indistinguishable but functionally distinct from the remainder of the esophageal circular muscular layer.
2. Stomach. In addition to the typical inner circular and outer longitudinal layers, the stomach contains an oblique innermost layer of smooth muscle.
3. Colon. The longitudinal muscle layer is concentrated in three bundles called the teniae coli. These fuse over the rectosigmoid forming a uniform layer over the rectum.
4. Anal canal. The internal anal sphinctor is formed by a thickened ring of smooth muscle in the inner circular layer. This sphinctor receives autonomic innervation. The external anal sphinctor is formed by striated muscle and receives somatic innervation.

NERVE SUPPLY OF THE GI TRACT. The GI system has both an extrinsic and an intrinsic nerve supply. These form an extensively integrated network.

Intrinsic (or intramural) nerves are composed of neurons contained within the wall of the GI system. These neurons receive input from the extrinsic nerve

Fig. 3-1. *Cross section of GI tract. (From P. R. Wheater, H. G. Burkitt, and V. G. Daniels,* Functional Histology, a Text and Colour Atlas *[1st ed.]. New York: Churchill Livingstone, 1979. P. 182.)*

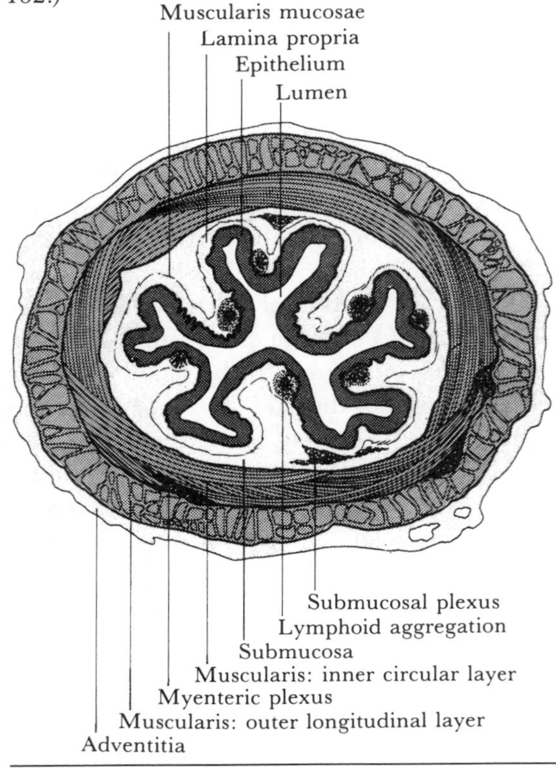

supply of the gut as well as afferent (sensory) input from the GI mucosa. The intrinsic nerves of the GI tract can mediate a variety of local reflexes in which both the afferent and efferent limbs are contained within the gut wall. These reflexes can, therefore, take place in the absence of extrinsic neural input. The intramural neurons are arranged in two distinct layers: (1) submucosal plexus (Meissner's plexus) is mostly involved in regulation of secretion, and (2) myenteric plexus (Auerbach's plexus), located between the circular and longitudinal layers of smooth muscle, is more important in regulation of GI motor function.

Extrinsic nerve supply of the GI tract includes both sympathetic and parasympathetic nerves. Parasympathetic stimulation promotes GI motility, relaxes sphinctors, and stimulates exocrine secretion by GI glands. Sympathetic stimulation, on the other hand, inhibits GI secretion and motility.

Sympathetic functions are mediated directly through the action of norepinephrine on smooth muscle and glandular cells and indirectly through the action of norepinephrine on presynaptic (alpha-2) receptors on cholinergic nerve endings causing a decreased parasympathetic output.

PHYSIOLOGICAL PROPERTIES OF GASTROINTESTINAL SMOOTH MUSCLE. There are two types of electrical activity in GI smooth muscle cells: slow waves and spike potentials.

In contrast to neurons and skeletal muscle where there is a stable resting membrane potential, intestinal smooth muscle cells have a "background" oscillation of membrane potential which is an inherent property to the muscle cells and is not significantly influenced by neural or endocrine regulation. These oscillations, occurring at a frequency of 3 to 12 cycles per minute, are called the "slow waves." The slow waves can be thought of as the "pacesetter potential," because, although they do not result in contraction of smooth muscle, they set the rate for spike potentials which in turn cause smooth muscle contraction. Slow waves are not an all-or-none phenomenon. They represent a "graded potential," i.e., their amplitude may vary.

Spike potentials are a group of one to a few prototypic electrical responses that occur in an all-or-none fashion and are triggered when slow waves depolarize the cell beyond a certain threshold. They are thought to be mediated by an influx of Ca^{2+} ions through voltage-dependent calcium channels in the smooth muscle cell membrane. These spike potentials cause smooth muscle contraction.

In response to elevation of intracellular Ca^{2+}, gastrointestinal smooth muscle cells display two different patterns of contraction.

1. Phasic contractions. The cycle of contraction — relaxation takes place over a relatively short period of time.
2. Tonic contractions. Contraction is sustained for minutes to hours. This pattern is more typical of smooth muscle cells that form sphinctors.

PERISTALTIC REFLEX. Distention of a segment of the GI tract which causes a wave of contraction slightly proximal, and a wave of relaxation slightly distal, to the site of distention. This reflex, called the peristaltic reflex, is independent of the extrinsic nerve supply of the GI tract and is thought to play some role in ensuring the movement of an ingested food bolus in the correct direction.

QUESTIONS

1. The electrical slow wave is an inherent property of GI smooth muscle, and its frequency is largely independent of neural regulation. The resting membrane potential, however, is found to be influenced by autonomic neural input. Given what you know about the overall effect of sympathetic and parasympathetic nerves on GI motility, predict their influence on resting membrane potential.
2. Metoclopramide is a drug with cholinergic properties. It promotes the effect of acetylcholine on GI smooth muscle. Which of the following conditions could be improved by administration of metoclopramide?
   1. Diarrhea
   2. Esophageal reflux
   3. Peptic ulcer disease
   4. Diabetic-gastroparesis–increased gastric emptying time

   Answer: A. 1, 2, and 3
   B. 1 and 3
   C. 2 and 4
   D. 4 only
   E. 1, 2, 3, and 4
3. Gastrointestinal motility is increased by
   1. Adrenergic stimulation
   2. Mucosal irritation
   3. Colonic bacteria
   4. Distension of gut lumen

   Answer: A. 1, 2, and 3
   B. 1 and 3
   C. 2 and 4
   D. 4 only
   E. 1, 2, 3, and 4

SECRETION

SALIVARY SECRETIONS. The salivary glands secrete 1000 to 1500 ml of saliva per day. Although there are many small buccal salivary glands, more than 90 percent of saliva is secreted by the paired parotid, submandibular, and sublinqual salivary glands. The salivary secretory unit is composed of acini and ducts. The acinar cells elaborate a primary secretion that contains enzymes, electrolyes at concentrations similar to plasma, and mucoproteins. The cells lining the proximal salivary ducts, called striated ducts, modify the electrolyte composition of this primary acinar secretion by actively reabsorbing Na^+, secreting K^+ and HCO_3^-, and passively reabsorbing Cl^-. The final composition of saliva is dependent on the salivary flow rate (see Fig. 3-2).

At slow rates, when there is ample time for active transport processes of the striated ducts, the ratio of Na^+ to K^+ concentration in saliva is similar to that of intracellular fluids, which are relatively K^+ rich. At faster rates the ratio of Na^+ to K^+ concentration approaches that of extracellular fluids, which are relatively Na^+ rich. Aldosterone influences Na^+-K^+ exchange in the striated ducts in the same way that it does in the distal convoluted tubule of the nephron: it increases Na^+ reabsorption and K^+ secretion.

There are two historically distinguishable types of salivary acini. Serous acini produce a watery secretion rich in an α-amylase called ptyalin. Mucinous acini elaborate a more viscous secretion containing a variety of mucoproteins. The parotid glands are almost entirely composed of serous acini, while the sublingual glands are predominantly composed of mucinous acini. The submandibular glands are mixed salivary glands. They contain both serous and mucinous acini.

The most important function of saliva is the role it plays in oral hygiene. Retained food particles between teeth serve as nutrients for bacteria. The acid that results from bacterial metabolism can dissolve the enamel of teeth. The flow of saliva protects the enamel by buffering and washing away this acid and also by removing retained food particles. Saliva contains gamma globulin A (IgA), antimicrobial proteins such as lysozyme, and thiocyanate ions, all of which act to protect the oral cavity against bacteria.

Saliva also plays a role in digestion. Ptyalin, an α-amylase capable of breaking internal α-1, 4 glycosidic linkages, is secreted in saliva. The breakdown of carbohydrates is initiated in the mouth by salivary ptyalin and is continued until inactivation of ptyalin by gastric acid. Although ptyalin is important in carbo-

Fig. 3-2. *Concentrations of major ionic constituents in the saliva as a function of saliva secretion rate. (From L. R. Johnson [ed.].* Gastrointestinal Physiology *[3rd ed.]. St. Louis: Mosby, 1985, P. 58.)*

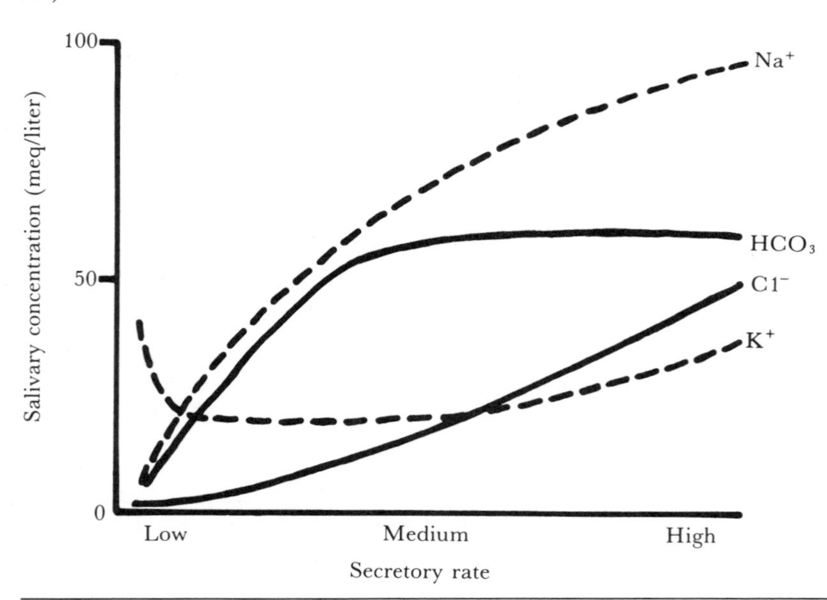

hydrate digestion, it is a nonessential enzyme because the pancreas secretes amylases with the same function as ptyalin.

Salivary secretion is under neural regulation. Salivary glands are innervated by both sympathetic and parasympathetic nerves. Signals such as taste, sight, smell, or thought of food are conveyed via afferent nerves to the salivatory nuclei in the brainstem which in turn causes efferent parasympathetic nerves to discharge. Parasympathetic stimulation can increase the rate of secretion by as much as 20-fold over baseline. Unlike other physiologic functions under autonomic regulation, in which sympathetic and parasympathetic nerves have antagonistic effects, in the regulation of salivary secretion, sympathetic stimulation is complementary to parasympathetic stimulation. Although both branches of the autonomic nervous system increase salivary secretion, the parasympathetic system is the predominant regulator of salivary glands.

GASTRIC SECRETIONS. The stomach secretes 1800 to 2000 ml of gastric juices per day. Functionally, the stomach can be thought of as having two distinct areas. The oxyntic gland area, corresponding to the proximal 80 percent of the stomach is where the oxyntic glands are located. These glands are made of oxyntic or parietal cells which produce acid and intrinsic factor and chief cells which secrete pepsinogen and mucous cells. The pyloric gland area which makes up the distal 20 percent of the stomach mainly consists of mucous cells, but it also contains some chief cells. Endocrine cells, called G cells, which secrete the hormone gastrin, are also located in the pyloric area.

Gastric secretions contain H^+, electrolytes, pepsinogen, intrinsic factor, and mucus.

The pH of gastric secretions can be as low as 1.0. This represents a millionfold concentration of H^+ ions by the parietal cells.

Several models have been proposed to explain the mechanism of H^+ secretion by gastric parietal cells. One of these is shown in Figure 3-3. The essential features of parietal-cell H^+ ion secretation are

1. The existence of an H^+-K^+ ATPase on the luminal surface of parietal cells capable of secreting H^+ ions against a millionfold concentration gradient.

Fig. 3-3. *Model for gastric acid secretion.*

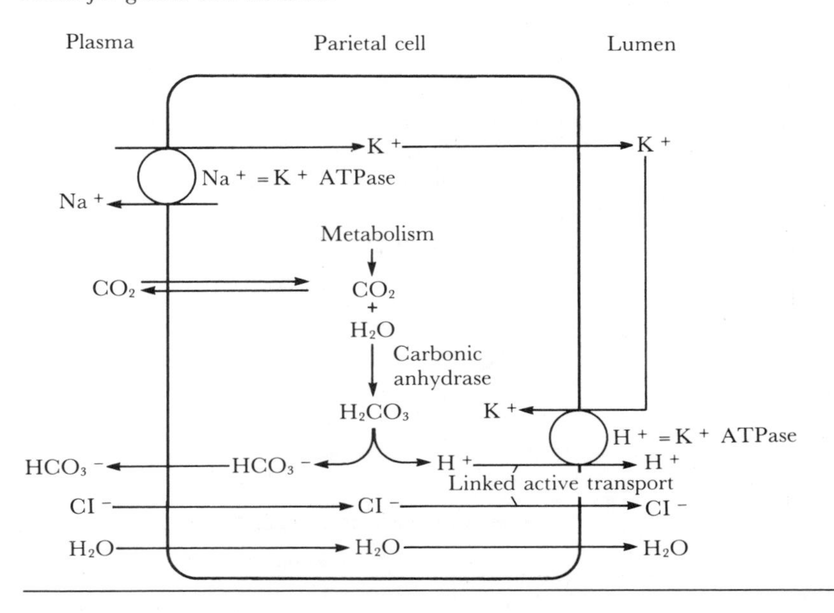

2. Active transport of Cl^- on the luminal surface of parietal cells. This active transport mechanism is thought to be linked to the active transport of H^+ ions.
3. The addition of an HCO_3^- ion to the plasma for each H^+ ion secreted into gastric juice. This explains the phenomenon of "alkaline tide" which refers to periods of active gastric secretion in which gastric venous blood has a high concentration of HCO_3^- and, therefore, a more alkaline pH.
4. Actively transported H^+ and Cl^- ions are osmotically active and passively draw water into the gastric lumen. Gastric secretions are, therefore, isotonic with plasma.

The electrolyte composition of gastric secretions is dependent on the secretion rate as shown in Figure 3-4. There is an inverse relation between Na^+ and H^+ ion concentrations in gastric secretions. As the secretion rate increases, the H^+ ion concentration increases while Na^+ ion concentration decreases. At all secretion rates, however, gastric secretions contain a higher concentration of K^+, Cl^-, H^+, and a lower concentration of Na^+ than plasma.

Pepsinogen is an inactive zymogen secreted by gastric chief cells. Once in the gastric lumen, pepsinogen is converted to pepsin by the action of gastric acid. Pepsinogen can, in turn, activate other pepsinogen molecules. As a proteolytic enzyme capable of cleaving internal peptide linkages, pepsin initiates the digestion of proteins in the stomach.

Intrinsic factor is a peptide secreted by parietal cells. Its function is to form a complex with vitamin B_{12}. This complex is then absorbed in the distal ileum. In the absence of intrinsic factor vitamin B_{12} cannot be absorbed.

Gastric secretions contain two chemically distinct types of mucus: soluble mucus and insoluble mucus. Soluble mucus, secreted in response to vagal stimulation, is similar to mucous secretions elsewhere in the GI tract. It serves as a lubricant and protects the gastric mucosa from mechanical injury. Insoluble mucus forms a gel on the surface of gastric mucosal cells and is one component of the gastric mucosal barrier which protects the gastric mucosa from digestion by acid and pepsin.

Gastric acid secretion is controlled by the interplay between three physiological secretagogues on the one hand and a number of gastric and intestinal inhibitory factors on the other. The secretagogues are acetylcholine from vagal stimulation, the hormone gastrin, and histamine. These three factors each potentiate the effects of the other and, therefore, function in a synergistic fashion in stimulating

Fig. 3-4. *Ionic composition of gastric secretions as a function of secretion rate. (From T. J. Sernka, and E. D. Jacobson.* Gastrointestional Physiology, the Essentials. *[2nd ed.]. P. 96 © 1983 The Williams & Wilkins Co., Baltimore.)*

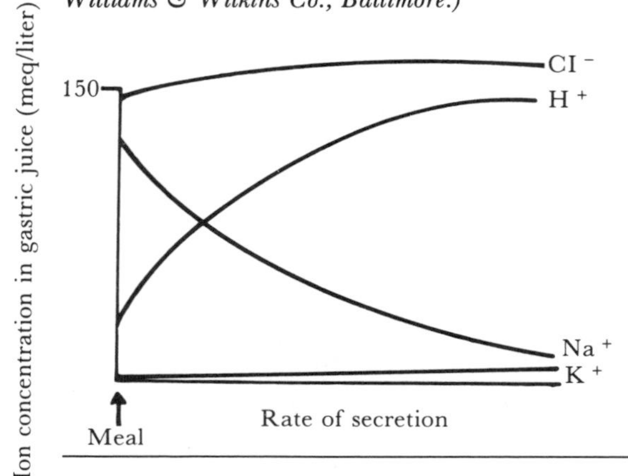

acid secretion. The inhibitory factors include low gastric juice pH, cholecystokinin-pancreozymin (CCK-PZ), secretin, gastric inhibitory polypeptide (GIP), and other as yet undefined factors. Gastric secretion can be considered as having three phases: cephalic, gastric, and intestinal.

During the cephalic phase, prior to the entrance of food into the stomach, signals such as taste, smell, sight or thought of food can activate the vagal nuclei in the medulla, causing vagal stimulation of the stomach. This enhances gastric secretion in two ways. First, acetylcholine released from vagal nerve endings directly stimulates parietal and chief cells to increase their secretion. Second, vagal stimulation causes release of gastrin from G cells in the antral mucosa. Gastrin, as mentioned earlier, is also a potent stimulus for acid secretion.

During the gastric phase, the presence of food in the stomach is sensed by mechanoreceptors that are activated by distension and chemoreceptors that are activated primarily by proteins and products of protein digestion. Activation of these receptors results in increased gastric secretion through the following mechanisms:

1. Vagovagal reflexes. Afferent fibers from gastric mechanoreceptors are carried to medullary vagal nuclei by the vagus nerve. Activation of these fibers can in turn increase vagal outflow to the stomach.
2. Local reflexes. Mechanical distention can activate local reflexes that have both their afferent and efferent fibers located within the gastric wall. Activation of these reflexes also increase gastric secretion.
3. Mechanical distention in the antrum can increase the release of gastrin.
4. Some of the chemical constituents of food, principally proteins and products of their digestion, stimulate gastric G cells to release gastrin.
5. Other food constituents such as caffeine can directly stimulate parietal cells.

During the intestinal phase, presence of food in the intestine is a minor stimulant of gastric secretion. The effect is mediated through release of gastrin from endocrine cells in the mucosa of the upper small intestine.

Several factors serve to decrease gastric secretion of acid. These include

1. Gastric acid itself. When gastric secretions have a pH of 2 or less, release of gastrin is completely inhibited.
2. Enterogastric reflex. Mechanical distention caused by presence of chyme in the small intestine decreases vagal output to the stomach and, thereby, decreases acid secretion.
3. Presence of fat, protein, or acid in the small intestine causes the release of a number of hormones, including cholecystokinin, secretin, and gastric inhibitory peptide, all of which inhibit gastric secretion.

PANCREATIC SECRETIONS. The pancreas secretes approximately 1000 ml of pancreatic juice into the duodenum per day. The basic secretory unit of exocrine pancreas is composed of acini and ducts. The acinar cells secrete digestive enzymes while the cells lining the small ductules secrete HCO_3^- ions. Figure 3-5 shows the proposed mechanism for HCO_3^- ion secretion. Carbonic acid, formed from carbon dioxide and water by action of carbonic anhydrase dissociates into H^+ and HCO_3^- ions. Bicarbonate ions are actively transported against an electrochemical gradient on the luminal side. Hydrogen ions are also actively transported in exchange for Na^+ ions on the plasma side. The electrolyte composition of pancreatic juice is dependent on the rate of secretion as shown in Figure 3-6.

There is an inverse relationship between Cl^- and HCO_3^- ion concentrations. The HCO_3^- ion secreted in pancreatic juice has two functions: (1) it neutralizes gastric acid and, therefore, protects the duodenal mucosa from injury by acid,

Fig. 3-5. *Model for pancreatic secretion of HCO_3^- ion.*

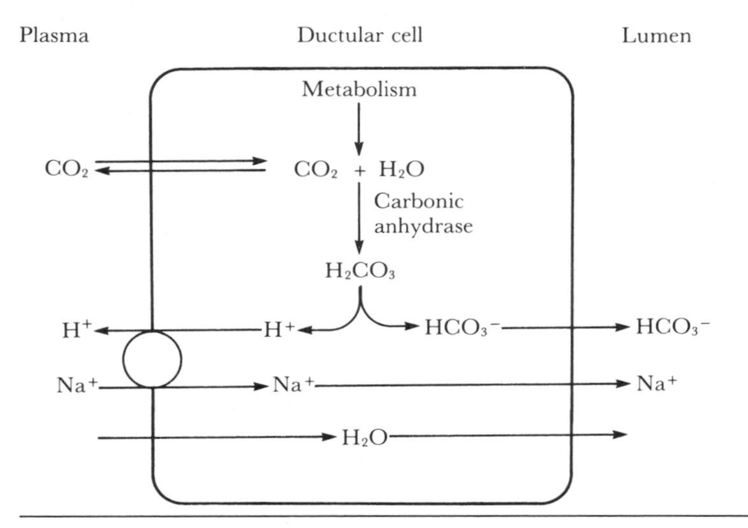

and (2) it raises the pH of duodenal fluids to a range that is optimal for the activity of pancreatic digestive enzymes.

The acinar cells secrete enzymes for the digestion of proteins, carbohydrates, fats, and nucleic acids. Unlike the digestive enzymes of saliva and the stomach, those of the pancreas are essential to digestion and absorption of nutrients. To prevent autodigestion of pancreatic tissues, the potent proteolytic enzymes of the pancreas are secreted in inactive zymogen form. These include trypsinogen, chymotrypsinogen, and procarboxypeptidase. Once in the duodenum, trypsinogen is converted to the active enzyme trypsin by the action of enterokinase, which is an enzyme present in the duodenal brush border. Trypsin can itself activate trypsinogen, thus setting up a positive feedback loop for activating trypsinogen. To prevent premature activation of trypsinogen in the pancreatic ducts, the acinar cells secrete a substance called trypsin inhibitor.

In addition to proteolytic enzymes, the pancreatic acini secrete ribonuclease and deoxyribonuclease, which are involved in digestion of nucleic acids. Other enzymes found in pancreatic secretions include those involved in fat digestion (lipases, phospholipases, and cholesterol esterase), as well as pancreatic amylase, which is involved in carbohydrate digestion.

Pancreatic secretions are under both neural and hormonal regulation. During the cephalic phase, prior to entry of chyme into the duodenum, vagal stimulation increases enzyme secretion by acinar cells. Later, as chyme enters the duodenum, the hormone cholecystokinin (CCK) is released from the duodenal mucosa. Peptides, amino acids, and fatty acids are particularly potent signals to the release of CCK. The CCK stimulates the acinar cells to increase enzyme secretion.

Presence of acid in the duodenum is a potent signal to the release of secretin from the upper small intestinal mucosa. Secretin stimulates the ductular lining cells to secrete a watery, HCO_3^--rich, secretion.

GENERAL PHYSIOLOGY OF BILE. Bile is continuously secreted by hepatocytes. Between meals, the sphinctor of Oddi is contracted, providing resistance to the flow of bile into the duodenum. Bile, following the path of least resistance, is stored in the gall bladder. In addition to its function in storage of bile, the gall bladder mucosa modifies the chemical content of bile by actively transporting sodium with either chloride or bicarbonate. Water is passively reabsorbed thus concentrating bile from 3- to 15-fold.

The secretion of bile is essential to the digestion and absorption of fat. It also

Fig. 3-6. *Ionic composition of pancreatic juice as a function of secretion rate. (From L. R. Johnson.* Gastrointestinal Physiology *[3rd ed.]. St. Louis: Morby, 1985. P. 85.)*

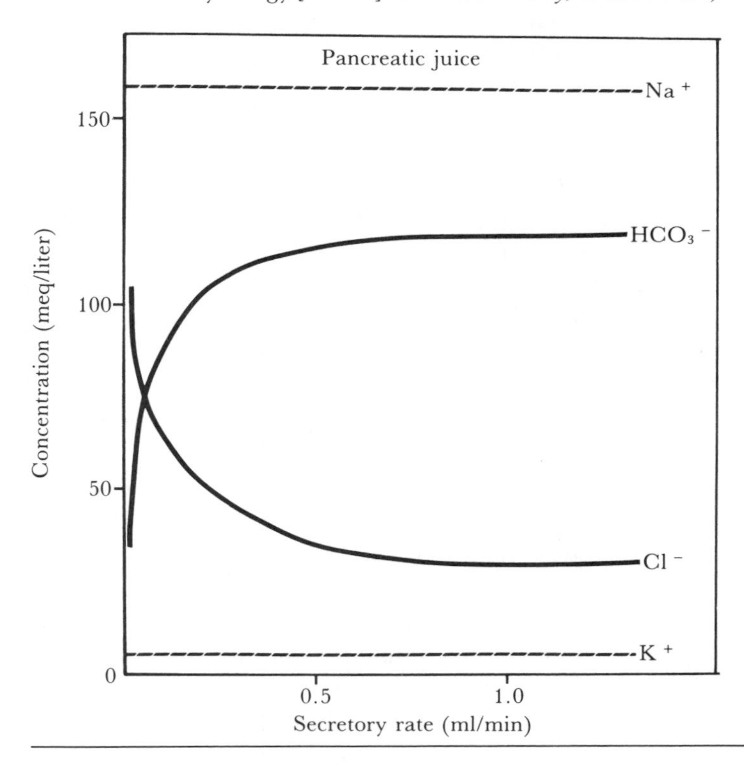

provides a route for elimination of endogenous waste products (e.g., bilirubin) and exogenous substances (e.g., certain drugs, toxins). The liver produces 600 to 1000 ml of bile per day. Constituents of bile include water (84%), bile salts (11%), lecithin (3%), cholesterol (0.5%), bilirubin (<1%), Na^+, K^+, Ca^{2+}, Cl^-, and HCO_3^-. Bile has an alkaline pH ranging from 7.5 to 8.5.

Bile Acids. Bile acids are amphiphilic molecules, i.e., they contain both hydrophilic and hydrophobic domains. In aqueous solution the hydrophobic domains of these molecules associate and exclude water molecules. The hydrophilic domains are externally directed and form the bile acid–water interface. This arrangement is called a micelle and is the most thermodynamically favorable state in which bile acids can exist in aqueous solution. These micelles can incorporate other amphiphilic (e.g., lecithin) or lipid-soluble (e.g., cholesterol, free fatty acids) molecules, thus solubilizing them in aqueous solution. These properties of bile acids form the basis for their physiologic role.

One such role is solubilization of other bile components, such as cholesterol and lecithin, by forming mixed micelles with them. This prevents cholesterol, which is otherwise insoluble, from precipitating out of solution and forming gallstones. Alterations in the composition of bile, such that it contains proportionately more cholesterol and less bile acids, is an important pathogenic mechanism in cholesterol gallstone disease.

The role of bile acids in digestion and absorption of dietary fat is also dependent on their amphiphilic nature. In the small intestine lumen, bile acids emulsify dietary lipids, making them available as substrates for pancreatic lipases.

The products of lipid digestion, such as fatty acids and monoglycerides, are then incorporated into mixed micelles with bile acids. Although these mixed micelles are not themselves directly absorbed, it is clear that incorporation into micelles facilitates the absorption of fatty acids and monoglycerides by the small intestinal mucosa.

Bile acids are synthesized from cholesterol by hepatocytes. The primary bile acids are those that are directly synthesized by the liver. In the intestinal lumen primary bile acids may be modified by bacterial enzymes. Products of these modifications are called secondary bile acids. The two main primary bile acids are cholic acid and chenodeoxycholic acid. Bacterial dehydroxylation of these results in the formation of the two main secondary bile acids: deoxycholic acid and lithocholic acid.

Bile acids may also be conjugated to either glycine or taurine in the liver. Conjugated bile acids have a lower pK_a and are more likely to be ionized at intestinal pH. Bile acids are truly amphiphilic only when they are ionized. Because of their lower pK_a at any given intestinal pH, a larger fraction of conjugated bile acids are ionized and, therefore, physiologically active. Intestinal bacterial enzymes can also deconjugate conjugated bile acids.

Only a small fraction of the bile acids released into the duodenum is lost in the feces. The remainder is reabsorbed from the intestine, taken up by the liver, and resecreted in bile. This is called the enterohepatic circulation of bile. There are two mechanisms for intestinal reabsorption of bile acids. First, un-ionized bile acids can be reabsorbed passively anywhere along the length of the intestine. Second, ionized bile acids can be reabsorbed via an active transport process in the distal ileum. The overall efficiency of the reabsorption process is approximately 95 percent. The total body pool of bile acids, including primary, secondary, conjugated, and unconjugated bile acids, is 2 to 4 g. This pool undergoes 6 to 10 cycles of secretion and reabsorption per day. Only 0.2 to 0.6 g of bile acids are lost in feces per day. In order to maintain steady state for the total body pool of bile acids, the rate of hepatic synthesis must be equal to the rate of fecal loss.

Bilirubin. Bilirubin is a waste product of heme degradation by the reticuloendothelial (RE) system. Eighty to ninety percent of the daily bilirubin production comes from the hemoglobin in senescent red blood cells, and the remainder comes from the breakdown of other heme-containing proteins. The average daily production of bilirubin in a normal adult is 250 to 300 mg per day. There is an enormous functional reserve in the capacity of the RE system to produce bilirubin. Faced with an increased load of hemoglobin degradation, e.g., in hemolytic anemia, the RE system can increase daily production of bilirubin to as high as 50 g per day. There is no known useful physiologic function for bilirubin.

Bilirubin is transported from its site of production in the RE system to the liver by plasma (Fig. 3-7). Because of its hydrophobicity and consequent water insolubility, bilirubin is transported in plasma bound to albumin.

Albumin-bound bilirubin is in equilibrium with a small amount of free or unbound bilirubin. Because of its lipid solubility, this unbound bilirubin freely permeates cell membranes. The hepatic uptake of bilirubin is dependent on several proteins within hepatocytes, called Y and Z ligandins, which have a high affinity for bilirubin. Binding to these intracellular proteins allows bilirubin to accumulate in hepatocytes.

Conjugation of lipid-soluble substances to hydrophilic moieties is a general mechanism by which hepatic metabolism renders a variety of endogenous (e.g., steroids, dilantin) and exogenous (e.g., drugs) substances water soluble and, therefore, excretable.

The principal hepatic metabolism of bilirubin is its conversion to bilirubin diglucuronide. The first step in this conversion is catalyzed by a system of enzymes, called glucuronyl transferases, located in the smooth endoplasmic reticulum. This enzyme system catalyzes the transfer of a glucuronic acid molecule from an activated donor called uridine diphosphate glucuronic acid (UDPGA) to bilirubin thus forming a bilirubin monoglucuronide. The site of the second glucuronidation reaction has not been well established. It may occur in

Fig. 3-7. *Bilirubin (BR) transport, conjugation, and excretion by the hepatocyte. (Reprinted by permission of the publisher from Bilirubin metabolism: State of the art, by R. Schmid.)* Gastroenterology *74:1307. Copyright 1978 by the American Gastroenterological Association.)*

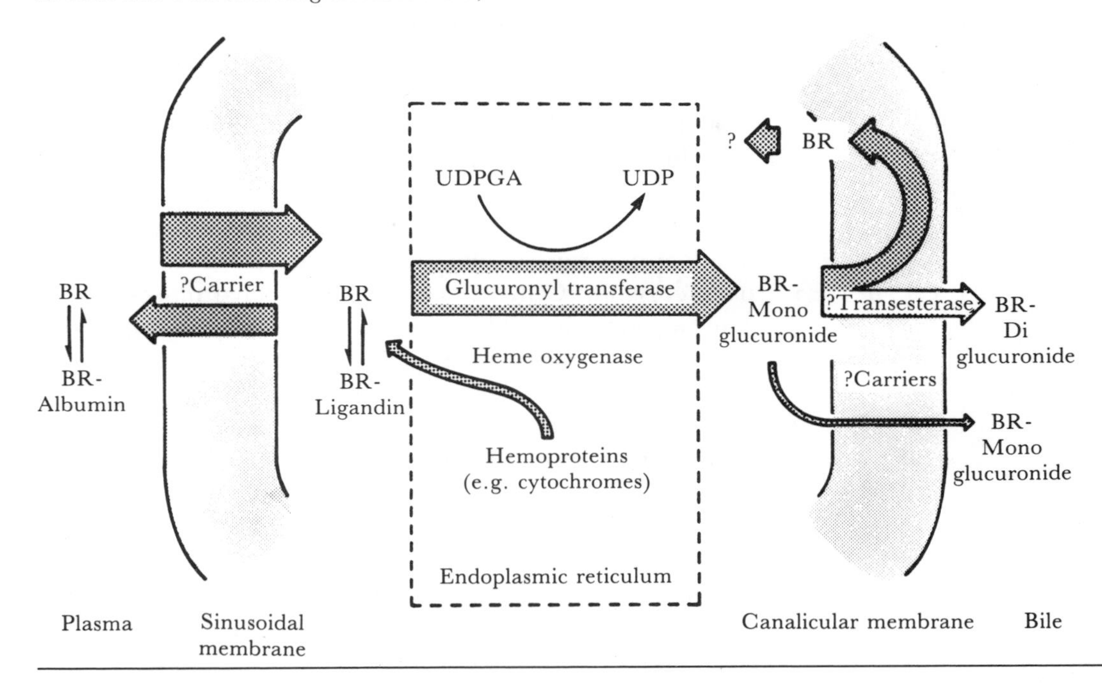

the smooth endoplasmic reticulum and be catalyzed by the same enzyme system that carried out the first glucuronidation reaction. Alternatively enzymes located in the bile canalicular membrane may be responsible for the second glucuronidation step in a process that may be intimately linked to the active transport of bilirubin from hepatocytes into bile canaliculi. Regardless of where the transfer of the second glucuronic acid to bilirubin occurs, it is the conversion of bilirubin to bilirubin diglucuronide which renders it water soluble. Only this water-soluble form is excreted in bile. The glucuronyl transferases involved in bilirubin metabolism have two important features. First, unlike other enzymes, they do not display strict substrate specificity. They are involved in the transfer of glucuronate groups from UDPGA to several other substrates as well. Second, they are inducible. A variety of drugs (e.g., phenobarbitol) have the capacity to induce their synthesis by the hepatocyte.

Transport of conjugated bilirubin out of hepatocytes across the bile canalicular membrane takes place against a concentration gradient. This active transport process is also involved in the transportation of a number of other organic anions into bile. Under normal circumstances only negligible amounts of conjugated bilirubin enter the plasma. If there is obstruction to the flow of bile (e.g., gallstones), or if the active transport process across the biliary canalicular membrane is interfered with, conjugated bilirubin can accumulate in plasma.

Once in the lower small intestine, bilirubin is converted to urobilinogens by gut bacteria. Most of the urobilinogen is lost in feces. Ten to twenty percent of the urobilinogen is reabsorbed, taken up by the liver, and excreted again in the bile, thus forming an enterohepatic circulation. A small proportion of the urobilinogen gets through the portal system and enters the systemic circulation where it is filtered by the kidney and excreted in urine.

Regulation of Bile Secretion. The factors that control the rate of hepatic secretion of bile include the size of the bile acid pool in the enterohepatic circulation, the hormone secretin, and hepatic blood flow.

Bile acids are potent choleretics. The greater the amount of bile acids in the enterohepatic circulation, the greater is the rate of hepatic bile secretion. The hormone secretin, released from the duodenal mucosa in response to the presence of acid in the duodenum, causes the secretion of a dilute but HCO_3^--rich bile presumably by stimulating a HCO_3^- pump in the cells lining the bile ductules. Reduction of hepatic blood flow reduces the rate of bile secretion.

While the hepatic secretion of bile is a continuous process, the entrance of bile into the duodenum is not. Between meals, bile is stored in the gallbladder as biliary pressures are not sufficient to drive the bile through a contracted sphinctor of Oddi. The presence of chyme, especially one rich in lipids in the duodenum, causes the release of a hormone called cholecystokinin-pancreozymin (CCK-PZ) from the duodenal mucosa. The CCK-PZ causes the contraction of the gallbladder while it relaxes the sphinctor of Oddi. This results in the expulsion of bile into the duodenum.

INTESTINAL SECRETION. The most proximal segment of duodenum contains submucosal glands called Brunner's glands which secrete an alkaline mucus. The function of these glands is to protect the duodenal mucosa by neutralizing gastric acid as it enters the duodenum. The daily secretion of Brunner's glands is less than 100 ml.

The small-intestinal crypts produce approximately 2 liters of secretions per day, the composition of which is very similar to that of interstitial fluids. In addition there are goblet cells in the intestinal mucosa that secrete mucus.

The colon secretes less than 100 ml of mucus per day which has a high concentration of K^+ and HCO_3^- ions.

Intestinal secretion is stimulated by presence of chyme in the intestine, through neural reflexes mediated by the vagas nerve and through direct mechanical stimulation of the mucosal lining cells.

QUESTIONS

4. Gastric acid secretion is increased by all of the following except:
   A. Sight, smell, and thought of food
   B. Distension of gastric antrum
   C. Gastrin
   D. Pepsin
   E. Histamine
5. Vagotomy affects gastric acid secretion by
   A. Disturbing gastric emptying
   B. Increased parietal cell threshold to gastrin or histamine
   C. Decreased release of glucagon by pancreas
   D. Increased sympathetic stimulation of stomach
6. Gastric emptying is slowed by
   1. Increased osmolality of material entering the duodenum
   2. Decreased pH of material entering the duodenum
   3. Free fatty acids in duodenum
   4. Recumbant position

   Answer: A. 1, 2, and 3
   B. 1 and 3
   C. 2 and 4
   D. 4 only
   E. 1, 2, 3, and 4
7. For each numbered statement, choose a lettered response. Each letter may be used more than once or not at all.
   A. Salivary secretions
   B. Gastric secretions

C. Both
D. Neither
1. Na+ concentration increases at faster secretion rate
2. Contain enzymes that are essential in digestion
3. Increase rate of secretion with thought, smell, and of food
4. Sympathetic stimulation increase secretion rate

8. Which of the following are first released as proenzymes?
A. Chymotrypsin, amylase, lipase
B. Chymotrypsin, secretin, elastase
C. Trypsin, ribonuclease, carboxypeptidase A
D. Trypsin, elastase, carboxypeptidase B

9. Bicarbonate ion in pancreatic secretions
A. Is formed by ultrafiltration of plasma bicarbonate
B. Is stimulated primarily by CCK-PZ
C. Increases in concentration as flow rates increase
D. Shifts pK_a of bile salts to neutral range

10. Vomiting is a symptom commonly seen in intestinal obstruction. Based on your understanding of gastrointestinal physiology, predict which of the following would be associated with the highest volume of vomiting:
A. Esophageal obstruction
B. Gastric outlet obstruction
C. Obstruction in the first part of the duodenum
D. Obstruction in the third part of the duodenum
E. Colonic obstruction

11. A variety of clinical conditions can result in accumulation of bilirubin in plasma leading to jaundice. Based on your knowledge of bilirubin metabolism, fill in the blanks in the following table with N for normal, O for absent, I for increased, and D for decreased

	Serum bilirubin			Urobilinogen	
	Indirect*	Direct*	Urine bilirubin	Stool	Urine
Hemolytic jaundice (breakdown of red cells)					
Obstructive jaundice (mechanical obstruction of bile flow, e.g., stones, tumor)					
Hepatocellular jaundice (principal defect is excretion of conjugated bilirubin across the bile canalicular membrane					
Neonatal jaundice (enzyme system for bilirubin conjugation is underdeveloped)					

12. Secondary bile salts are formed by
A. Oxidation of primary bile salts in the GI tract
B. Reduction of primary bile salts in the GI tract

* The van den Bergh reaction is a laboratory test for detection of bilirubin. Conjugated bilirubin directly reacts with the test reagent and gives a positive result. Unconjugated bilirubin does not react directly, and alcohol must be added to the mixture for unconjugated bilirubin to give a positive result. Conjugated bilirubin, therefore, is called direct reacting bilirubin, and unconjugated bilirubin is called indirect reacting bilirubin.

C. Degradation of cholesterol
D. Conjugation of primary bile salts with glycine and taurine
E. Deconjugation of conjugated bile salts by intestinal bacteria

13. Which of the following is true of bile secretion?
   A. Chyme rich in lipid in the duodenum causes release of gastrin
   B. Chymotrypsin causes gallbladder contraction
   C. CCK-PZ causes gallbladder contraction
   D. Hepatic secretion of bile occurs only after meals
   E. Entrance of bile into the duodenum is continuous
14. The rate-limiting step (and also the step that is most sensitive to hepatic injury) in the process by which hepatocytes handle bilirubin is
   A. Transport
   B. Uptake
   C. Conjugation
   D. Excretion
   E. Reabsorption

DIGESTION

Digestion, the chemical breakdown of foods to absorbable components, can be divided into luminal and membrane digestion. Luminal digestion is carried out by enzymes that are present in salivary, gastric, and pancreatic secretions. Membrane, or brush border digestion, is carried out by enzymes that are synthesized by enterocytes and are bound to the small-intestinal brush border.

Although some digestion occurs in the mouth and stomach, the vast majority of protein and carbohydrate digestion, as well as all fat digestion, takes place in the small intestine.

Absorption, the transfer of digestive products from the intestinal lumen to blood, takes place almost exclusively in the small intestine. (A few drugs and some lipid soluble substances such as ethanol are absorbed in the stomach.) The absorptive surface area of the small intestine is increased by three superimposed levels of mucosal folding. First, there are circularly arranged folds of mucosa and submucosa called plica circulares or Kerckring's valves that increase the absorptive surface area by three- to fourfold. Second, the plica circulares are covered by finger-like projections of the mucosa called villi. The existence of these intestinal villi increases the absorptive surface area by an additional ten-fold. Third, the luminal membrane of the enterocytes lining the villi are thrown into numerous folds called microvilli which increase the absorptive area by another 20 fold. Thus, the combination of these structural adaptations increase the area of the absorptive small intestinal mucosa by 600 to 800 fold.

WATER AND ELECTROLYTE ABSORPTION. Typically 1.0 to 1.5 liters of fluid are ingested per day. The various GI secretions contribute another 8 to 9 liters of fluids so that on the average 10 liters of fluids enter the small intestine per day (Table 3-1). Approximately 90 percent of this fluid is absorbed in the small intestine, resulting in the passage of about 1 liter of fluid into the colon where all but 100 to 200 ml is reabsorbed.

The upper small intestine is the site of osmotic equilibration of luminal contents. Depending on the osmolality of chyme, water diffuses either in or out of the gut lumen, making luminal contents isoosmotic to plasma. Chyme remains isotonic to plasma throughout the remainder of the intestine. Movement of water across the GI mucosa is by passive diffusion and follows the transport of osmotically active solutes such as Na^+, glucose, and amino acids.

Table 3-1

Summary of Fluid and Electrolyte Absorption in the Gastrointestinal Tract

	Water ml/24 hr	Na^+ meq/24 hr	K^+ meq/24 hr
Entering small intestine (diet and secretions)	10,000	775	140
Entering colon	1000	75	5
In stool	100	2–5	10–20
Absorbed in small intestine	9000	700	130
Absorbed in colon	900[1]	70	−15[2]

[1]Maximum reabsorptive capacity of colon is 3 to 4 liters per day.

[2]Note that K^+ is secreted in the colon.

Sodium ions are absorbed actively throughout the small and large intestine. The intestinal epithelial cells are functionally polarized by division of their cell membrane into mucosal and basolateral membranes. On the mucosal side Na^+ enters the enterocytes through three separate processes (Fig. 3-8).

1. Diffusion by itself down its concentration gradient
2. Coupled to glucose or amino acid transport
3. Coupled to Cl^- transport

The Na^+-K^+ ATPase which actively transports Na^+ out of and K^+ into enterocytes against an electrochemical gradient is limited to the basolateral membrane. This asymmetric distribution of the sodium pump causes a net transfer of Na^+ from gut lumen to blood.

Intestinal transport of K^+ is by passive diffusion. Passive movement of K^+ down its electrochemical gradient in the jejunum and ileum results in the net transfer of this ion from gut lumen to blood. In the colon, however, K^+ is secreted from plasma into the lumen. This secretion also represents a passive process, the driving force being the high transepithelial potential difference (15–50 mv lumen negative) in the colon.

The mechanism of transport for Cl^- and HCO_3^- is uncertain and may be interrelated. There is evidence for both active and passive transport. Both Cl^- and HCO_3^- are absorbed in the jejunum. In the ileum and colon Cl^- is absorbed, but HCO_3^- is secreted. The mechanism of HCO_3^- "absorption" in the

Fig. 3-8. *Intestinal absorption of Na^+. A. Passive movement of Na^+ down its concentration gradient. B. Glucose cotransport. C. Cl^- cotransport.*

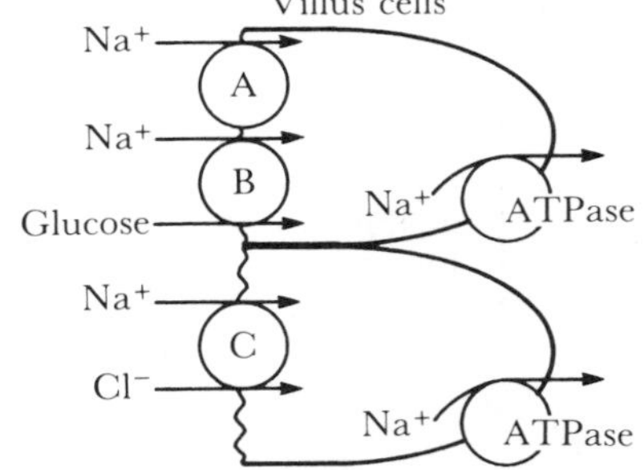

jejunum may not be absorption of the HCO_3^- ion but may be due to secretion of H^+ into the lumen which titrates the HCO_3^- to water and carbon dioxide. Exchange diffusion of Cl^- and HCO_3^- has been suggested to explain the direction of transport of these ions in the ileum and colon.

CARBOHYDRATE DIGESTION AND ABSORPTION. The main digestible dietary carbohydrates include polysaccharides such as starch and glycogen, disaccharides such as sucrose and lactose, and monosaccharides such as glucose and fructose (Table 3-2). Cellulose, a carbohydrate of plant origin, is a linear polymer of glucose molecules joined by β-1, 4 linkages, Because humans do not have the enzymatic capability of breaking these linkages, cellulose is not digestible and, therefore, contributes to dietary fiber.

Prior to absorption, carbohydrates must be digested to their component monosaccharides because only monosaccharides can be absorbed. Digestion of carbohydrates can be divided into (1) luminal digestion carried out by enzymes present in salivary and pancreatic secretions that break down polysaccharides to oligo- and disaccharides, and (2) brush border digestion by enzymes that are synthesized in small-intestinal cells and incorporated in the microvillous membranes of these cells. These brush border enzymes hydrolyze the products of luminal digestion into their component monosaccharides. Luminal digestion of carbohydrates begins in the mouth by the action of ptyalin, the salivary α-amylase. Prior to its inactivation by gastric acid, ptyalin hydrolyzes 20 to 40 percent of ingested starch. Pancreatic amylase carries out the remainder of starch hydrolysis. Salivary and pancreatic α-amylase can only hydrolyze internal 1, 4 glycosidic linkages; therefore, the products of starch digestion by α-amylases are maltose, maltotriose, and isomaltose (α-limit dextrin) (Fig. 3-9).

The small intestinal brush border contains the enzymes maltase, isomaltase, lactase, and sucrase. The action of these enzymes completes the digestion of carbohydrates resulting in their breakdown to their component monosaccharides which can then be absorbed.

The rate limiting step in carbohydrate assimilation is the transport of digestive products across the brush border membrane. Several different transport processes have been identified. Glucose and galactose are actively transported by a common Na^+-dependent carrier system. The energy for this active transport process is provided by the Na^+-K^+ ATPase on the basolateral membrane of these intestinal cells. Fructose is transported by facilitated diffusion. Digestion

Table 3-2 *Principal Digestible Dietary Carbohydrates*

Carbohydrates	Source	Total Carbohydrate (%)	Chemical Structure
Starch	Plants, grains	50	
Amylose			Straight-chain polymer of glucose molecules in α-1,4 linkage
Amylopectin			Branched-chain polymer of α-1,4 linked glucose chains connected by α-1,6 branching points
Glycogen	Animal muscle and liver	2	Similar to amylopectin but more highly branched
Sucrose	Table sugar	30	Glucose α-1,2 linked to fructose (disaccharide)
Lactose	Milk	10	Galactose β-1,4 linked to glucose (disaccharide)

Fig. 3-9. *Products of starch hydrolysis by α-amylase (From L. R. Johnson. [ed.].* Gastrointestinal Physiology *[3rd ed.]. St. Louis: Mosby, 1985. P. 109.)*

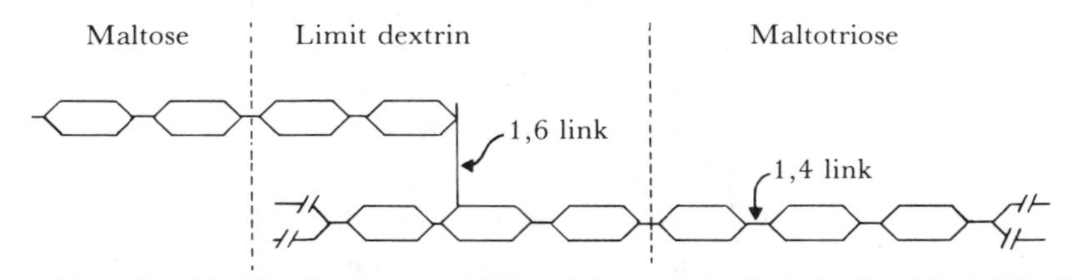

and absorption of carbohydrates is usually completed in the proximal jejunum (Fig. 3-10).

FAT DIGESTION AND ABSORPTION. Triglycerides account for over 90 percent of dietary fat. Phospholipids, cholesterol esters, cholesterol, and fat-soluble vitamins make up the rest.

The presence of chyme (especially one that is rich in lipids and proteins) and acid in the duodenum causes the release of two peptide hormones, cholecystokinin (CCK) and secretin from the upper small intestinal mucosa. The combined effect of these hormones results in pancreatic secretion of HCO_3^- and digestive enzymes. The CCK also causes contraction of the gallbladder and flow of bile into the duodenum. Through the action of these hormones, the presence of lipids in the duodenum brings about the conditions that are necessary for fat emulsification, digestion, and absorption.

Because lipolytic enzymes function only at the lipid-water interface, dietary lipids must first be emulsified before being digested. Emulsification takes place in the duodenum and requires two conditions: (1) a neutral pH and (2) presence of bile salts.

Pancreatic secretions contain several enzymes involved in lipid digestion which are secreted in the active form. These include

Fig. 3-10. *Summary of carbohydrate digestion and absorption.*

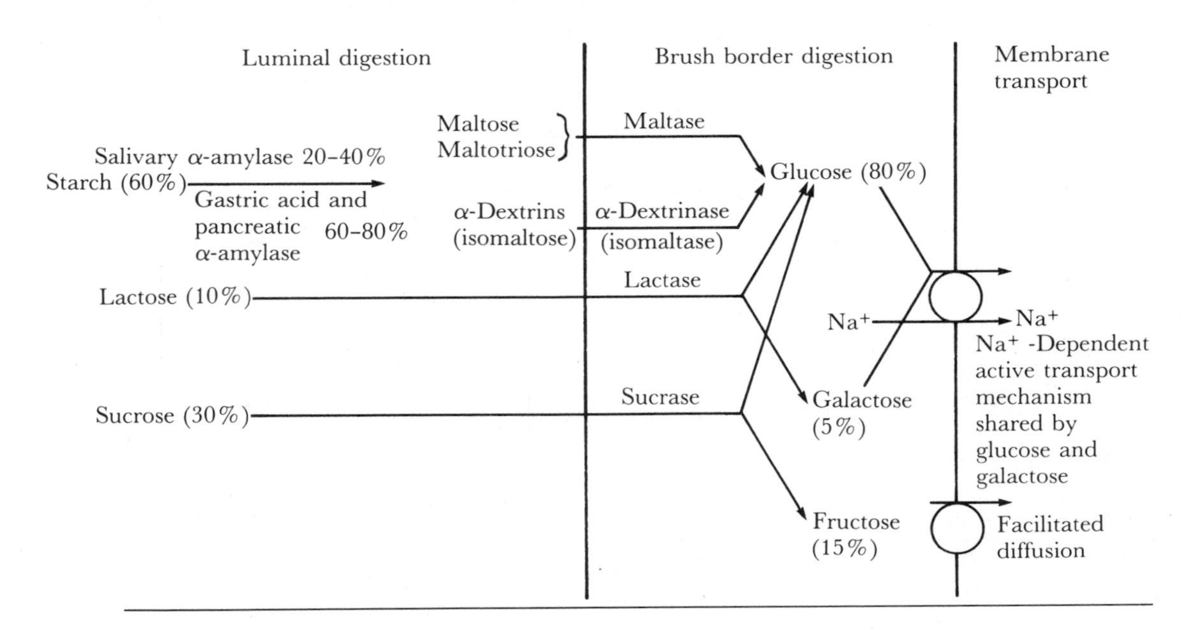

1. Pancreatic lipase. This enzyme catalyzes the hydrolysis of ester linkages in the 1 and 3 positions of triglycerides, resulting in the formation of fatty acids and two monoglycerides. For optimum function, pancreatic lipase requires the presence of a peptide called colipase which is also secreted by the pancreas. Colipase has two functions. First, the optimum pH for lipase is changed from 8 to 6 by binding to colipase; therefore, lipase is made more active at the physiological pH of small intestine. Second, colipase facilitates the action of lipase by attaching to the lipid-water interface of emulsified lipids, allowing fixation of lipase to this interface. The formation of lipase-colipase–bile-salt–lipid complex is the rate limiting step in fat digestion and absorption.
2. Phospholipase. Several phospholipases with different specificities are present in pancreatic secretions (Fig. 3-11). These enzymes convert lecithin to lysolecithin.
3. Cholesterol esterase. This enzyme converts cholesterol esters to cholesterol and free fatty acids.

To be absorbed, the products of lipid digestion must first traverse a layer of water which is adjacent to the microvillous membrane and glycocalyx of small-intestinal cells. This layer is not mixed with luminal chyme regardless of the degree of peristalsis. It is, therefore, called the "unstirred water layer" (UWL), and movement of solute molecules through it is by diffusion. With the exception of short and medium chain fatty acids, the products of lipid digestion are insoluble in the UWL and must be incorporated into micelles in order to reach the microvillous membrane of enterocytes. Formation of micelles into which fatty acids, cholesterol, lysolecithin, and fat-soluble vitamins can be incorporated is another function of bile acids. These micelles are not themselves absorbed by intestinal cells. In the UWL and near the brush border membrane of intestinal cells an equilibrium exists between free lipid digestion products and lipid digestion products in micelles. It is the free form of lipid digestion products that penetrates the intestinal cell membrane by passive diffusion.

Once inside the small-intestinal cell, the products of lipid digestion are reconstituted into triglycerides, phospholipids, and cholesterol esters by reesterification. They are then incorporated into chylomicra, which, like micelles, serve to solubilize them in aqueous solution. Chylomicra are composed of triglycerides (84%) and cholesterol (2%) which form the core, and lipoproteins (1%) and phospholipids (13%) which form the outer coat. These chylomicra then leave the basolateral membrane of intestinal cells by exocytosis and enter lymph through

Fig. 3-11. *Specificity of various phospholipases.*

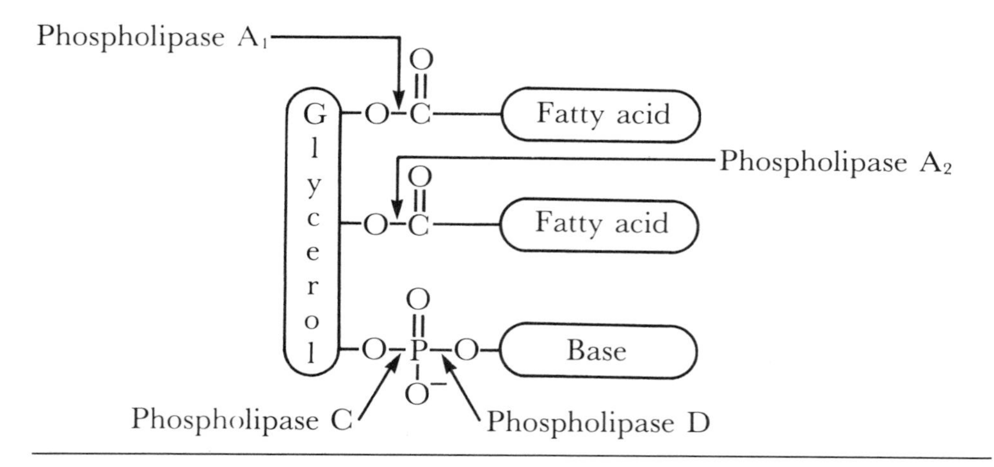

lacteals in the intestinal villi. Although cholesterol is present in chylomicra, the main route of entry for cholesterol is by incorporation in another kind of lipoprotein complex called the very low density lipoprotein (VLDL).

Short and medium chain fatty acids are not reesterified in intestinal cells but instead move across the cell by passive diffusion and end up in the portal venous system.

PROTEIN DIGESTION AND ABSORPTION. In addition to the proteins ingested in the diet, there are several endogenous proteins that enter the GI tract and are thus subject to the same digestive and absorptive processes as dietary proteins. These include proteins in GI secretions, proteins in sloughed mucosal cells of the GI tract, and plasma proteins. Protein digestion begins in the stomach by the action of pepsin and gastric acid. Both of these, however, are dispensable, and in their absence proteins are readily digested and absorbed. The principle enzymes involved in protein digestion are secreted by the pancreas as inactive zymogens. These include trypsinogen, chymotrypsinogen, proelastase, and procarboxypeptidases A and B. Trypsinogen is activated to trypsin by a duodenal brush border peptidase called enterokinase. Trypsin, in turn, activates other pancreatic proteases. Luminal digestion of proteins by these proteases results in free amino acids and small peptides. The small peptides are further broken down by intestinal brush border di- and polypeptidases.

Absorption of amino acids by small-intestinal cells is dependent on several active transport processes all of which are specific for L-amino acids. Groups of amino acids share common active transport carriers. Separate carrier systems are known to exist for neutral, basic, and acidic amino acids. The amino acids proline and hydroxyproline are also known to have a separate active transport carrier. All of these active transport systems are Na^+ dependent, and the Na^+-K^+ ATPase in the basolateral membrane of intestinal cells provides the energy for transport of amino acids against a concentration gradient.

In addition to active transport of amino acids, the small-intestinal mucosa contains systems for the active transport of di- and tripeptides. Through the process of pinocytosis, a quantitatively negligible amount of protein may be absorbed without prior digestion to peptides and amino acids. This is not a well-understood mechanism, but it may be involved in the pathogenesis of food allergy.

QUESTIONS

15. A 20-year-old black man complained of watery diarrhea after drinking milk. The patient probably has
    1. Deficiency of sucrase that is essential for digestion of the sugar in milk
    2. Osmotic diarrhea caused by retention of the undigested disaccharide and its breakdown products released by bacterial action
    3. Severe disacchariduria with frequent urination
    4. Deficiency of a specific brush border β-galactosidase

    Answer: A. 1, 2, and 3
    B. 1 and 3
    C. 2 and 4
    D. 4 only
    E. 1, 2, 3, and 4
16. Patients with the absence of aminooligopeptidase would be expected to have which of the following nutrients in the lower intestinal lumen?
    1. Intact exogenous proteins, amino acids, and oligopeptides
    2. Oligopeptides
    3. Amino acids, intact exogenous proteins
    4. Tetrapeptides

    Answer: A. 1, 2, and 3
    B. 1 and 3

C. 2 and 4
D. 4 only
E. 1, 2, 3, and 4

17. Which of the following enzymes is both produced and functions in the small intestine?
    A. Ribonuclease
    B. Phospholipase A
    C. Trypsin
    D. Elastase
    E. Lactase
18. Micelle formation is prerequisite to the efficient absorption of the following except:
    A. Vitamin A
    B. Cholesterol
    C. Linolenic acid
    D. Vitamin K
    E. Glycerol
19. Na^+ absorption in normal human gut
    1. Requires a special entry mechanism
    2. Is enhanced by transport of amino acids and glucose
    3. Requires ATPase in the basolateral membrane
    4. Establishes a transcellular electrical potential gradient

    Answer: A. 1, 2, and 3
    B. 1 and 3
    C. 2 and 4
    D. 4 only
    E. 1, 2, 3, and 4
20. Which of the following is not true in the colon?
    A. Na^+ is actively absorbed
    B. Chloride and bicarbonate are transported in opposite directions by an ion exchange mechanism
    C. K^+ is actively secreted
    D. Water is absorbed by osmosis
21. For each numbered statement choose a lettered response.
    A. Gastrectomy
    B. Resection of ileum
    C. Both
    D. Neither
    1. Vitamin B_{12} deficiency
    2. Increased rate of bile acid synthesis

GASTROINTESTINAL HORMONES

Important GI hormones have already been discussed in the context of regulation of motility, secretion, digestion, and absorption. Table 3-3 summarizes the structure, function, and regulation of the well-known GI hormones.

A variety of other peptides found in the GI mucosa as well as other sites, such as the central nervous system, are known to influence GI function when they are present in pharmacologic concentrations. The physiologic role of these peptides remains to be established. Some of these "candidate hormones" include

1. Motilin. A peptide released from the upper small-intestinal mucosa that may be involved in the regulation of gastric and intestinal motility in the interdigestive period.
2. Vasoactive intestinal peptide (VIP). A peptide of gastrointestinal origin that inhibits gastric acid and pepsin secretion, stimulates pancreatic and intestinal

Table 3-3
Properties of Gastrointestinal Hormones

	Gastrin	CCK-PZ	Secretin	GIP
Structure	Peptide	Peptide	Peptide	Peptide
Location of endocrine cells	Antrum of stomach	Small intestine	Small intestine	Small intestine
Stimuli for hormone release	Amino acids or peptides in stomach; parasympathetic nerves (ACh); distension of gastric antrum	Amino acids or fatty acids in small intestine	Acid in small intestine	Fat (to a lesser extent, carbohydrate) in small intestine
Stimuli for hormone inhibition	Acid in lumen of stomach; secretin			
Target-cell responses*				
Stomach				
Acid secretion	Stimulates		Inhibits	Inhibits
Antral contraction	Stimulates		Inhibits	Inhibits
Pancreas				
Bicarbonate secretion		Potentiates secretion action	Stimulates	
Enzyme secretion		Stimulates	Potentiates CCK-PZ actions	
Liver				
Bicarbonate secretion	Potentiates secretion action	Stimulates		
Gallbladder contraction		Stimulates		
Sphincter of Oddi		Relaxes		
Small-intestine motility	Stimulates ileum; inhibits ileocecal sphincter			
Large intestine motility		Stimulates mass movement		
Growth (trophic) of stomach and small intestine		Exocrine pancreas	Exocrine pancreas	

*Only the target-cell responses that are known to occur at physiological concentrations of the hormones are listed.

CCK-PZ = cholecystokinin-pancreozymin; GIP = gastric inhibitory polypeptide; ACH = acetylcholine.

From A. J. Vander, J. H. Sherman, and D. S. Luciano. *Human Physiology: The Mechanisms of Body Function* (4th ed.). New York: McGraw-Hill, 1985. P. 478.

secretion, and causes vasodilation. Pancreatic tumors that release large amounts of VIP into the circulation are associated with a profuse watery diarrhea that is thought to be caused by the influence of VIP on intestinal function.

3. Pancreatic polypeptide. A polypeptide produced in the pancreas which may inhibit pancreatic exocrine secretion and cause relaxation of the gallbladder.

QUESTIONS

22. Secretin stimulates the pancreas to secrete
    A. Hyperosmolar solution containing HCO_3^- and Cl^-
    B. Isoosmolar solution with increased enzyme and HCO_3^- content
    C. Isoosmolar solution with increased Cl^- and reciprocal HCO_3^- concentration
    D. Isoosmolar solution with relatively high HCO_3^- and low Cl^- concentration

23. The most potent stimulus for the secretion of secretin is
   A. Fat
   B. Carbohydrate
   C. Protein
   D. Acid
   E. Ca^{2+}
24. Gastrin
   1. Shares the same C-terminal amino acids with CCK-PZ
   2. Its action on gastric acid secretion is antagonized by secretin
   3. Potentiates the action of acetylcholine and histamine on gastric parietal cells
   4. Its secretion is inhibited by low gastric pH

   Answer: A. 1, 2, and 3
   B. 1 and 3
   C. 2 and 4
   D. 4 only
   E. 1, 2, 3, and 4

ANSWERS

1. Sympathetic stimulation decreases GI motility; therefore, it must hyperpolarize gastrointestinal smooth muscle cells. The opposite is true for parasympathetic stimulation (Fig. 3-12).
2. C. Esophageal reflux can be treated with metoclopramide, because increased cholinergic activity increases the tone of the lower esophageal sphinctor. Increased cholinergic activity also decreases gastric emptying time by promoting contractions of gastric smooth muscle.
3. C. Mucosal irritation causes increased gastrointestinal motility through activation of local intrinsic neural reflexes. The peristaltic reflex is also a local intrinsic neural reflex that promotes GI motility in response to local distension of gut wall.
4. D.
5. B. The three principal secretagogues of gastric H^+, acetylcholine, gastrin, and histamine, interact in a synergistic fashion.

Fig. 3-12. *Autonomic influence on gastrointestinal smooth muscle membrane potential. (From A. C. Guyton.* Textbook of Medical Physiology *[7th ed.]. Philadelphia: Saunders, 1986. P. 755.)*

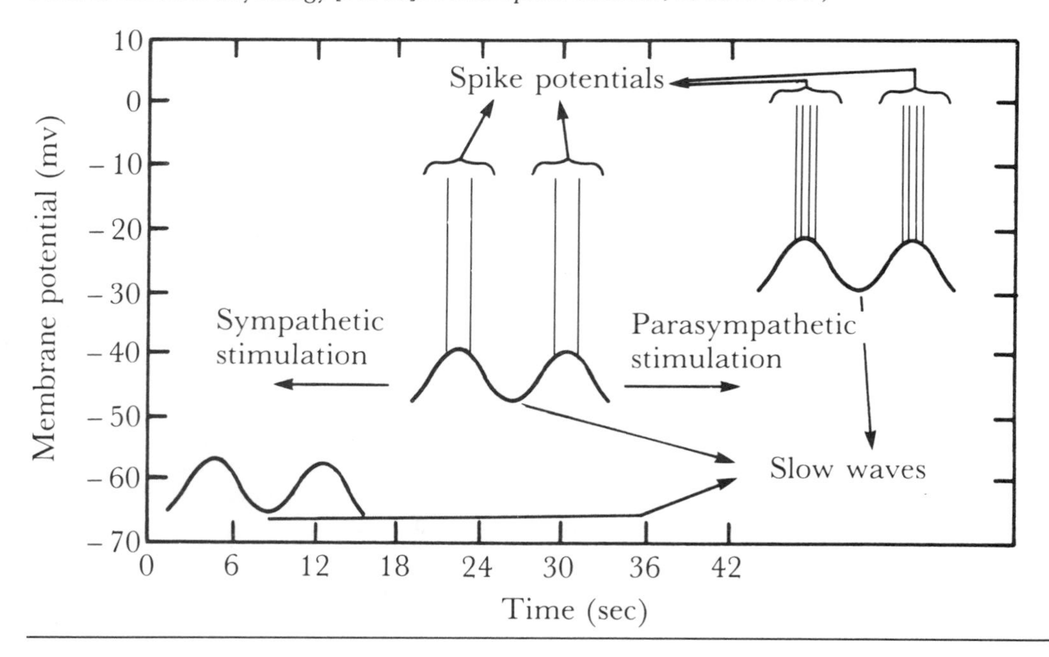

6. A. The rate of gastric emptying is regulated by small-intestinal receptors that are responsive to the osmolarity, acidity, and chemical content of the chyme that exits from the stomach. Increased osmolarity and acidity in chyme cause a slowing of gastric emptying. Carbohydrates generally empty faster than proteins, which in turn empty faster than fats. Gravity does not influence gastric emptying significantly. The effector mechanisms through which small-intestinal receptors bring about changes in gastric emptying are not well understood and probably include both neural and hormonal mechanisms.
7. 1 — A
   2 — D. Complete digestion could be carried out in the total absence of salivary ptyalin and gastric pepsin.
   3 — C
   4 — A
8. D. Peptidases are all secreted in inactive proenzyme form. Amylase, nucleases, and lipases are secreted as active enzymes by the pancreas.
9. C. As pancreatic flow rate increases, there is an increase in HCO_3^- concentration. The HCO_3^- is primarily stimulated by the hormone secretin. The pK_a of bile salts is an intrinsic property of their chemical structure and is not altered by the environment in which they are found.
10. D. Obstruction at this level is distal to the ampulla of Vater, the point at which pancreatic and biliary secretions enter the duodenum. Salivary, gastric, pancreatic, and biliary secretions would, therefore, accumulate proximal to the obstruction leading to increased volume of vomitus. In colonic obstruction, upper gastrointestinal secretions can be absorbed in the small intestine, and vomiting is not a major part of the clinical presentation.
11.

	Serum bilirubin			Urobilinogen	
	Indirect	Direct	Urine bilirubin	Stool	Urine
Hemolytic jaundice (indirect breakdown of red cells)	I	N	O	I	I
Obstructive jaundice (mechanical obstruction of bile flow, e.g., stones, tumor)	I	I	I	D	D
Hepatocellular jaundice (principal defect is the excretion of conjugated bilirubin across the bile canalicular membrane	I	I	I	D	I
Neonatal jaundice (enzyme system for bilirubin conjugation is underdeveloped)	I	N	O	N	N

N = Normal; O = Absent; I = Increased; D = Decreased.

In hemolytic states an increased bilirubin load is presented to the liver. Jaundice does not occur until production of bilirubin exceeds the tremendous reserve in the capacity of the liver to conjugate bilirubin (which can be as high as 50 g per day). Once this capacity has been overcome, unconjugated bilirubin begins to accumulate in plasma. An increased amount of conjugated bilirubin is being formed, but because the hepatic and biliary systems are normal, all of this added conjugated bilirubin is properly excreted, and none accumulates in plasma. Increased urobilinogen in stool and urine results from the increased excretion of conjugated bilirubin. Urine contains bilirubin only when plasma contains increased amounts of conjugated bilirubin. Unconjugated bilirubin, being bound to albumin, is not filtered at the glomerulus even when its plasma concentration is high.

In obstructive jaundice, the hepatic conjugation of bilirubin is normal, but conjugated bilirubin cannot be properly excreted. Eventually both conjugated and unconjugated bilirubin accumulate in plasma causing bilirubinuria.

Although excretion of conjugated bilirubin is the step which is most sensitive to hepatic injury, all other steps, including uptake and conjugation are also affected in hepatocellular jaundice. Both conjugated and unconjugated bilirubin accumulate as expected. The total amount of urobilinogen in the enterhepatic circulation is decreased. Because of decreased hepatic uptake, most of the urobilinogen is excreted in urine.

12. B. Secondary bile salts are formed by bacterial dehydroxylation (reduction) of primary bile salts.
13. C.
14. D.
15. C. Lactase deficiency is the most common disaccharidase deficiency. It is seen predominantly in blacks.
16. C. Intact exogenous proteins are substrates of pancreatic proteolytic enzymes and, therefore, would not be found intact in the lower intestine.
17. E. The other enzymes are produced by pancreatic acinar cells.
18. E. Monoglycerides, cholesterol, fat-soluble vitamins, and long-chain fatty acids are absorbed by the enterocyte only after they are incorporated into micelles. Glycerol and short chain fatty acids are more water soluble and can be absorbed directly from aqueous solution.
19. E.
20. C. The K^+ ion is secreted in the colon down its electrochemical gradient, not by active transport.
21. 1—C. Intrinsic factor produced by gastric parietal cells is required for efficient absorption of vitamin B_{12} in the distal ileum. Both gastrectomy and resection of ileum would lead to B_{12} deficiency over time if this vitamin were not supplemented parenterally.
21. 2—B. The distal ileum is a site of active reabsorption of bile acids. In the absence of the ileum, bile acids would be lost from the enterohepatic circulation, and their rates of synthesis would have to be increased by hepatocytes.
22. D.
23. D.
24. E. Gastrins are a group of peptides of varying molecular weight that share the same five C-terminal amino acids. The peptide hormone CCK-PZ also has this same C-terminal sequence, which may account for its mild stimulator effect on gastric acid secretion. Secretin, on the other hand, inhibits gastric acid secretion.

BIBLIOGRAPHY

1. Davenport, H. W. *Physiology of the Digestive Tract* (5th ed.). Chicago: Yearbook, 1982.
2. Greenberg, N. J. *Gastrointestinal Disorders. A Pathophysiologic Approach* (3rd ed.). Chicago: Yearbook, 1986.
3. Guyton, A. C. *Textbook of Medical Physiology* (7th ed.). Philadelphia: Saunders, 1986.
4. Johnson, L. R. (ed.). *Gastrointestinal Physiology* (3rd ed.). St. Louis: Mosby, 1985.
5. Sernka, T. J., and Jacobson, E. D. *Gastrointestinal Physiology, the Essentials* (2nd ed.). Baltimore: Williams & Wilkins, 1983.
6. Vander, A. J., Sherman J. H., and Luciano D. S., *Human Physiology — The Mechanisms of Body Function* (4th ed.). New York: McGraw-Hill, 1985.
7. West, J. B. (ed.). *Best and Taylor's Physiologic Basis of Medical Practice* (11th ed.). Baltimore: Williams & Wilkins, 1985.
8. Wheater, P. R., Burkitt, H. G., and Daniels, V. G. *Functional Histology. A Text and Colour Atlas.* New York: Churchill Livingstone. 1979. P. 182.

4 Cardiovascular Physiology

Benjamin Hsu

CARDIAC ELECTROPHYSIOLOGY

ACTION POTENTIAL. As with most cells in the body, the resting membrane potential of cardiac cells is about −80 mv, the inside negative with respect to the outside. The ionic basis of the resting potential in cardiac cells is similar to that in neurons and skeletal muscle fibers (see Chap. 1). There are fundamental differences, however, with the cardiac potential.

A typical action potential of a cardiac muscle fiber may be divided into five phases (Fig. 4-1):

Phase 0: Rapid upstroke. When the membrane is depolarized to a threshold potential, voltage-sensitive Na^+ channels open resulting in rapid increase in permeability to Na^+. This phase is analogous to the rapid depolarization seen in nerve and muscle fibers.

Phase 1: Partial repolarization. During this phase, the fast Na^+ channels are inactivated, producing a rapid decrease in permeability to Na^+. These channels are refractory to further stimulation until they are "reset" by repolarization. There is also a small increase in potassium permeability during this phase contributing to the repolarization.

Phase 2: Plateau. A unique feature of the cardiac action potential, the plateau is due to (1) the lack of a large rapid increase in permeability to K^+ as seen in nerve and muscle fibers and (2) a slow inward current arising from an increase in Ca^{2+} permeability. The Ca^{2+} that enters is involved in excitation-contraction coupling.

Phase 3: Rapid repolarization. The ionic basis for this phase includes (1) a gradual increase in K^+ permeability which was occurring during the plateau and (2) the inactivation of the slow inward Ca^{2+} channels. Fast Na^+ channels are reset during this phase, and the cell becomes subject to premature activation.

Phase 4: Diastolic depolarization. In myocardial fibers, this phase represents the

Fig. 4-1. *A typical cardiac action potential recorded in ventricular myocardium showing ERP (effective refractory period) and RRP (relative refractory period).*

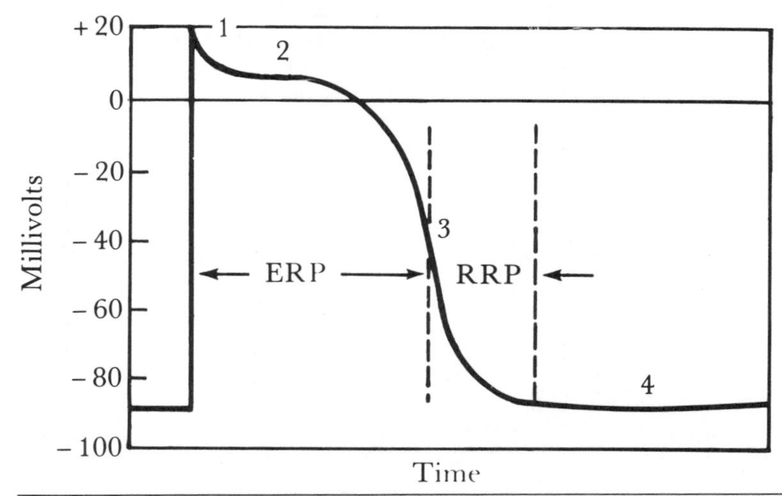

interval between rapid repolarization and rapid upstroke. These myocardial fibers are predominantly permeable to K^+ between action potentials, and their resting membrane potential is determined by the Nernst potential for K^+. Phase 4 is called diastolic depolarization because in the sinoatrial (SA) and atrioventricular (AV) nodes, and to some extent in Purkinje fibers, a spontaneous depolarization occurs (see below).

The action potential of cells in the SA and AV nodes (Fig. 4-2), called a "slow response" action potential, differs from that of myocardial and Purkinje fibers ("fast response"). It has a slow upstroke velocity, a smaller magnitude of action potential, and a brief plateau. In SA and AV nodal cells, there are no fast Na^+ channels and the slower action potential upstroke is mediated by Ca^{2+} channels. A spontaneous, gradual depolarization in the resting membrane potential occurs during phase 4, which is due to a decrease in K^+ permeability in these nodal, or pacemaker, cells. When the membrane potential is depolarized to threshold, an action potential is initiated. The rate at which this spontaneous depolarization brings the membrane potential to threshold determines the rate at which the heart beats. It follows, then, that the slope of the phase 4 depolarization, the level of the resting potential, and the level of the threshold potential influences the rate at which these cells fire an action potential. Acetylcholine decreases the slope of phase 4 depolarization while catecholamines increase the slope. Cells in the rest of the myocardium retain a high K^+ permeability during phase 4 and thus show little or no depolarization. Phase 4 is called diastolic depolarization because relaxation takes place during this phase while muscle contraction (systole) takes place mainly during the plateau (phase 2), when Ca^+ is entering into the cells.

The interval from the beginning of the action potential (phase 0) to the point at which the cell is able to conduct another action potential of any magnitude (middle of phase 3) is called the effective (or absolute) refractory period (ERP) (see Fig. 4-1). A second action potential cannot be elicited during this period. The relative refractory period (RRP) is the interval during which an action potential may be elicited but not of full magnitude. An action potential of full magnitude cannot be elicited until the membrane is completely repolarized (phase 4). Because the ERP of the cardiac action potential is so long, the contractile response of the myocardial fiber is more than half over by the time a second response can be elicited. Thus the refractory period allows time for the myocardial fibers to relax and the ventricles to become refilled with a sufficient volume of blood before the next contractile response.

EXCITATION-CONTRACTION COUPLING. Excitation-contraction (EC) coupling in cardiac muscle is similar to that in skeletal muscle (see Chap. 1). The Ca^{2+} that

Fig. 4-2. *Slow response cardiac action potential.*

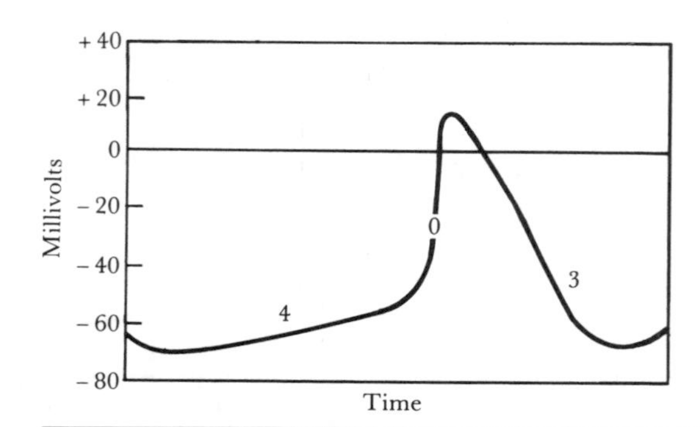

enters during the plateau phase mediates the coupling. In cardiac fibers, the Ca^{2+} that enters from the extracellular fluid (ECF) induces the release of more Ca^{2+} from the sarcoplasmic reticulum (SR) and directly interacts with troponin as well. In skeletal muscle fibers, Ca^{2+} comes almost exclusively from the SR. Catecholamines increase the Ca^{2+} influx from the ECF to bring about a more forceful contraction.

CONDUCTION. An action potential propagates down a single myocardial fiber by a mechanism similar to that seen in nerve (see Chap. 1). Unlike skeletal muscle, in which fibers are excited via direct innervation from motor neurons, fibers in the heart are stimulated by excitatory impulses from neighboring fibers which spread by means of gap junctions. The origin of excitation is the SA node. Cardiac excitation is conducted in the following order: the SA node, internodal (atrial) pathways, the AV node, the bundle of His, the Purkinje system, and the ventricular muscle. All of these structures possess automaticity except ventricular muscle, but the SA node discharges with the highest rate, and depolarization spreads from it to the rest of the myocardium before any other region is able to discharge spontaneously. Thus the SA node is the normal pacemaker of the heart. The AV nodal cells discharge with the second highest rate, followed by the His-Purkinje fibers.

Conduction velocity increases with the rate of change of potential (dV/dt) during phase 0 and with the amplitude of the action potential. It also increases when the level of the resting membrane potential is more negative. Impulses are then propagated with the highest speed in the Purkinje system, with intermediate speed in the internodal pathways, His bundle, and ventricular muscle, and with the slowest speed in the SA and AV nodes.

ELECTROCARDIOGRAM (ECG). The normal sequence and abnormal events in cardiac excitation can be seen in the ECG. The initial discharge from the SA node (located in the right atrial wall) is too small to be seen in the ECG. The excitation is spread by atrial tracts throughout the atrial myocardium, depolarization of which produces the P wave (Fig. 4-3). Excitation enters the AV node (located in the interatrial septum) via internodal pathways as the atrial myocardium completes its depolarization. From the AV node, the excitation moves to the bundle of His, located in the interventricular septum. A conduction delay in the AV node and His bundle allows time for ventricular filling. The His bundle divides into right and left bundle branches which eventually run into the Purkinje system, fibers of which ramify into the myocardium of both ventricles. The depolarization of the ventricular myocardium produces the QRS complex. The PR interval represents the conduction delay through the AV node and the His bundle. Ventricular repolarization produces the T wave. Mechanical contraction of the ventricles takes place during the QT interval.

Surface ECG waves depend upon extracellular currents produced when electrical excitation propagates down any excitable fiber. The magnitude of the current that is produced is proportional to the rate of change of the membrane potential of the fiber. In ventricular myocardial cells, this rate of change is greatest during phase 0 (rapid depolarization) and phase 3 (rapid repolarization) of the action potential. Hence, these phases correlate with the QRS complex and the T wave, respectively (Fig. 4-3). In the normal heart repolarization proceeds in the opposite sequence to depolarization, thus accounting for the fact that R waves (upright portion of the QRS complex) and T waves are both upright (or both inverted) in the same leads despite opposite charges.

In a first-degree heart block, all atrial impulses reach the ventricles, but are slowed, producing a long PR interval. In a second-degree block, not all atrial

Fig. 4-3. *Electrocardiogram showing the important deflections and intervals. A ventricular myocardial action potential is shown above it to demonstrate the relative timing between the two.*

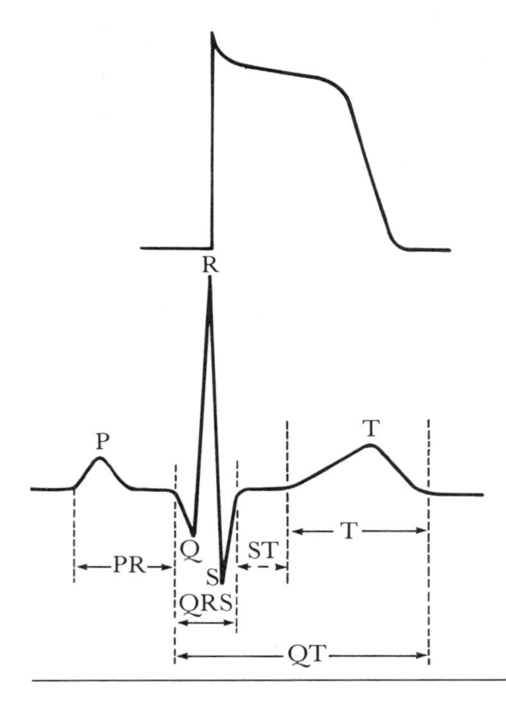

impulses reach the ventricles. A third-degree block is a complete block, in which the ventricles beat independent of and usually at a slower rate than that of the atria. Ventricular tachycardia and fibrillation are more serious than their atrial counterparts because cardiac output is compromised. Atrial fibrillation is more benign because atrial pumping accounts for only a small fraction of ventricular filling.

QUESTIONS

1. What are the four important electrophysiological differences between cardiac and skeletal muscle fibers?
2. Arrange the following structures in order of increasing conduction velocity: right ventricular muscle, the Purkinje system, and the AV node.
3. Which of the following is (are) true of the QRS complex?
   A. It always precedes ventricular contraction
   B. It represents ventricular repolarization
   C. It is caused mainly by current flow through fast Na^+ channels
4. Drug X specifically acts on K^+ channels in the SA node to open them. Describe the effects this drug would have on the heart.
5. Arrange the following in order of decreasing rate of spontaneous discharge: AV node, myocardium, SA node, and His-Purkinje fibers.
6. What is the ECG manifestation of SA node depolarization?
7. Describe what happens in a third-degree (complete) heart block.
8. During which phase of the cardiac action potential is calcium entry important?

CARDIAC CYCLE, OUTPUT, AND MECHANICS

CARDIAC CYCLE. The cardiac cycle, with its corresponding valvular events, heart sounds, pressures and volumes, and the ECG, is shown in Figure 4-4. Systole is usually defined as ventricular contraction and diastole as ventricular relaxation. Atrial events are reflected in the jugular pulse pressure curve, because no valves

Fig. 4-4. *The cardiac cycle, showing left atrial, aortic, and left ventricular pressure pulses correlated in time with aortic flow, ventricular volume, heart sounds, venous pulse, and electrocardiogram. (From R. M. Berne, and M. N. Levy.* Cardiovascular Physiology *[4th ed.]. St. Louis: Mosby, 1981.)*

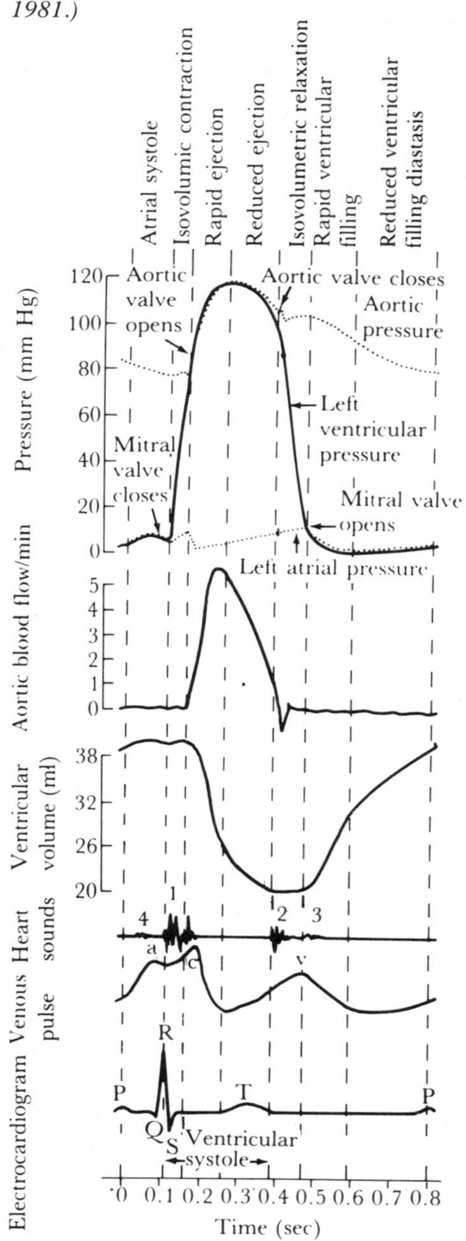

separate the superior vena cava from the right atrium. The a wave in the jugular curve is produced by atrial contraction. When the ventricles contract, the AV valves close and bulge into the atria. The blood coming into the atria suddenly accumulates and the c wave appears. Venous delivery into the atria continues during ventricular systole and the progressive increase in atrial pressure gives rise to the v wave.

Ventricular diastole begins with the closure of the semilunar (aortic and pulmonic) valves. These valves close when the pressures within the ventricles fall below those in the aorta and pulmonary artery, respectively. The first part of the diastole is isovolumic relaxation, during which the ventricles relax, the ventricular pressures drop rapidly, and the ventricular volumes remain constant (because the AV valves are still closed). When the ventricular pressures fall below those of the atria, the AV valves open, marking the end of isovolumic relaxation. A rapid diastolic filling from the atria then occurs, followed by a slower filling during mid-diastole. At the end of diastole, atrial contraction transiently increases the ventricular volume and pressure curves to produce the presystolic a wave.

Ventricular systole begins with closure of the AV valves. These valves close when the ventricles begin to contract and raise ventricular pressures above atrial pressures. The first part of ventricular systole is isovolumic contraction during which the ventricles contract, the ventricular pressures rise rapidly, and the ventricular volumes remain constant (because the semilunar valves are still closed). When the ventricular pressures exceed those in the great arteries the semilunar valves open, isovolumic contraction ends, and systolic ejection begins. Ejection continues until ventricular pressures fall below those of the great arteries. Systole ends with the closure of the semilunar valves. The "rebound" of these valves produces the dicrotic notch, or incisura, in the downstroke of the pressure curves of the great arteries.

The systolic blood pressure is the peak pressure (120 mm Hg) in the aorta during systolic ejection. The diastolic blood pressure is the lowest aortic pressure (80 mm Hg) during diastole. The difference between the systolic blood pressure and the diastolic blood pressure is the pulse pressure. The pulse pressure is narrower in aortic stenosis due to a slower rate of rise in pressure during ejection and wider in aortic insufficiency due to a faster rate of rise in pressure and a lower diastolic pressure. The mean blood pressure is approximately equal to the diastolic blood pressure plus one-third of the pulse pressure.

The first heart sound (S1) is due to closure of the AV valves, the mitral closing before the tricuspid. The second heart sound (S2) comes from closure of the semilunar valves, the aortic closing before the pulmonic. The opening of normal valves is silent. In children, a third sound (S3) may be heard during early diastole because of rapid filling of the ventricles. A fourth heart sound in late diastole may be heard in abnormally stiff ventricles as the atrium contracts to propel blood into the ventricle.

CARDIAC OUTPUT AND MECHANICS. The cardiac output can be measured by the Fick method. This method assumes that the amount of any substance taken up by an organ (or by the whole body) per unit time equals the difference between the aterial and venous concentrations of the substance multiplied by the blood flow through that organ. This can be applied using the body as the organ and oxygen as the substance in order to calculate the cardiac output (CO):

$$\begin{aligned} CO &= \frac{\text{oxygen consumption rate by the body}}{\text{arterial oxygen content} - \text{venous oxygen content}} \qquad (4\text{-}1) \\ &= \frac{250 \text{ ml } 0_2\text{/min}}{190 \text{ ml } 0_2\text{/liter blood} - 140 \text{ ml } 0_2\text{/liter blood}} \\ &= 5 \text{ liter blood/min} \end{aligned}$$

This method is accurate only at constant and normal or relatively low cardiac outputs. At low outputs, blood perfuses the tissues for longer times resulting in a greater arteriovenous oxygen difference (the limiting factor for accuracy).

The CO is equal to the heart rate (HR) times the stroke volume (SV) (the volume of blood ejected in one beat).

$$CO = HR \times SV \qquad (4\text{-}2)$$

In a resting man, the SV is about 80 ml and the heart rate about 70 beats per min. The CO is then about 5.6 liter per min. Factors which affect either the heart rate or the stroke volume control the CO. The heart rate is regulated mainly by the autonomic nervous system. The stroke volume is a function of (1) contractility, (2) preload, and (3) afterload.

Contractility, or the contractile state of the heart, refers to the strength of the heart muscle, or the force which it is capable of developing as it contracts. Contractility can be measured by the maximum rate of pressure (or force) development (max dP/dt) within the ventricle. The stroke volume is equal to the ventricular end-diastolic volume minus the ventricular end-systolic volume. The ejection fraction is equal to the stroke volume divided by the ventricular end-diastolic volume. A more forceful contraction propels more blood into the arteries leaving the ventricles a smaller end systolic volume. Hence stroke volume and ejection fraction increase with contractility. Factors which affect the contractile state of the heart are called inotropic factors. Positive inotropic factors include sympathetic stimulation, catecholamines, increasing the extracellular Ca^{2+} concentration, decreasing the extracellular Na^{+} concentration, digitalis administration, and increasing the heart rate. Many of these effects are ultimately mediated by an increased Ca^{2+} influx into the myocardial fibers. Sympathetic stimulation and catecholamines exert their inotropic effect by turning on the adenylate cyclase–cAMP system, which activates a protein kinase, which in turn opens Ca^{2+} channels to augment the "slow inward" current during the action potential plateau.

The mechanism for digitalis' inotropic effect can be explained by means of a carrier in the cardiac cell membrane which exchanges Ca^{2+} (in) for Na^{+} (out). Digitalis inhibits the Na^{+}-K^{+}–activated ATPase pump in the cell membrane (the same pump found in nerve and skeletal muscle fibers; see Chap. 1). This causes Na^{+} to accumulate in the cell and stimulates the Na^{+}-Ca^{2+} exchange system. Raising the ECF Ca^{2+} or lowering the ECF Na^{+} also stimulates this exchange system. Negative inotropy results from beta-adrenergic blockage (i.e., pharmacologically blocking the catecholamine receptors), heart failure, parasympathetic discharge (to some extent), acidosis, hypoxia, and hypercapnea.

The preload is the ventricular end-diastolic volume or pressure. The relationship between the preload and the stroke volume is described by the Frank-Starling law, which states that the force, or tension, developed in a muscle fiber depends on the degree to which the fiber is stretched. The length of the fiber is determined by the average sarcomere length within the fiber. There is an optimal initial sarcomere length, and thus fiber length, from which the most forceful contraction is obtained. At this length, actin and myosin filaments are most ideally situated to interact with the maximum number of force-generating sites. The heart normally operates at initial (end-diastolic) fiber lengths which are less than this optimal length; therefore an increase in the initial fiber length up to the optimal length produces a more forceful contraction. Applying this to cardiac muscle mechanics, the initial fiber length is equivalent to the left ventricular end-diastolic volume (LVEDV). A ventricular function curve (Fig. 4-5), plotting stroke volume against LVEDV, illustrates Frank-Starling's law. Stroke volume increases with ventricular filling, or preload, up to a certain point. When the muscle fibers are stretched beyond their physiologic limits, cross-bridges between actin and myosin filaments cannot form and contractile force decreases as shown by the dashed line (descending limb of Starling's curve). This effect is not seen under stable physiologic conditions. The LVEDV increases with venous return (blood volume returning to the heart per unit time). Exercise, overtransfusion, and sympathetic venoconstriction increase venous return, while hemorrhage, diuretics, and venodilatation decrease venous return. The LVEDV fails to

Fig. 4-5. *Ventricular function curve, where LVEDV is left ventricular end-diastolic volume and the dashed line represents the descending limb of Starling's curve.*

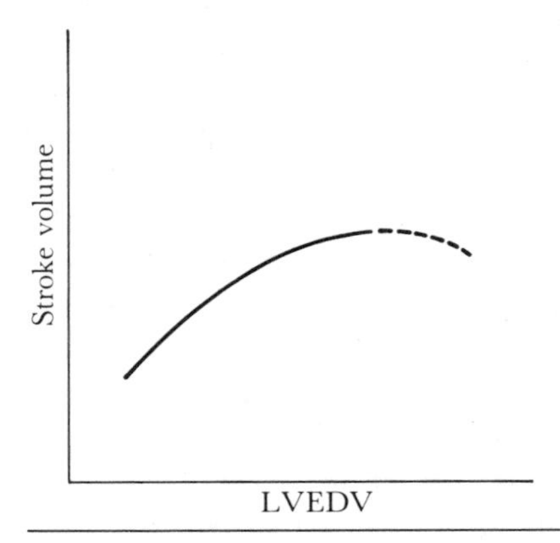

increase normally with decreasing compliance. Thus a hypertophied ventricle, which may have very little compliance, tends to limit ventricular filling.

The ventricular function curve also illustrates the relationship between contractility and the stroke volume (Fig. 4-6). A single curve seen in Figure 4-6 represents a given contractile state of the heart. When contractility is increased (e.g., sympathetic stimulation), the curve is shifted upward, and when contractility is decreased (e.g., heart failure), the curve shifts downward. For a constant preload, the stroke volume can be changed by changing contractility. Similarly, for a given contractile state, the stroke volume can be changed by changing the preload.

The sequence of events in heart failure and the body's compensatory response are shown in Figure 4-7. The first event is a decrease in the contractile state of the heart (the failure itself), which shifts the Starling curve downward. One response is fluid retention (mediated by the kidney), which increases venous return and the LVEDV (preload). By the Frank-Starling mechanism, this increases the stroke volume; but it does so only a little, because the Starling curve is quite flat for failing hearts at high LVEDVs. The body also compensates by eliciting an increased sympathetic discharge. Flooded with catecholamines, the heart pumps in a higher contractile state, and the Starling curve shifts upward to a safer position.

The third factor that determines stroke volume is the afterload, or pressure load, i.e., the total peripheral resistance. This is the load which the heart must

Fig. 4-6. *Ventricular function curves and different contractility states.*

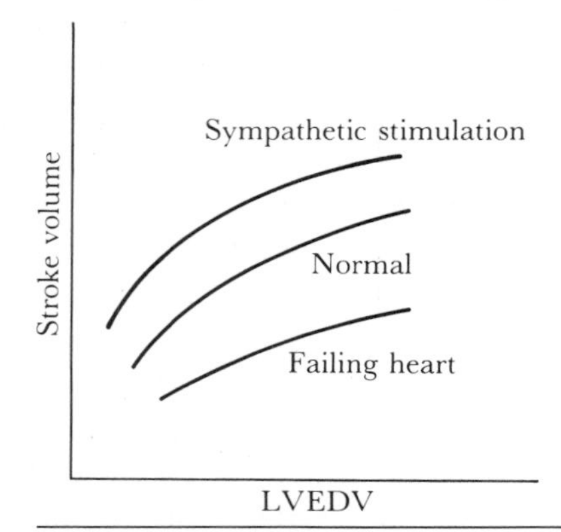

Fig. 4-7. *Heart failure and compensatory steps. A. Pump failure. B. Fluid retention. C. Sympathetic stimulation.*

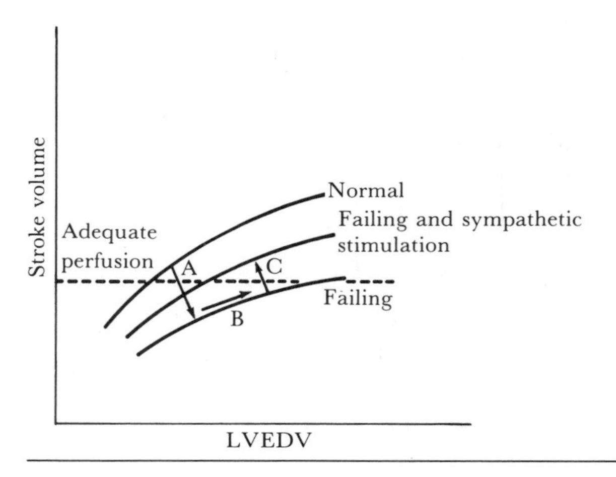

pump against in order to eject blood, and its magnitude is best represented by the diastolic blood pressure. At constant preload and contactility the stroke volume is inversely related to the afterload. By decreasing the peripheral resistance (e.g., with a vasodilator) the stroke volume can be increased.

The changes in ventricular pressure and volume during the cardiac cycle can also be seen in a pressure-volume diagram (Fig. 4-8). An increase in the contractile state is indicated by dashed lines. Compliance is defined as the ratio of volume change to pressure change. The heart is compliant if it takes little pressure to fill the ventricle to a given volume. When the ventricular walls are hypertrophied, the heart is less compliant, or stiffer. Work is done by the heart during the ejection phase and is represented by the area under the ejection phase of the curve (approximated by stroke volume multiplied by systolic pressure). In aortic stenosis, ejection must be carried out at higher pressures (pressure overload), requiring more work (greater area within the pressure-volume loop, Fig. 4-9) and a greater oxygen demand. In aortic regurgitation, blood leaks through the aortic valve into the left ventricle during diastole and

Fig. 4-8. *Pressure-volume loop of left ventricle for a single cardiac cycle showing MC (mitral closure), AO (aortic opening), AC (aortic closure), and MO (mitral opening). The dashed line represents the loop in a state of increased contractility.*

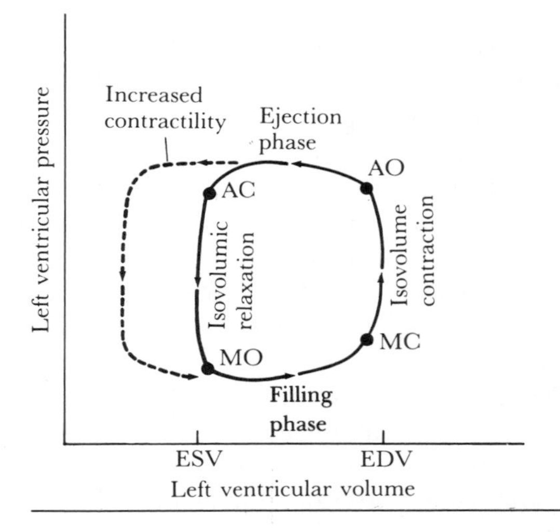

Fig. 4-9. *Pressure-volume loops for normal and pathological states.*

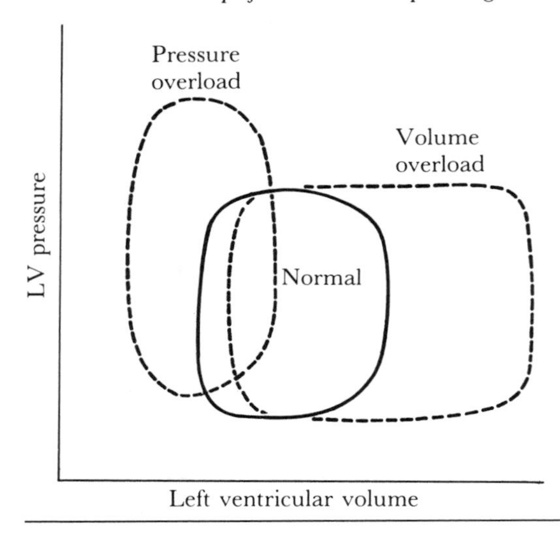

causes an increase in LVEDV. The stroke volume is increased and the work performed by the ventricle is increased due to the volume overload.

QUESTIONS

9. What valvular events mark the beginning and termination of ventricular systole?
10. During which phase does the greatest rate of pressure generation in the ventricles occur?
    A. Isovolumic relaxation
    B. Early diastolic filling
    C. Late diastolic filling
    D. Isovolumic contraction
    E. Systolic ejection
11. Explain the source of heart sounds S1 and S2.
12. A *normal* systolic/diastolic blood pressure of 22/14 is most likely to be measured in the
    A. Left ventricle
    B. Right ventricle
    C. Aorta
    D. Pulmonary artery
13. What valvular events correspond to the maximum and minimum left ventricular volume?
14. The QRS is timed closest to
    A. Mitral closure
    B. Aortic opening
    C. Aortic closure
    D. Mitral opening
15. How does pulse pressure change with aortic stenosis and aortic insufficiency?
16. If the brain consumes oxygen at a rate of 45 ml/min, the carotid arterial oxygen content is 190 ml/liter blood, and the internal jugular venous content is 130 ml/liter blood, what is the cerebral blood flow (total blood flow through the brain)?
17. How is the stroke volume changed by the following:
    A. Digitalis administration
    B. Increasing venomotor tone
    C. Increasing arteriomotor tone

D. Sympathetic stimulation of the heart
E. Hemorrhage
F. Administration of an arteriole vasodilator
G. Uncompensated congestive heart failure
H. Perfusing the heart with a highly concentrated saline solution

18. During exercise, the venous return to the heart is increased. This *alone* allows the heart to eject more blood because of
   A. An increase in contractility
   B. The Frank-Starling mechanism
   C. Decreased total peripheral resistance
   D. Decreased preload
   E. Increased afterload
19. Stroke volume is (are)
   A. Left ventricular end-diastolic volume minus left ventricular end-systolic volume
   B. Ejection fraction times left ventricular end-diastolic volume
   C. The area inside the pressure-volume loop
   D. Ejection fraction times cardiac output
   E. Cardiac output / heart rate
20. Match the letter in the left column with the appropriate number in the right column.

A. Contractile state	1. LVEDV
B. Preload	2. Left ventricular max dP/dt
C. Afterload	3. Aortic systolic pressure
	4. Aortic diastolic pressure
	5. Stroke volume

21. How does the body compensate in acute congestive heart failure?
22. In which case is more work required
   A. Right ventricular pumping
   B. Left ventricular pumping

HEMODYNAMICS

The pressure, velocity of flow, cross-sectional area, and capacity of blood vessels of the systemic circulation are shown in Figure 4-10.

POISEUILLE-HAGEN FORMULA. One important feature of Figure 4-10 is the pressure drop across the arterioles. The arterioles are the major source of resistance to blood flow. The relationship between flow, driving pressure, and resistance is described by a formula analogous to Ohm's law

$$\Delta P = Q \times R \qquad (4\text{-}3)$$

in which ΔP is the difference in pressure between the two ends of the tube, Q is the flow, and R is the resistance. The resistance can be defined in terms of the caliber (radius r) and length (L) of the blood vessel and the viscosity (η) of the blood.

$$R = \frac{8\,\eta\,L}{\pi r^4} \qquad (4\text{-}4)$$

Putting the above two equations together, we describe the Poiseuille-Hagen formula as

$$\Delta P = Q \times \frac{8\,\eta\,L}{\pi r^4} \qquad (4\text{-}5)$$

Fig. 4-10. *Pressure, velocity of flow, cross-sectional area, and capacity of the blood vessels of the systemic circulation. (From R. M. Berne, and M. N. Levy.* Cardiovascular Physiology *[4th ed.]. St. Louis: Mosby, 1981.)*

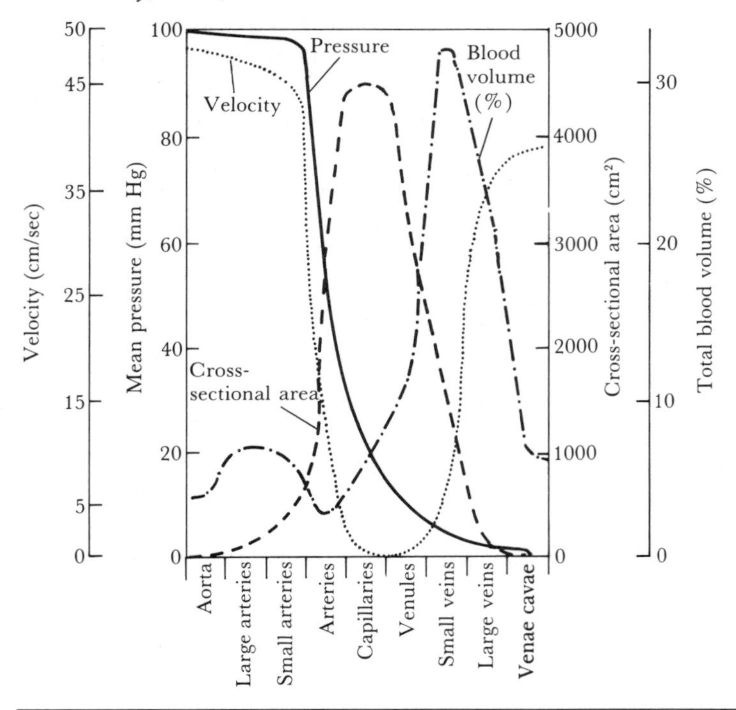

A decrease in the lumen caliber results in a greater pressure drop. Arterioles are the major source of resistance in the systemic circulation because they are the first vessels to impose a significant reduction in lumen caliber. Because the pressure change is inversely proportional to the fourth power of the radius, a small degree of vasoconstriction is sufficient to produce a dramatic change in the pressure. The arterioles contain smooth muscle which is regulated to constrict or dilate; therefore, the arterioles play a role not only as the major resistance factor, but also as the major site of regulation of the systemic arterial blood pressure.

Resistance is also affected by blood viscosity. Viscosity may be elevated in severe polycythemia, macroglobulinemia, and hereditary spherocytosis (red cells abnormally rigid). Viscosity is decreased in severe anemia.

CONTINUITY EQUATION. Figure 4-10 also shows an inverse relationship between the total cross-sectional area and the velocity of blood flow. This relationship is described by the continuity equation

$$\text{Flow} = V_1A_1 = V_2A_2 \tag{4-6}$$

in which V is velocity (cm/sec) and A is cross-sectional area (cm^2). Because the total flow is constant through the various series components of the circulatory system, V and A are inversely proportional. The total cross-sectional area is greatest at the capillary level.

BERNOULLI'S PRINCIPLE. For nonviscous fluids moving in a solid tube, a relationship between the velocity of flow, the transmural pressure (the lateral pressure against the blood vessel wall), and the height of the fluid (with respect to an arbitrarily chosen height, i.e., potential energy = 0 when h = 0) is given by the following equation:

$$E = \frac{1}{2}\rho v^2 + \rho gh + P \qquad (4\text{-}7)$$

in which E is energy density (energy per unit volume), ρ is density of the fluid, v is flow velocity, g is acceleration due to gravity, h is the relative height of the fluid, and P is transmural pressure. Although blood is viscous and blood vessels are distensible, this formula may be applied to a rough approximation. It must be emphasized that the variables in the equation describe properties at a single point in the liquid. Conservation of energy dictates that the energy density be constant. For a given fluid height, a decrease in flow velocity would increase the transmural pressure against the wall. An example of this is an aneurysm of the aorta (dilatation due to weakness in the wall). At the aneurysm, the cross-sectional area is enlarged. Consequently according to the continuity equation, the velocity of blood flow must be slower at the aneurysm since flow must be constant at every point along the aorta (see Equation 4-6). If all other variables in Bernoulli's equation are constant, the transmural pressure at this lesion must increase. The increased pressure pushes the wall out further, dilating the vessel even more, and decreasing further the flow velocity. This vicious cycle may ultimately rupture the aortic wall. Another application of Bernoulli's principle can be seen in the effect of gravity on blood. If all other variables in Equation 4-7 are held constant, and the height of the fluid varied, the transmural pressure changes accordingly. This explains why blood pressures above the level of the heart are lower than those below.

LAW OF LAPLACE. The transmural pressure pushes against the wall and distends the blood vessel. The wall tension is the force which opposes this pressure to keep the vessel intact. Their relationship is described by

$$T = Pr \qquad (4\text{-}8)$$

in which P is the transmural pressure, T is the wall tension, and r is the radius of the vessel lumen. In a blood vessel with a smaller radius, less wall tension is needed to balance the distending pressure. For this reason, capillaries with only endothelial walls do not normally rupture.

In a heart chamber, the Laplace relationship is:

$$T = PR/2h \qquad (4\text{-}9)$$

in which T is the intramyocardial tension, P is the intraventricular pressure, R is the radius of the ventricle, and h is the thickness of the ventricular wall. When the heart is dilated, wall tension must be increased to maintain the same systolic pumping pressure. However, if the ventricular walls are hypertrophied, the thicker walls (increased h) help to allieviate the wall tension in a dilated chamber.

STARLING HYPOTHESIS. The forces which drive fluid across the capillary wall are the osmotic pressure and the hydrostatic pressure. They are related by Starling's equation

$$\text{Flow} = k\,[(P_c + \pi_i) - (P_i + \pi_p)] \qquad (4\text{-}10)$$

in which P_c = capillary hydrostatic pressure, p_i = interstitial fluid hydrostatic pressure, π_p = plasma protein oncotic pressure, π_i = interstitial fluid oncotic pressure, and k = filtration constant for the capillary wall. The capillary hydrostatic pressure pushes fluid from vessel to the interstitial (tissue) space and is increased with arterial or venous blood pressure. It drops along the capillary due to flow of fluid outward. Typical interstitial fluid hydrostatic pressure (5 mm Hg) pushes fluid back into the vessel and is constant throughout the length of the capillary. The plasma protein oncotic pressure (osmotic pressure), about 25 mm Hg, pulls water across the semipermeable capillary membrane (see Chap. 1) from the tissue space into the vessel. The interstitial fluid oncotic pressure (1 mm Hg) comes from the small amount of albumin that escapes from the capillary and pulls water out of the capillary. The net result is filtration of fluid into the interstitial space at the arterial end of the capillary and reabsorption of fluid into the vessel at venous end.

LYMPHATIC CIRCULATION. The lymphatic circulation functions to remove tissue fluid that has not been reabsorbed in the capillaries and to bring this extra fluid back into the circulation. Lymph flow is due to four factors: (1) the pumping action of skeletal muscle movement, (2) the negative intrathoracic pressure during inspiration, (3) the suction effect of high-velocity flow of blood in the great veins into which the lymphatics drain, and (4) the rhythmic contractions of the walls of the large lymph ducts. As veins do, lymph vessels also have valves that prevent backflow.

VENOUS CIRCULATION. The veins are called capacitance vessels because of the large volume of blood they can store without much of a rise in the venous pressure (Fig. 4-10). The venous return, as previously mentioned, is an important determinant of stroke volume and is, in fact, equal to the cardiac output under steady state conditions. Transfusions, fluid retention, venoconstriction, and skeletal muscle pumping of venous flow all augment venous return. Hemorrhage, diuresis, venodilatation, and standing upright (pooling in lower extremities) all lower venous return. The intrathoracic pressure becomes more negative during inspiration and increases venous return whereas expiration decreases venous return.

QUESTIONS

23. Use the following information to calculate the systemic vascular (total peripheral) resistance and pulmonary vascular resistance. A unit of resistance is mm Hg min/liter.
    Mean aortic pressure = 101 mm Hg
    Mean right atrial pressure = 1 mm Hg
    Mean pulmonary artery pressure = 10 mm Hg
    Mean left atrial pressure = 5 mm Hg
    Cardiac output = 5 liters/min
24. A medical student draws a blood sample from a patient, taking a long time to fill the syringe. The venous pressure is 5 mm Hg and the pressure inside the syringe is 1 mm Hg. Can you think of some ways of taking the sample 4 times as quickly?
25. In which type of blood vessel is
    A. The total cross-sectional area greatest
    B. Total blood volume greatest
    C. The pressure greatest
26. Which of the following would cause increased interstitial fluid volume (edema)?
    A. Nephrosis leading to proteinuria
    B. A large burn resulting in loss of capillary membrane selective permeability
    C. Arteriolar dilatation
    D. Heart failure that increases venous pressure

E. Venular constriction
F. Inadequate lymph flow after a radical mastectomy
G. Liver cirrhosis in which hepatic synthesis of plasma proteins is depressed
H. An accumulation of osmotically active substances in the interstitial space
I. An increase in the hydrostatic pressure in the interstitial space

27. In which blood vessel would you expect wall tension to be greatest?

CARDIOVASCULAR REGULATION

The cardiovascular system is regulated mainly by the autonomic nervous system, i.e., the parasympathetic and sympathetic nervous systems. The parasympathetic system controls primarily the heart rate, while the sympathetic system controls the heart rate, heart contractility, arteriolar resistance, and venous capacitance. The source of these neural controls lies within specific nuclei in the medulla, whose activities are modulated by peripheral baroreceptors and chemoreceptors and by inputs from higher brain centers as well.

The parasympathetic innervation of the heart acts predominantly on the SA node, atria, and AV node. Increased activity of vagal fibers slows the heart rate. The cardioinhibitory center in the medulla initiates a tonic vagal discharge at rest; therefore, the rate of the SA node in a denervated heart is higher than the rate observed in a heart with intact innervation. Less important parasympathetic effects on the heart include a decrease in atrial contractility and in the conduction velocity in the AV node.

The sympathetic regulation of the heart is more extensive. The cardiac receptors which mediate the sympathetic effects are of the beta-1-adrenergic type. The major transmitter is norepinephrine from sympathetic nerve terminals although epinephrine in the bloodstream from the adrenal medulla also plays a role. The responses to sympathetic stimulation include

1. SA node — increased heart rate
2. Atria — increased contractility and conduction velocity
3. AV node and His-Purkinje system — increased automaticity and conduction velocity
4. Ventricles — increased contractility, conduction velocity, and automaticity

Clearly, the effect of sympathetic stimulation to the heart is a rise in cardiac output due to a faster heart rate and a larger stroke volume.

While there is very little parasympathetic innervation of the vascular system, nearly all vessels of the body are supplied with sympathetic nerve fibers. The postsynaptic adrenergic receptor which mediates vasoconstriction is of the alpha-1 type, and the principal transmitter is norepinephrine. There is also a tonic discharge from these vasoconstrictor fibers maintaining a partial state of constriction in blood vessels. This continuous discharge is initiated by the vasomotor center in the medulla. The result of increased sympathetic activity is increased arteriolar constriction and a rise in blood pressure. Veins are also innervated by vasoconstrictor fibers, and venoconstriction results in a depletion of blood volume in the venous reservoirs and a greater venous return.

In the coronary and skeletal muscle arterioles, there are beta-2-adrenergic receptors which respond to sympathetic stimulation by vasodilatation. Sympathetic stimulation occurs by means of circulating epinephrine or sympathetic vasodilator fibers which innervate these arterioles and release norepinephrine. However, more important than adrenergic stimulation are the vasodilator effects of local metabolites produced or depleted in working muscle. A lowered oxygen tension, and an increase in adenosine, various kinins, and histamine all serve to dilate arterioles in the heart and skeletal muscle.

The medullary cardiovascular centers receive inputs from special receptors in the periphery to complete the feedback loop. Arterial baroreceptors are located in the carotid sinus and aortic arch. These receptors are sensitive to stretch and send afferent fibers via cranial nerves IX and X to the medulla. An elevation in arterial pressure increases the rate of discharge from the receptors. This produces an increase in parasympathetic discharge, which slows the heart rate, and a decrease in sympathetic discharge to the vessels, causing the arterioles and veins to dilate. The net result is a reduction in cardiac output, peripheral resistance, and blood pressure. A decline in the arterial blood pressure produces the opposite effects: decreased parasympathetic discharge, a faster heart rate, increased sympathetic discharge, and arterial and venous constriction.

The aortic and carotid bodies are chemoreceptors sensitive to oxygen tension, carbon dioxide tension, and pH. These receptors play a more important role in the control of respiration, but the oxygen tension also regulates the cardiovascular system in that its decrease triggers a reflex rise in heart rate and blood pressure. This is reasonable because blood is now carrying less oxygen and tissues require more blood per unit time to get the same amount of oxygen.

Higher brain centers can directly activate the medullary cardiovascular centers. Excitement, fear, anger, and painful stimuli accelerate the heart rate and raise the blood pressure. Other stimuli such as grief slow the heart rate and lower the blood pressure. Several other important circumstances also affect the heart and blood vessels. Onset of inspiration increases the heart rate and blood pressure; expiration does the opposite. A rise in intracranial pressure reduces blood flow to the medullary vasomotor center which increases its rate of discharge. The resultant vasoconstriction raises the blood pressure which in turn slows the heart rate. An increase in body temperature is sensed by the hypothalamus, which activates the sympathetic system to dilate the skin blood vessels. Table 4-1 provides a summary of factors affecting the caliber of arterioles, including factors not previously discussed.

The integration of these regulatory mechanisms can be seen when someone rises from the supine to the upright position. Gravity causes blood to pool in the venous capacitance vessels of the lower extremities, which reduces venous return. This reduces stroke volume and thus cardiac output, which in turn lowers the blood pressure. The initial drop in blood pressure is sensed by baroreceptors in the carotid sinus and aortic arch. The compensatory responses are an increased heart rate and arteriolar constriction. The increase in heart rate

Table 4-1. *Summary of Factors Affecting the Caliber of the Arterioles*

Constriction	Dilatation
Increased noradrenergic discharge	Decreased noradrenergic discharge
Circulating catecholamines (except epinephrine in skeletal muscle and liver)	Circulating epinephrine in skeletal muscle and liver
Circulating angiotensin II	Activation of cholinergic dilators in skeletal muscle
Locally released serotonin	Histamine
Decreased local temperature	Kinins
	"Axon reflex"
	Decreased oxygen tension
	Increased carbon dioxide tension
	Decreased pH
	Lactic acid, potassium, adenosine
	Increased local temperature

Reproduced with permission from W. F. Ganong. *Review of Medical Physiology* (12th ed.). Los Altos, CA: Lange, 1985.

is not quite enough to balance the diminished stroke volume, so the cardiac output is still somewhat decreased. Arteriolar constriction increases the total peripheral resistance to maintain arterial blood pressure. Without such compensatory responses, the reduction in cardiac output would compromise cerebral blood flow which might result in fainting. If the individual moves around (e.g., walks), the pumping action of muscle assists the venous return.

Exercise also illustrates cardiovascular regulation. During exercise, the heart increases its output both by speeding up its rate and by pumping out a greater stroke volume. The increase in heart rate is due to decreased parasympathetic discharge and increased sympathetic discharge. The larger stroke volume may be attributed to a higher contractile state as a result of sympathetic stimulation and an increased venous return, due to sympathetic venoconstriction and the pumping action of muscle during exercise. The increase in heart rate accounts for most of the increase in cardiac output. In the periphery, the cardiac output is redistributed to favor working muscles. Local metabolites vasodilate arterioles in working muscle to increase blood flow. In nonworking organs, sympathetic vasoconstriction reduces blood flow. For isotonic (dynamic) exercise, the net result is that systolic blood pressure is raised modestly while diastolic blood pressure remains relatively constant. The mean blood pressure is constant or slightly elevated and the systemic vascular resistance is significantly lowered. The muscle tissue also extracts a greater amount of oxygen per unit blood volume during exercise, producing a wider arteriovenous oxygen difference. The body is capable of increasing oxygen consumption up to 20 times during exercise. To meet this demand, the heart rate may increase by a factor of 4, the stroke volume by 1.5, and the arteriovenous oxygen difference by 3.3.

QUESTIONS

28. In a transplanted (denervated) heart, why does cardiac output increase during exercise?
29. What are some of the immediate cardiovascular compensatory reactions to a hemorrhage?
30. During jogging, determine whether the following parameters increase, decrease, or remain the same:
    A. Heart rate
    B. Contractility
    C. Sympathetic tone
    D. Parasympathetic tone
    E. Mean arterial blood pressure
    F. Total peripheral resistance
    G. Blood flow through the superior mesenteric artery
    H. Lumen caliber in arterioles in the grastrocnemius muscle
    I. Blood volume in the venous circulation
    J. Arteriovenous oxygen difference
    K. Stroke volume
31. How does increased sympathetic activity affect the following structures?
    A. SA node
    B. Atria
    C. AV node and His-Purkinje system
    D. Ventricles
32. What controls vascular resistance in active skeletal muscle?
33. Which of the following occur when the carotid sinus is stimulated?
    A. Decreased norepinephrine release at the SA node
    B. Increased acetylcholine release at the SA node
    C. Decreased total peripheral resistance
    D. Decreased arterial blood pressure

34. In the normal adult at rest, which autonomic system predominantly influences the heart rate?
35. In rising from a supine to standing position, determine whether the following increases, decreases, or remains unchanged:
    A. Heart rate
    B. Cardiac output
    C. Stroke volume
    D. Transmural pressure across the femoral vein
    E. Total peripheral resistance
    F. Venous blood volume.

CIRCULATION THROUGH SPECIFIC ORGANS

The distribution of the cardiac output to various organs and their oxygen consumption rates are shown in Table 4-2. On a per gram basis, the kidneys receive by far the largest blood flow of any organ, and the heart consumes oxygen at the highest rate. Blood flow and oxygen consumption rate per gram of tissue is exceptionally high in four organs: heart, kidney, liver, and brain. This reflects the fact that such organs are metabolically highly active.

CEREBRAL CIRCULATION. The blood flow through the brain is about 750 ml per minute. Because the brain is extremely sensitive to ischemia, its circulation is regulated in such a way that the total cerebral blood flow (CBF) is maintained at a constant despite changes in blood pressure. Using Equation 4-4, the overall CBF can be defined in terms of the mean arterial pressure (MAP), the internal jugular pressure (IJP), and the cerebral vascular resistance (CVR).

$$CBF = \frac{MAP - IJP}{CVR} \tag{4-11}$$

The MAP depends on the cardiac output and the total peripheral resistance. The CVR is a function of intracranial pressure, the caliber of cerebral arterioles, and blood viscosity (see Equation 4-4). An increase in the intracranial pressure compresses the cerebral vessels which increases the CVR. The vascular diameter is regulated on a regional level by local factors. Regional blood flow correlates well with regional metabolic activity. A local increase in carbon dioxide tension or a local decrease in oxygen tension or pH causes vasodilation. The role of sympathetic innervation is minimal in cerebral vessels.

Table 4-2.
Distribution of Cardiac Output to Various Organs and Their Oxygen Consumption Rates

		Blood Flow		Arteriovenous O_2 Difference	O_2 Consumption		Resistance in R units		Percentage of Total	
Region	Mass (kg)	ml/min	ml/100 g/min	(ml/liter)	ml/min	ml/100 g/min	Absolute	per kg	Cardiac Output	O_2 Consumption
Liver	2.6	1500	57.7	34	51	2.0	3.6	9.4	27.8	20.4
Kidneys	0.3	1260	420.0	14	18	6.0	4.3	1.3	23.3	7.2
Brain	1.4	750	54.0	62	46	3.3	7.2	10.1	13.9	18.4
Skin	3.6	462	12.8	25	12	0.3	11.7	42.1	8.6	4.8
Skeletal muscle	31.0	840	2.7	60	50	0.2	6.4	198.4	15.6	20.0
Heart muscle	0.3	250	84.0	114	29	9.7	21.4	6.4	4.7	11.6
Rest of body	23.8	336	1.4	129	44	0.2	16.1	383.2	6.2	17.6
Whole body	63.0	5400	8.6	46	250	0.4	1.0	63.0	100.0	100.0

* Reproduced with permission from P. Bard. (ed.). *Medical Physiology* (11th ed:). St. Louis: Mosby, 1961.

A special feature of the cerebral circulation is autoregulation to maintain a constant overall CBF. The MAP is normally 90 mm Hg. It can fall to 60 mm Hg without any compromise in the overall CBF as a result of vasodilatation. The MAP can rise up to 150 mm Hg with no change in the CBF because of vasoconstriction. The mechanism for this autoregulation is not clear. Another unique feature of the cerebral circulation is the blood-brain barrier, which excludes hydrogen ion, bicarbonate, mannitol, proteins, and any charged or highly polar substances from the brain by means of nonporous capillaries and the neuroglia. The function of this barrier is to maintain the necessary ionic composition in the cerebral spinal fluid, the ECF that bathes the brain.

PULMONARY CIRCULATION. While a constant flow is maintained in the cerebral circulation, the pulmonary circulation maintains a constant pressure. The pulmonary blood flow is the cardiac output (because the right heart output must equal the left heart output), which obviously cannot remain constant. Unlike the systemic vessels, the pulmonary vessels have thin walls, little smooth muscle, and little neural regulation. The pulmonary circulation is a low pressure system (systolic 20 mm Hg, diastolic 10 mm Hg) and a low resistance system (2 resistance units compared to 20 resistance units in the systemic vasculature). The vessels are capable of autoregulation in that resistance can be decreased in the face of an increased cardiac output. This mechanism is so effective that there is no change in pulmonary artery pressure until cardiac output is increased 2½ times. This is accomplished by recruiting more of the pulmonary capillary bed, not by arteriolar dilatation. In addition, local regulation occurs in order to optimize the ventilation-perfusion ratio (see Chap. 6). A fall in the alveolar oxygen tension or arteriolar pH causes vasoconstriction, just opposite of what happens in systemic arterioles. Passive factors also play a role. Decreased lung volume, increased intrathoracic pressure, and increased alveolar pressure all decrease pulmonary blood flow. For example, during expiration, intrathoracic pressure increases, and lowers the pressure gradient from the systemic veins to the intrathoracic cavae, thus causing a decreased venous return to the right heart. The right ventricular output decreases, and so does pulmonary blood flow. The opposite occurs with inspiration. Bernoulli's equation (see Equation 4-7) holds true for the pulmonary circulation in that the pulmonary artery pressure at the apices of the lung (in the erect man) is about 10 mm Hg lower than at the level of the heart. At the bases the pressure is about 10 mm Hg higher. Blood flow at the bases of the lungs is greater than at the apices. Applying Equation 4-4 to the pulmonary circulation, the pulmonary flow (right ventricular cardiac output [CO]) can be related to the pulmonary artery pressure (PAP), the left atrial pressure (LAP), and the pulmonary vascular resistance (PVR).

$$\mathrm{CO} = \frac{\mathrm{PAP} - \mathrm{LAP}}{\mathrm{PVR}} \tag{4-12}$$

Rearranging gives

$$\mathrm{PAP} = \mathrm{CO} \times \mathrm{PVR} + \mathrm{LAP} \tag{4-13}$$

Equation 4-13 indicates that pulmonary hypertension (increased PAP) may be caused by an increase in either the CO, PVR, or LAP. An increase in PVR can result from pulmonary arteriolar vasoconstriction (e.g., secondary to alveolar hypoxia), fibrosis in vessel walls due to diffuse lung disease, or pulmonary emboli. A common cause of pulmonary hypertension is a rise in the LAP as a result of left heart failure or mitral stenosis. In left ventricular failure, the left ventricular end-diastolic volume and pressure increase, which is transmitted to the left atrium and then the pulmonary capillary. The increase in pulmonary capillary pressure causes fluid to be moved into the interstitial space, including the alveolar space. The result is called pulmonary edema and its consequences include poor gas exchange and dyspnea.

CORONARY CIRCULATION. The blood vessels that supply the myocardium are compressed when it contracts. In the left ventricle, ventricular pressure is slightly greater than aortic pressure during systole but much less than the aortic pressure during diastole. As a result, flow through the arteries that supply the subendocardial portion of the left ventricle occurs only during diastole. Because diastole is shortened to a much greater degree than systole when the heart rate is increased, left ventricular coronary flow is reduced during tachycardia. The left ventricular subendocardium is vulnerable to ischemia and is the most common site of myocardial infarction. In the right ventricle and atria, aortic pressure is greater than right ventricular and atrial pressure during systole as well as diastole. Coronary flow in those parts of the heart continues throughout the cardiac cycle.

The heart extracts large amounts of oxygen from the blood and yields the greatest arteriovenous oxygen difference compared to any other organ. Oxygen consumption increases with heart rate, contractile state, intraventricular pressure or volume, myocardial wall tension, and total muscle mass. Oxygen consumption by the heart cannot be significantly increased by further extraction of oxygen from the blood but only by increasing coronary flow. As previously mentioned, the most important factors affecting the coronary arteriole lumen caliber are local metabolites. A drop in the oxygen tension or pH, or a rise in carbon dioxide, potassium, lactate, or adenosine bring about vasodilatation. The coronary arterioles also receive neural innervation from noradrenergic vasoconstrictive fibers, but autonomic regulation is of minor importance. The lumen may be occluded by atherosclerosis and, less commonly, by coronary artery vasospasm. The former is the major cause of myocardial infarctions. Angina pectoris is a paroxysmal chest pain caused by a transient ischemia of the myocardium (without infarction) due to coronary artery atherosclerotic disease. Variant (i.e., nonexercise related) angina arises from coronary artery vasospasm.

QUESTIONS

36. Through which organ (excluding the lungs) is
    A. The percentage of cardiac output greatest
    B. The blood flow per gram tissue greatest
    C. The oxygen consumption per gram tissue greatest
37. During vigorous exercise, an athlete's cardiac output increases fivefold. How much does his cerebral blood flow increase?
38. When cardiac output doubles, what happens to pulmonary vascular resistance and pulmonary arterial pressure?
39. During ventricular systole, what happens to coronary flow in the right and left ventricular subendocardium?
40. What factors are most important in determining myocardial oxygen consumption rate?

ANSWERS

1. A. The cardiac action potential has a plateau and consequently a longer refractory period.
   B. The resting membrane potential of some cardiac cells (nodal cells) spontaneously depolarizes to produce automaticity.
   C. Cardiac muscle fibers are stimulated by excitation from neighboring muscle cells whereas skeletal muscle fibers are stimulated directly by motor neurons.

D. Excitation-contraction coupling in myocardial cells is dependent on extracellular calcium but not in skeletal muscle cells.

2. AV node, right ventricular muscle, then the Purkinje system.
3. A and C
4. It would produce an increased K^+ conductance which would slow the rise in resting membrane potential during phase 4 causing bradycardia.
5. SA node, AV node, His-Purkinje fibers, and myocardium
6. There is none.
7. In a complete heart block, none of the atrial impulses are transmitted to the ventricles. The ventricles beat at an independent and slower rate than the atria and the P wave is dissociated from the QRS complex.
8. During phase 2, the plateau.
9. Atrioventricular valve closure marks the beginning and semilunar valve closure the termination.
10. Isovolumic contraction.
11. S1 is due to atrioventricular valve closure and S2 to semilunar valve closure.
12. D. Pulmonary artery. The systolic ventricular pressures are equal to the systolic pressures in the corresponding great arteries. The diastolic ventricular pressures are significantly less than the corresponding diastolic pressures in the great arteries.
13. Mitral valve closure (end-diastolic volume) and aortic valve closure (end-systolic volume).
14. A. Mitral closure.
15. Decreases and increases.
16. Using Equation 4-1,

$$\text{Cerebral blood flow} = \frac{45 \text{ ml } 0_2\text{/min}}{(190 \text{ ml } 0_2\text{/liter blood}) - (130 \text{ ml } 0_2\text{/liter blood})}$$

$$= 0.75 \text{ liter blood/min}$$

17. A. Increase
    B. Increase
    C. Decrease
    D. Increase
    E. Decrease
    F. Increase
    G. Decrease
    H. Decrease
18. B
19. A, B, and E
20. A. 2
    B. 1
    C. 4
21. By retaining fluids to increase venous return, which increases LVEDV, which in turn slightly increases stroke volume, and by sympathetic stimulation of the myocardium to increase contractility.
22. A. Left ventricular pumping, because it involves pumping against a higher pressure.
23. Using Ohm's law, which states that $R = \Delta P/Q$,

    systemic vascular resistance = (101 mm Hg − 1 mm Hg)/5 liters/min = 20 units

    pulmonary vascular resistance = (10 mm Hg − 5 mm Hg)/5 liters/min = 1 unit

24. According to Equation 4-5, the flow $Q = \Delta P \pi r^4 / 8 \eta L$. Anything that increases the flow through the needle by a factor of 4 will speed up the blood

drawing fourfold. One way is to use a needle with a radius larger by a factor of $\sqrt{2}$. This increases the r term by a factor of 4. Another way is to shorten the needle length by a factor of 4. A third way is to raise the venous pressure to 17 mm Hg by lowering the arm (increasing the P term fourfold) or using a vein in the lower extremity (ouch!), if the arm is too short. Lastly, though somewhat drastic, the medical student can cause an anemia in the patient until the blood viscosity is reduced by a factor of 4.

25. A. Capillaries
    B. Veins
    C. Aorta
26. All except I. See Equation 4-10.
27. The aorta, because lumen radius is greatest there. See Equation 4-8.
28. The heart rate and contractility is increased by epinephrine in the bloodstream. Because venous return also increases during exercise, the Frank-Starling mechanism assures a greater stroke volume.
29. The reduced blood volume results in less venous return and less cardiac output. Blood pressure decreases and the arterial baroreceptors are stretched to a lesser degree. A decreased parasympathetic discharge produces tachycardia. An increased sympathetic outflow produces vasoconstriction and venoconstriction. The increased cardiac output and systemic vascular resistance raises the blood pressure so that blood perfuses the brain adequately.
30. A. Increased
    B. Increased
    C. Increased
    D. Decreased
    E. Unchanged
    F. Decreased
    G. Decreased
    H. Increased
    I. Decreased
    J. Increased
    K. Increased
31. A. Increased heart rate
    B. Increased contractility and conduction velocity
    C. Increased automaticity and conduction velocity
    D. Increased contractility, conduction velocity, and automaticity
32. Local metabolites.
33. All of them.
34. The parasympathetic system.
35. A. Increased
    B. Decreased
    C. Decreased
    D. Increased
    E. Increased
    F. Increased
36. A. Liver
    B. Kidneys
    C. Heart muscle
37. Very little.
38. Pulmonary vascular resistance is reduced and pulmonary arterial pressure remains unchanged.
39. It decreases in the right ventricle and stops in the left ventricle.
40. Heart rate, contractile state, intraventricular pressure and volume, and total muscle mass.

BIBLIOGRAPHY

Berne, R. M., and Levy, M. N. *Cardiovascular Physiology* (4th ed.). St. Louis: Mosby, 1981.

Ganong, W. F. *Review of Medical Physiology* (12th ed.). Los Altos, CA: Lange, 1985.

Perlroth, M. G. (ed.). *Cardiovascular Physiology Syllabus.* Lecture handout, Stanford University School of Medicine, 1985.

Petersdorf, R. G., et al. (eds.). *Harrison's Principles of Internal Medicine* (10th ed.). New York: McGraw-Hill, 1983. Pp. 1311–1352.

5 Renal Physiology

Charles H. Tadlock

The composition and constancy of the body fluids are regulated primarily by kidney function. Fluctuations in fluid balance, electrolyte concentrations, acid-base status, and blood pressure affect renal function in maintaining homeostasis. Losses from skin, lungs, and intestine are important (Table 5-1); however, the responsibility for adjusting solute and water excretion is borne by the kidney. By the process of filtration, secretion, and reabsorption, the kidney directly regulates the quantity and composition of the urine (Table 5-2) and thereby indirectly regulates that of the other body fluid compartments. These processes result in the reabsorption of glucose, amino acids, bicarbonate, sodium, potassium and chloride and the elimination of urea, uric acid, creatinine, phosphates, sulfates, and hydrogen ions (Table 5-3). The mechanisms that enable the kidney to respond to daily alterations in fluid intake and diet, and to excrete waste products in amounts that precisely balance the quantities acquired by ingestion and metabolic transformation, are the subject matter of renal physiology.

QUESTIONS

1. Match the substances in the right column with the processes in the left column.

A. Entirely reabsorbed by the kidneys	Urea
B. Primarily reabsorbed by the kidneys	Glucose
C. Primarily excreted by the kidneys	Na^+
	Uric acid
	H^+
	K^+
	Phosphates
	Sulfates
	Cl^-
	HCO_3^-
	Creatinine
	Water

Table 5-1. *Usual Daily Sources of Water Intake and Loss for the Average Adult*

Source	Volume (ml)
Intake	
Fluid intake	1200–1800
Water content of ingested food	700–1000
Water of oxidation	250–300
Total	2000–3000
Output	
Urine	1500–2000
Insensible loss	
Skin	300–600
Lungs	200–400
Gastrointestinal loss (water content of feces)	100
Total	2000–3000

Reproduced with permission from W. G. Walker, and A. Whelton. Disorders of Water and Electrolyte Balance. In A. M. Harvey, R. J. Johns, V. A. McKusick, A. H. Owens, and R. S. Ross (eds.), *The Principles and Practice of Medicine* (21st ed.). Norwalk, CT: Appleton-Century-Crofts, 1984. P. 45.

Table 5-2. *Composition of Urine*

Characteristic	Range
Volume	400 ml/day–> 4 liters/day
Osmolality	50–1200 mosm/liter
pH	4.5–7.8
[Na^+]	0–300 meq/liter
[K^+]	20–70 meq/liter

2. Estimate normal urine flow in milliliters per minute during
   A. A water conserving state
   B. A water excess state
   C. What would be a reasonable estimate of the osmolality of the urine in A and B?

ENDOCRINE FUNCTIONS OF THE KIDNEY

The kidneys function not only as excretory organs but also as endocrine organs. Functional kidney tissue is necessary for the activation of erythropoietin from its protein precursor in the blood and for the conversion of circulating 25-hydroxycholecalciferol to 1,25-dihydroxycholecalciferol (vitamin D). Furthermore, the hormone renin is manufactured and released by modified smooth muscle cells in the walls of the afferent arterioles. Acting via the renin-angiotensin-aldosterone system, renin affects every aspect of water and electrolyte balance. Prostaglandins are also produced by the kidneys; their exact role(s) have yet to be clarified.

QUESTION

3. True or false. Renal failure can cause
   A. Anemia
   B. Ca^{2+} deficiency and bone disease
   C. Hyperkalemia
   D. Metabolic acidosis
   E. Uremia

FUNCTIONAL ANATOMY OF THE KIDNEY

MACROSCOPIC ANATOMY OF THE KIDNEY. The kidney is divided into the outer cortex and inner medulla (Figs. 5-1, 5-2). Extensions of renal cortex or columns divide the medulla into distinct pyramidal-shaped regions, the renal pyramids. The apex of each of the renal pyramids is the papilla and opens into a

Table 5-3. *Renal Handling of Water, Na^+, K^+, HCO_3^-, Glucose and Solute*

	Filtered	Excreted	Reabsorbed	Filtered Load Reabsorbed (%)
Water (liters/day)	180	1.5	178.5	99.2
Na^+ (meq/day)	25,000	150	24,850	99.4
K^+ (meq/day)	900	70*	900	100
HCO_3^- (meq/day)	4300	2	4298	99.9
Cl^- (meq/day)	18,000	150	17,850	99.2
Glucose (mmoles/day)	900	0.5	899.5	99.9+
Solute (mosm/day)	52,000	800	51,200	98.5

* By secretion, rate highly variable.
Adapted with permission from H. Valtin. *Renal Function: Mechanisms Preserving Fluid and Solute Balance in Health* (2nd ed.). Boston: Little, Brown, 1983. P. 8.

Fig. 5-1. *A. Superficial cortical and juxtamedullary nephrons and their vasculature. The glomerulus plus the surrounding Bowman's capsule are known as the renal corpuscle. There is some overlapping nomenclature; for example, the loop of Henle consists of the proximal straight tubule, the descending and ascending thin limbs, and the ascending thick limb, even though the first and last parts are also considered to belong to the proximal and distal tubules, respectively. The beginning of the proximal tubule, the so-called urinary pole, lies opposite the vascular pole, where the afferent and efferent arterioles enter and leave the glomerulus. The ascending thick limb of the distal tubule is always associated with the vascular pole belonging to the same nephron; the juxtaglomerular apparatus is located at the point of contact (see also Fig. 5-19). B. Capillary networks have been superimposed on the nephrons illustrated in A. (From H. Valtin.* Renal Function: Mechanisms Preserving Fluid and Solute Balance in Health *[2nd ed.]. Boston: Little, Brown, 1983. P. 4.)*

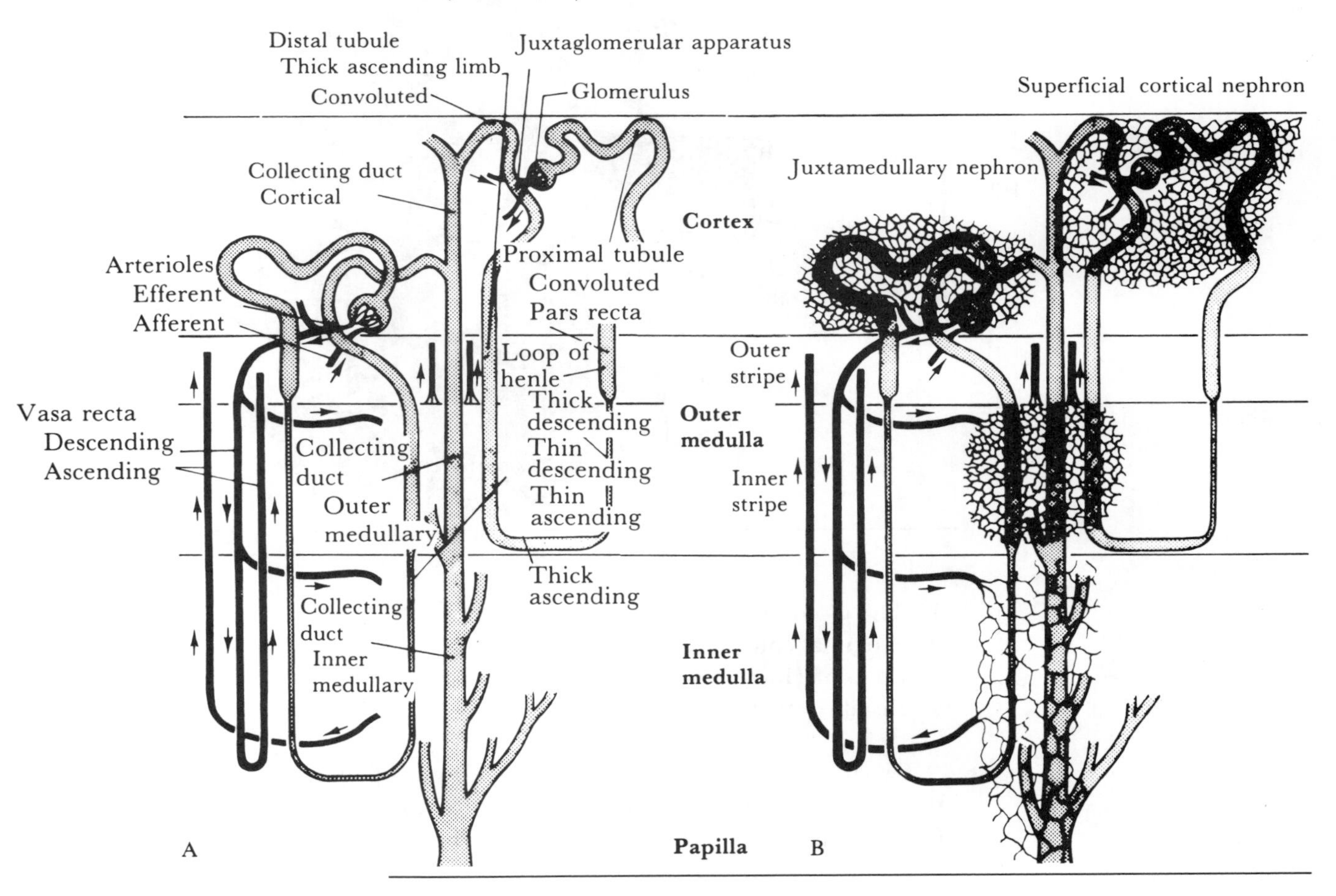

minor calyx. The minor calyces join together to form the funnel-shaped renal pelvis, which is the dilated proximal portion of the ureter into which the urine flows.

MICROSCOPIC ANATOMY OF THE KIDNEY. The functional and structural unit of the kidney is the nephron. Each human kidney consists of approximately one million nephrons all of which ultimately contribute fluid to the collecting system. The distal tubules of the nephrons merge into the collecting system forming sequentially larger collecting ducts until ultimately only 10 to 25 open at the level of the papilla to empty into each of the minor calyces.

The nephron is divided into several histologically and functionally distinct segments, each of which occupies a characteristic location in the cortex or medulla.

Fig. 5-2. *Sagittal section of a human kidney showing the major gross anatomical features. The renal columns are extensions of cortical tissue between the medullary areas. (From J. B. West.* Best and Taylor's: Physiological Basis of Medical Practice *[11th ed.]. P. 451. © 1985 The Williams & Wilkins Co., Baltimore.)*

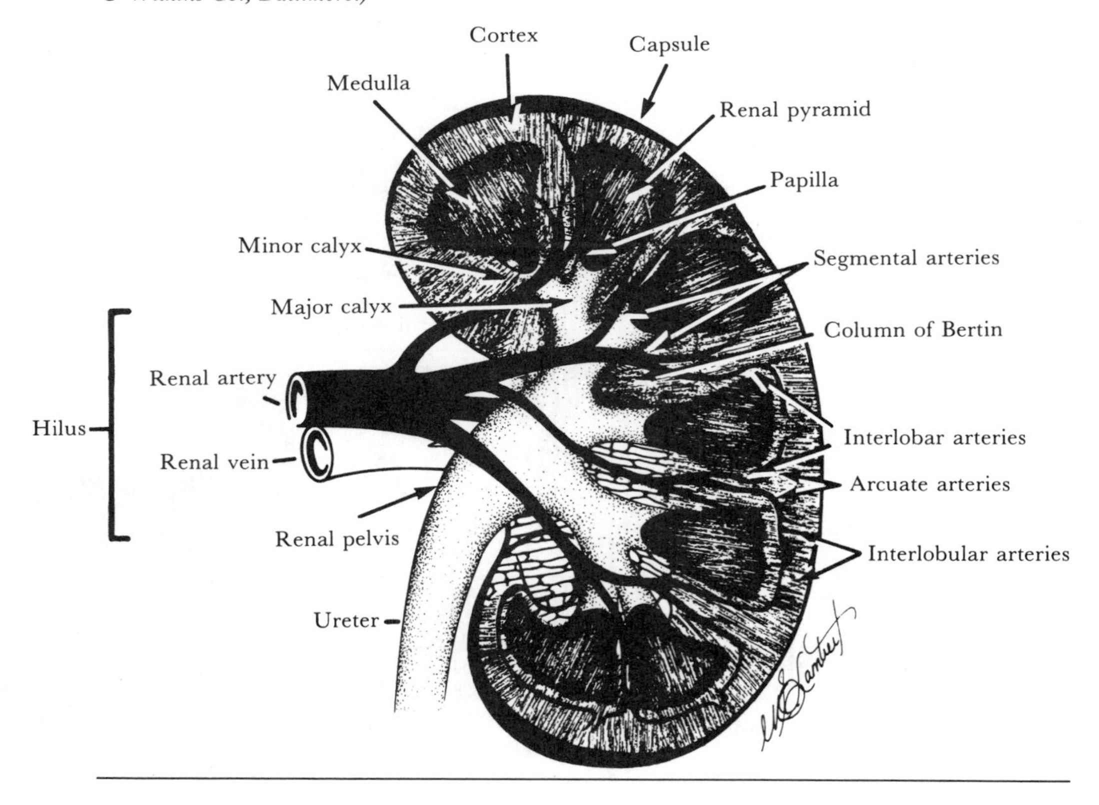

1. The glomerulus
2. Proximal convoluted tubule
3. Loop of Henle
4. Distal convoluted tubule

Two distinct classes of nephrons are recognized on the basis of the location of their glomeruli in the cortex: outer cortical and juxtamedullary (see Fig. 5-1). Outer cortical nephrons are distinguished by

1. The location of their glomeruli in the outer portion of the cortex
2. A short loop of Henle extending only into the outer medulla
3. Efferent arterioles that divide to form the peritubular capillary plexus

Juxtamedullary nephrons are distinguished by

1. The location of their glomeruli in the cortex near the corticomedullary junction
2. A long loop of Henle extending deep into the medullary substance with thin descending and thin ascending limbs
3. Efferent arterioles that form the vasa recta in addition to contributing to the peritubular plexus

Numerically, the short-looped outer cortical nephrons are the overwhelming majority composing seven-eighths of the total, the remainder being juxtamedullary.

The collecting system consists of cortical, medullary, and papillary collecting ducts.

The formation of urine begins in the glomerulus in which the ultrafiltrate of plasma is extruded into Bowman's space, the proximal end of the nephron. After passing through and being modified by the various tubule segments, the fluid in its final form, the urine, passes into the calyceal system. No further modification of the urine is now possible. The urine then flows through the minor and major calyces, into the renal pelvis, ureter, bladder, urethra, and out of the urethral orifice.

THE RENAL CIRCULATION. The renal artery divides successively into the interlobar, arcuate, and interlobular arteries and finally into the afferent arterioles. The afferent arteriole enters the glomerulus and forms the glomerular tuft of capillaries and then reforms to exit the glomerulus as the efferent arteriole. The term glomerulus, in strict usage, refers only to the glomerular tuft of capillaries. Efferent arterioles of cortical nephrons then form a second capillary bed around the cortical tubules (i.e., the peritubular capillary plexus). Efferent arterioles exiting from juxtamedullary nephrons contribute to the peritubular plexus and also form an additional capillary bed, the descending vasa recta, which descends straight down into the medulla (see Fig. 5-1 B). Both of these capillary beds drain into a common venous plexus. The thin limbs of the loop of Henle are surrounded by this plexus. The alignment of capillary beds in series and their arrangement about the tubular system makes possible the reabsorption of the essential substrates, including water, that were first filtered out in the glomerular capillary bed. The venous drainage system is named in the same manner as the arterial system with interlobular, arcuate, interlobar, and renal veins.

THE RENAL NERVES. The renal nerves are primarily composed of sympathetic efferents from the mesenteric plexus. A few afferent nerves and some cholinergic innervation by vagal fibers are also present but are of unknown significance. Stimulation of renal nerves results in the following:

1. A marked decrease in renal blood flow as a result of vasoconstriction primarily of the afferent and efferent arterioles
2. Increased renin secretion by the juxtaglomerular apparatus (β-2 receptor mediated)
3. Increased reabsorption of the Na^+ and water by the renal tubules, an effect independent of either the renin-angiotensin system or prostaglandins

QUESTIONS

4. Describe the course the fluid ultimately composing the urine takes in passing from the blood into the bladder.
5. A. How does the arrangement of these capillary beds in series and their arrangement around the renal tubular system contribute to reabsorption of essential substrates? Describe these relationships in terms of the Starling forces.
   B. Precipitous decreases in hydrostatic pressure occur at two places in the renal circulation. Name them.
6. Which of the following cause renal vasoconstriction resulting in a decrease in renal blood flow (RBF)?
   1. Exercise, rising from a supine to a sitting position or administration of catecholamines
   2. Hypoxia with a PaO_2 of less than 50 percent normal
   3. A decreased discharge from baroreceptor fibers resulting from a decrease in systemic blood pressure

4. Stimulating the vasomotor center in the medulla, the cortex, or the anterior tip of the temporal lobe

   Answer: A = 1,2,3
   B = 1,3
   C = 2,4
   D = 4
   E = all of the above

7. Which of the following would be true following denervation of the kidney?
   1. Renal vasoconstriction and decreased RBF
   2. Normal physiological function of the kidney
   3. Inability of the kidney to regulate glomerular filtration rate (GFR)
   4. Renal vasodilatation and increased RBF

   Answer: A = 1,2,3
   B = 1,3
   C = 2,4
   D = 4
   E = all of the above

BODY FLUID COMPARTMENTS AND THEIR COMPOSITION

TOTAL BODY WATER (TBW). The percentage of the human body composed of water depends primarily upon the proportion of fat present because fat cells contain very little water (Table 5-4). The average, nonobese male is approximately 60 percent water. The average nonobese female is approximately 50 percent water due to the slightly greater proportion of body fat present. Neonates are 70 percent water by weight.

The volume of TBW as well as the volumes of its various compartments (Table 5-5) may be determined experimentally by using the indicator-dilution principle and calculating the volume of distribution of various markers using the following equation:

$$\text{Volume of compartment} = \frac{\text{Amount of marker given} - \text{amount of marker lost}}{\text{Concentration of the marker in the compartment}}$$

BODY FLUID COMPARTMENTS. The body fluid is divided into two major compartments, extracellular fluid (ECF) and intracellular fluid (ICF). The ECF may be further subdivided into four subcompartments (see Fig. 5-3). The percentage of the TBW found in each compartment and subcompartment is as follows:

ICF = 55%
ECF = 45%
Interstitial water = 20%
Bone and connective tissue water = 15%
Plasma water = 7.5%
Transcellular water (i.e., water located in the viscera) = 2.5%

Table 5-4. *Percent Water of Various Body Fluid Compartments*

ECF (interstitial)	99
Plasma	94
Whole blood	88
Cell (ICF)	75
Bone	25
Fat	20

Table 5-5. *Substances and Equations Used to Measure the Size of the Major Body Fluid Compartments*

Compartment	Substance	Equation
TBW	Antipyrine	Volume compartment = (Amount marker − Amount of marker lost)/[marker]
	Deuterium	
	Tritium	
ECF	Inulin	Same as above
	Raffinose	
	Sucrose	
	Mannitol	
	Thiocyanate	
	Radiochloride	
	Radiosodium	
	Radiobromide	
Plasma	Iodinated I-131 serum albumin	Same as above
	Evans blue, or T-1824	
	Chromicized Cr-51 erythrocytes	
Interstitial fluid	Not measured directly	Interstitial fluid = ECF − plasma
ICF	Not measured directly	ICF = TBW − ECF

Adapted with permission from H. Valtin. *Renal Function: Mechanisms Preserving Fluid and Solute Balance in Health* (2nd ed.). Boston: Little, Brown, 1983. P. 26.

The bone and connective tissue water compartment is frequently neglected in discussions of the ratio of the ECF : ICF because it is difficult to measure and requires equilibration times greater than those of other compartments. If one ignores the bone and connective tissue water, the ICF represents two-thirds of the TBW and the ECF one-third.

COMPOSITION OF THE FLUID COMPARTMENTS. Figure 5-4 illustrates the compositions of the extracellular and intracellular fluids. Table 5-4 lists the percentage of water found in various compartments. The ICF, for instance, is only 75 percent water. Knowledge of the relative proportions of Na^+, K^+, Cl^-, and HCO_3^- in one compartment compared with those in another is vital to understanding renal physiology (Table 5-6).

QUESTIONS

8. A 70-kg man is given 1 millicurie (mCi) of tritiated water (HTO) intravenously. During the period of equilibration, one-half percent of the dose administered is lost in the urine. After equilibration a sample of the man's plasma contains 0.024 mCi per liter plasma.
   A. Calculate the volume of distribution of the marker.
   B. What body fluid compartment does this correspond to?
9. The male patient in question 8 is now given 12.5 g mannitol intravenously. Assuming sufficient time has passed for equilibration to occur and that 1.5 g mannitol has been lost in the urine
   A. What is the volume of distribution of the marker given a serum mannitol concentration of 58.8 mg per 100 ml?
   B. What fluid compartment does this correspond to?
   C. Calculate the ICF volume.
10. The same man (70 kg) is now injected with 10 mg of T-1824 dye (Evan's blue) intravenously. A few minutes later, a blood sample has a concentration of 3.2 mg T-1824 per liter plasma.

Fig. 5-3. *Distribution of body water. Intracellular water constitutes the largest portion of the body water. The balance of the body water is defined as the extracellular fluid. The plasma is separated from the interstitial-lymph compartments by the capillary membrane (dashed line). It is highly permeable to water and solutes with the important exception of the plasma proteins. The three cavities (circular dashed lines) are potential extracellular fluid spaces. Dense connective tissue and bone compartments appear to be inaccessible to the commonly employed saccharide marker molecules.*

The transcellular fluid is exemplified by the fluid contained in the gastrointestinal and urinary tracts. (Modified from R. H. Maffly. The Body Fluids: Volume, Composition, and Physical Chemistry. In B. M. Brenner, and F. C. Rector [eds.], The Kidney *[2nd ed.]. Philadelphia: Saunders, 1981. P. 88.)*

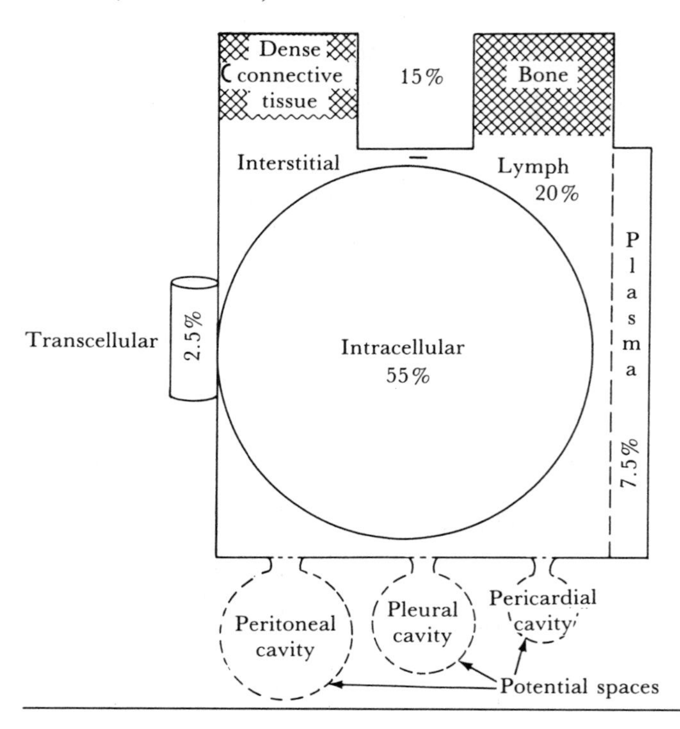

A. What is the volume of distribution of the marker?
B. What body fluid compartment does this correspond to?
C. Calculate the interstitial fluid volume.

11. Name the two major cations and two major anions in
    A. The ICF
    B. The ECF

PRINCIPLES OF WATER AND ELECTROLYTE BALANCE

EFFECTIVE AND INEFFECTIVE OSMOLES. Maffly's principle states that water goes where the osmoles go, i.e., water flows from a region of low osmolality to a region of higher osmolality. Because sodium is the main extracellular osmole, this means that water goes where the sodium goes in the ECF. Abnormalities in Na^+ balance are reflected, therefore, in the amount and distribution of extracellular body water (e.g., edema, see Fig. 5-5).

A corollary to Maffly's principle is that the movement of water across membranes is also governed by the distribution of osmoles. Only solutes which do not freely cross the membrane effect the movement of water across that membrane. Such solutes are called effective osmoles because the addition of an effective osmole to one side of a membrane osmotically obligates water to flow to that side of the membrane until equilibrium is established. Sodium is an effective osmole because it remains in the ECF and does not cross into the ICF; therefore,

Fig. 5-4.

Solute composition of the major body fluid compartments. The concentrations are expressed as chemical equivalents to emphasize that the compartments are made up primarily of electrolytes. Intracellular values are estimates only. (Modified from J. L. Gamble. Chemical Anatomy, Physiology and Pathology of Extracellular Fluid *[6th ed.]. Cambridge, MA. Harvard University Press, 1954. Reprinted by permission.)*

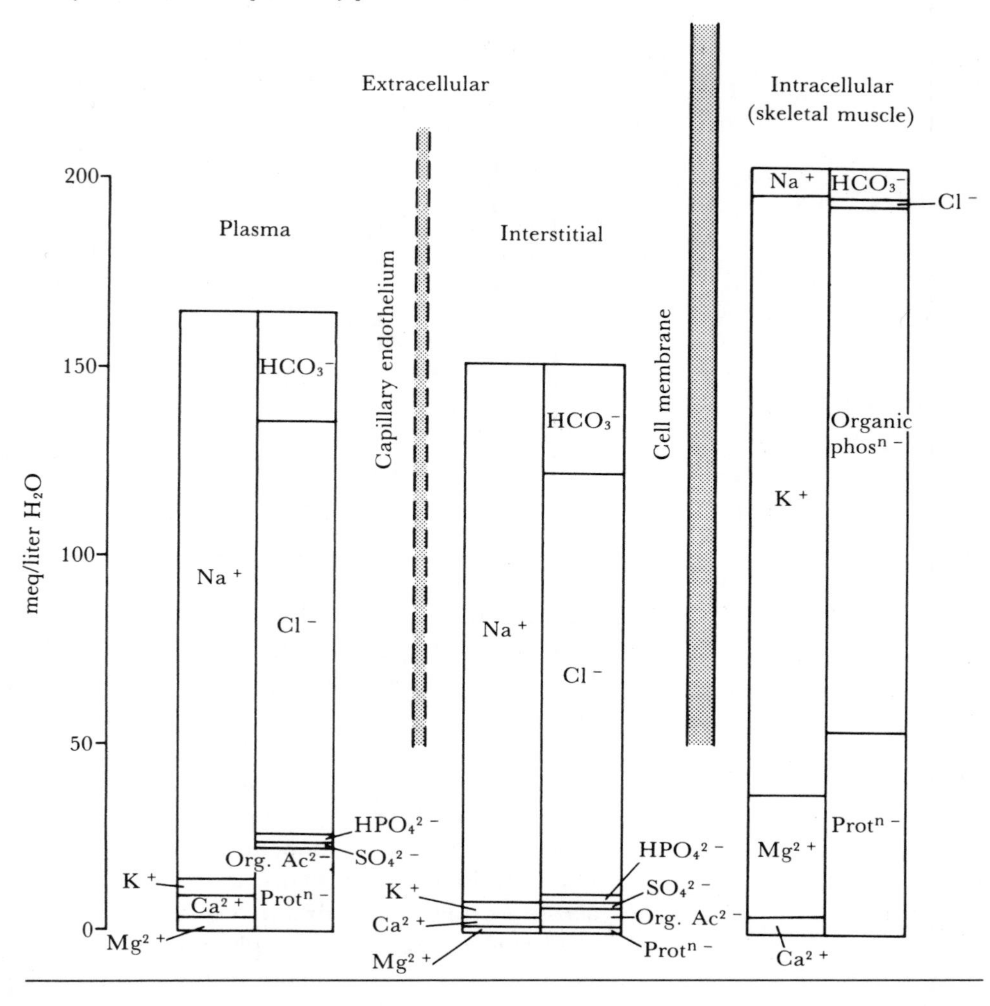

Table 5-6.

Concentrations of Important Ions in Intracellular and Extracellular Fluids

Ion	ECF (meq/kg H_2O)	ICF (meq/kg H_2O)
K^+	4–5	150–160
Na^+	144–150	5–10
Cl^-	110–114	5–10
HCO_3^-	24	15
Plasma pH	7.4 or 40 × 10^{-9} meq H^+/kg H_2O (the ICF is more acidic)	
Plasma protein	7 g/dl	

if sodium is added to the ECF, it draws water out of the cells until equilibrium is attained

$$[\text{Osmoles}]_{ECF} = [\text{Osmoles}]_{ICF}$$

Other ECF effective osmoles of importance include glucose, glycerol, and mannitol. Examples of ineffective osmoles are urea, alcohol, and any other osmole that moves rapidly across cell membranes and, therefore, does not

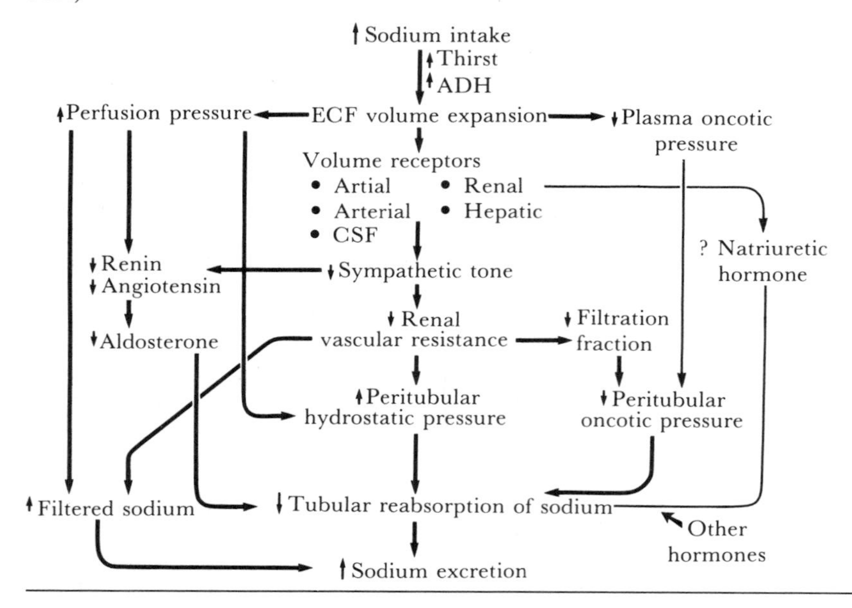

Fig. 5-5. *Schematic view of possible afferent and efferent mechanisms in the response to increases in sodium intake. (From J. F. Seeley, and M. Levy. Control of Extracellular Fluid Volume. In B. M. Brenner, and F. C. Rector [eds.]* The Kidney *[2nd ed.]. Philadelphia: Saunders, 1981. P. 399.)*

osmotically obligate water to follow it. In uremia the osmolality of the body's fluids measured in the laboratory is increased by its increased content of urea; however, because urea is an ineffective osmole no change in the EFC-ICF ratio occurs.

DETERMINATES OF PLASMA VOLUME. The volume of ECF is determined primarily by total body stores of sodium; however, the amount of the ECF in the plasma subcompartment also depends on the oncotic and hydrostatic pressures. The relationship between hydrostatic pressure, oncotic pressure, and fluid flow across a capillary membrane is given by the Starling equation.

The major solutes in the plasma are Na^+, HCO_3^-, Cl^-, glucose, and urea. Because Na^+ is the major cation in the ECF, the total concentration of electrolytes may be estimated by doubling the Na^+ concentration thereby accounting for the anions present in the ECF. One can then add the osmolalities of the two major nonelectrolytes and estimate the total osmolality. This is summarized in the following equation:

$$\text{Serum osmolality} = 2[Na^+] + [\text{glucose}] + [\text{urea}]$$

in which $[Na^+]$ is the concentration of sodium ions in meq per liter, [glucose] is the concentration of glucose molecules in mmole per liter, and [urea] is the concentration of urea molecules per liter. Note that Na^+ is a univalent ion; therefore, 1 meq per liter of Na^+ is equal to 1 mosm per liter Na^+. Similarly 1 mmole of glucose or urea is equal to 1 mosm of glucose or urea. Thus this equation estimates osmolality of the serum in mosm per liter. For the purposes of this text, 1 liter of serum or plasma is equivalent to 1 kg of water; therefore, these will be used interchangeably when discussing concentration. Serum is actually plasma with the plasma proteins removed. Plasma is, in fact, approximately 91 percent water and approximately 9 percent solute. This error cancels out because the activity of ions in plasma is also 91 percent, and it is activity rather than actual concentration that is measured clinically.

Glucose and urea are measured in mg per liter; a more useful form of the serum osmolality equation takes this into account. Given that the molecular weight of glucose is 180 and that of urea is 28, the following equation can be derived:

$$\text{Serum osmolality} = 2[Na^+] + \text{glucose}\ \frac{\text{mg/100ml}}{18} + \text{urea}\ \frac{\text{mg/100 ml}}{2.8}$$

Thus the normal serum osmolality can be calculated. Using normal values for $[Na^+]$ of 140 meq per liter, for [glucose] of 90 mg per dl, and for [urea] of 14 mg per dl

$$\begin{aligned}\text{Serum osmolality} &= 2(140) + \frac{90\ \text{mg/dl}}{18} + \frac{14\ \text{mg/dl}}{2.8}\\ &= 2(140) + 5 + 5\\ &= 290\ \text{mosm/kg}\ H_2O\end{aligned}$$

Osmolality of the body fluids is 290 milliosmoles per kilogram water everywhere except in the plasma (and, of course, in the transcellular water), in which, due to the presence of plasma proteins, primarily albumin, it is 291 mosm per kilogram water.

One, milliosmole per kilogram water net oncotic pressure (i.e., osmotic pressure due to proteins or other large molecules restricted to the plasma) is equivalent to roughly 20 mm Hg of hydrostatic pressure. Thus this oncotic pressure just balances the net capillary hydrostatic force of about 20 mm Hg (see Fig. 5-9).

DISTRIBUTION OF IONS IN BODY COMPARTMENTS. If the concentrations of a singly charged ion on either side of a semipermeable membrane are known, then the Nernst equation (see Equation 1-3) may be used to determine the electrical potential generated by the tendency of ions to move along their concentration gradients.

$$\text{Potential (mv)} = 61 \log (C_{in}/C_{out}) \text{ for univalent ions at } 37^\circ\ C \qquad (5\text{-}1)$$

in which C_{in} is the concentration of the ion in the cell and C_{out} is the concentration of the ion outside of the cell.

FLUID BALANCE. At equilibrium, input equals output; except in complete renal shutdown, what goes in must come out. In a stable chronic condition, the body *must* be in balance. Acute changes in body weight may be assumed to be due to gain or loss of body water. The exact composition of fluid depends on the nature of the insult (for example, excessive sweating would result in the loss of hypotonic fluid).

QUESTIONS

12. A 24-year-old businessman in a coma is brought into the emergency room. His serum sodium is 141 meq per liter, his BUN is 28 mg per dl, and his blood glucose is 180 mg per dl. The molecular weight of glucose is 180 and that of BUN is 28.
    A. Calculate his serum osmolality.
    B. His serum osmolality is 380. Why?
13. $[K^+]_{ICF} = 156$, $[K^+] = 4.0$. Calculate the size and direction of the electric potential.

14. Would the following increase, decrease, or have no effect on the hematocrit?
    A. Addition of hypotonic saline intravenously
    B. Addition of hyperosmotic mannitol or glucose
    C. Ingestion of excessive quantities of water
    D. End-stage renal disease
    E. Acute ingestion of excessive quantities of alcohol
15. Volume contraction may be accompanied by which of the following? What about volume expansion?
    A. Hyperosmolality
    B. Isosmolality
    C. Hyposmolality
16. Plasma Na^+ concentration reveals nothing about total body stores of Na^+. Similarly, both volume contraction and volume expansion can accompany states of hypernatremia, hyponatremia, and eunatremia. Plasma osmolality, in contrast, is ordinarily reflected by the plasma Na^+ concentration. Name the three parameters that determine plasma Na^+ concentration.
17. Because all the body fluid compartments are in equilibrium, the osmolalities of all the body fluid compartments are equal (ignoring the oncotic pressure due to plasma proteins). The plasma and ECF Na^+ concentrations are approximately equal to the ICF concentration of K^+, and in each case these cations constitute roughly half the osmolality of their respective compartments. Plasma Na^+, therefore, reflects both intracellular and extracellular osmolality and total body cation stores. Calculate the total body (exchangeable) cation in a 70-kg male with a plasma Na^+ concentration of 140. Do the same for a plasma Na^+ concentration of 120. What does the concentration of Na^+ in the plasma tell us about the concentration of K^+ in the ICF?
18. Loss of cation from one compartment causes water to leave that compartment and enter the other until $[osm]_{ECF}$ equals $[osm]_{ICF}$, complicating analysis of water and electrolyte imbalances. A 60-kg female volunteers to the following procedures. Calculate the ECF and ICF volumes assuming complete equilibration but negligible excretion.
    A. Infusion of 2 liters of 0.3 M urea
    B. Infusion of 1.5 liters of 0.15 M NaCl
    C. Infusion of 1.5 liters of a 0.3 M solution of urea to which has been added 0.15 M NaCl
    D. Infusion of 3 liters of 0.3 M NaCl
    E. Infusion of 0.3 M glucose
19. Match the appropriate letter to the Darrow-Yannet diagram to which it corresponds (Fig. 5-6). The first diagram illustrates the normal state. In each diagram, the healthy state is drawn in solid lines and the new steady-state in dashed lines.*
    A. Loss of pure water (perspiration with salt replacement)
    B. Gain of isosmotic fluid (normal saline IV)
    C. Loss of isosmotic fluid (severe diarrhea)
    D. Loss of NaCl only (adrenal insufficiency or perspiring in desert with only water replaced)
    E. Gain of NaCl only (potato chips PO bolus)
    F. Gain of pure water (syndrome of inappropriate antidiuretic hormone [SIADH], polydipsia).
20. If a normal person (70-kg male or 60-kg female) lost 2 percent of body

* Question and answer adapted from H. Valtin. *Renal Function: Mechanisms Preserving Fluid and Solute Balance in Health* (2nd ed.). Boston: Little, Brown, 1983. Pp. 38–39, 271–274.

weight by dehydration (net loss of body water only) what would the plasma osmolality and plasma [Na^+] be? Assume normal osmolality of 290 mosm /liter and [Na^+] of 140 meq/liter.

21. Do dextrans, high-molecular-weight polymers of glucose, also exert an oncotic pressure even though they may be uncharged?
22. A patient presented with a serum [Na^+] of 155 meq/liter due to acute dehydration, yet his skin turgor, blood pressure, and venous pressure appeared perfectly normal. What percent of his body water had he lost? Why were the parameters mentioned not obviously abnormal?

PHYSIOLOGY OF THE GLOMERULUS

THE GLOMERULUS. The mesangial cells, the endothelial cells of the glomerular capillaries and the epithelia of Bowman's capsule, together with their fused basement membranes, constitute the renal corpuscle. The glomerular tuft of capillaries (the glomerulus proper) is encapsulated by the renal corpuscle (Fig. 5-7). The terms renal corpuscle and glomerulus will hereafter be used interchangeably, as they frequently are, but strictly speaking they are not the same

Fig. 5-6. *Darrow-Yannet diagram of types of contraction and expansion of the body fluid compartments. The plasma and the interstitial fluid are depicted as a single extracellular compartment. In each diagram, the healthy state is drawn in solid lines and the new steady state in dashed lines. (From H. Valtin.* Renal Function: Mechanisms Preserving Fluid and Solute Balance in Health *[2nd ed.]. Boston: Little, Brown, 1983. P. 272.)*

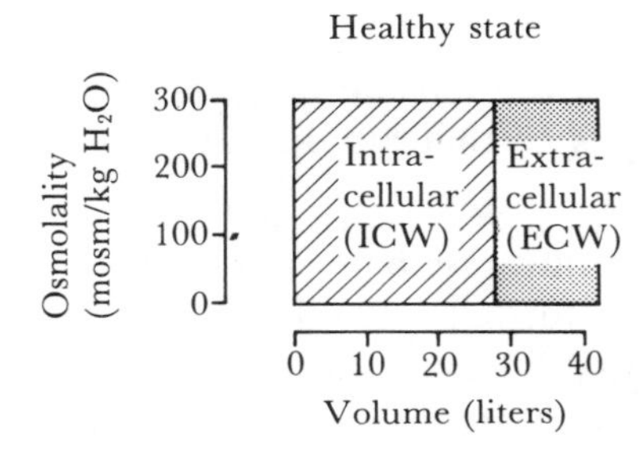

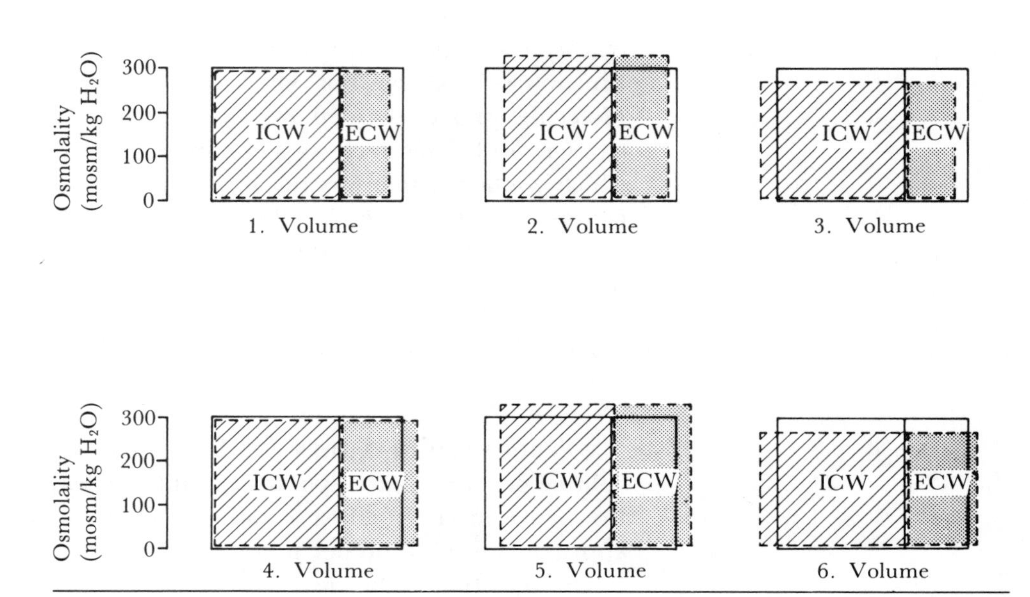

thing. The function of the renal corpuscle is to form an ultrafiltrate of plasma. Plasma water and small solutes move out of the glomerular capillaries propelled by hydrostatic pressure through fenestrae in the endothelial cells that form the capillary wall, filter through the basement membrane, and then pass through the slit pores between the podocyte endfeet to enter Bowman's space as an ultrafiltrate of plasma (Fig. 5-8). Mesangial cells provide structural support to the glomerular capillaries and also serve to clear the filtrate of any macromolecules that may have leaked across the filtration barrier. In addition, by virtue of the contractile elements they contain, it is thought that they may also serve to regulate intraglomerular blood flow.

Bowman's capsule is the most proximal portion of the nephron (see Fig. 5-7). It consists of parietal epithelial cells forming the outer spherical capsule and visceral epithelial cells whose foot processes interdigitate surrounding the glomerular tuft of capillaries and forming an inner spherical capsule. The space between these two capsules is Bowman's space. Bowman's space is continuous with the lumen of the proximal convoluted tubule at the urinary pole of the capsule.

THE GLOMERULUS FILTRATION BARRIER. The components of the glomerular filtration barrier (i.e., between the capillary lumen and Bowman's space) (Fig. 5-8) are

1. The capillary endothelial fenestrae which are 500–1000 angstroms (Å) in diameter
2. The fused basement membranes of the endothelial cells and the visceral epithelial podocytes
3. The podocytes with their foot processes separated by slit pores (250 Å wide) with diaphragms further limiting the actual pore size

The glomerular filtration barrier is charge selective, size selective, and, to an extent, shape (or deformability) selective. Because glomerular endothelium lacks the diaphragms that normally decrease the apertures of the fenestrae of other capillaries, they do not impose a barrier to the movement of proteins (colloids), although they do prevent the movement of erythrocytes. The basement membrane, on the other hand, is a continuous filamentous layer consisting of type IV and V collagen and negatively charged sialoglycoproteins. It is probably the size and charge selective component of the barrier. Finally, there are the visceral epithelial cell podocytes with their foot processes and slit pores whose exact function is unknown but which may also contribute to the filtration barrier. Molecules greater than 50 to 100 Å in diameter are unable to pass the filtration barrier. Molecules such as albumin that are somewhat smaller than this but which contain numerous negative charges are removed by a combination of charge and size selective filtration. Molecules that are unable to deform sufficiently are also excluded. (Perhaps they cannot squeeze through the right holes.) Overall, glomerular capillaries are significantly less permeable to proteins than are their systemic counterparts.

For many smaller substances, protein binding limits filtration. Only that portion of a substance or drug that is unbound to protein is subject to filtration.

GLOMERULAR FILTRATION RATE. The glomerular filtration rate (GFR) is the total volume per unit time of plasma ultrafiltrate leaving the capillaries and entering Bowman's space. In a healthy adult male, GFR is roughly 180 liters per day or approximately 120 ml per minute.

The renal blood flow is approximately 25 percent of the total cardiac output

Fig. 5-7. *The renal corpuscle. The upper part shows the vascular pole, with afferent and efferent arterioles and the macula densa. Note the juxtaglomerular cells in the wall of the afferent arteriole. Podocytes cover glomerular capillaries. Their nuclei protrude on the cell surface while their processes line the outer surface of the capillaries. Note the flattened cells of the parietal layer of Bowman's capsule. The lower part of the drawing shows the urinary pole and the proximal convoluted tubule. (From L. C. Junqueira, and J. Carneiro.* Basic Histology *[4th ed.]. P. 400. © 1983 by Lange Medical Publications, Los Altos, CA.)*

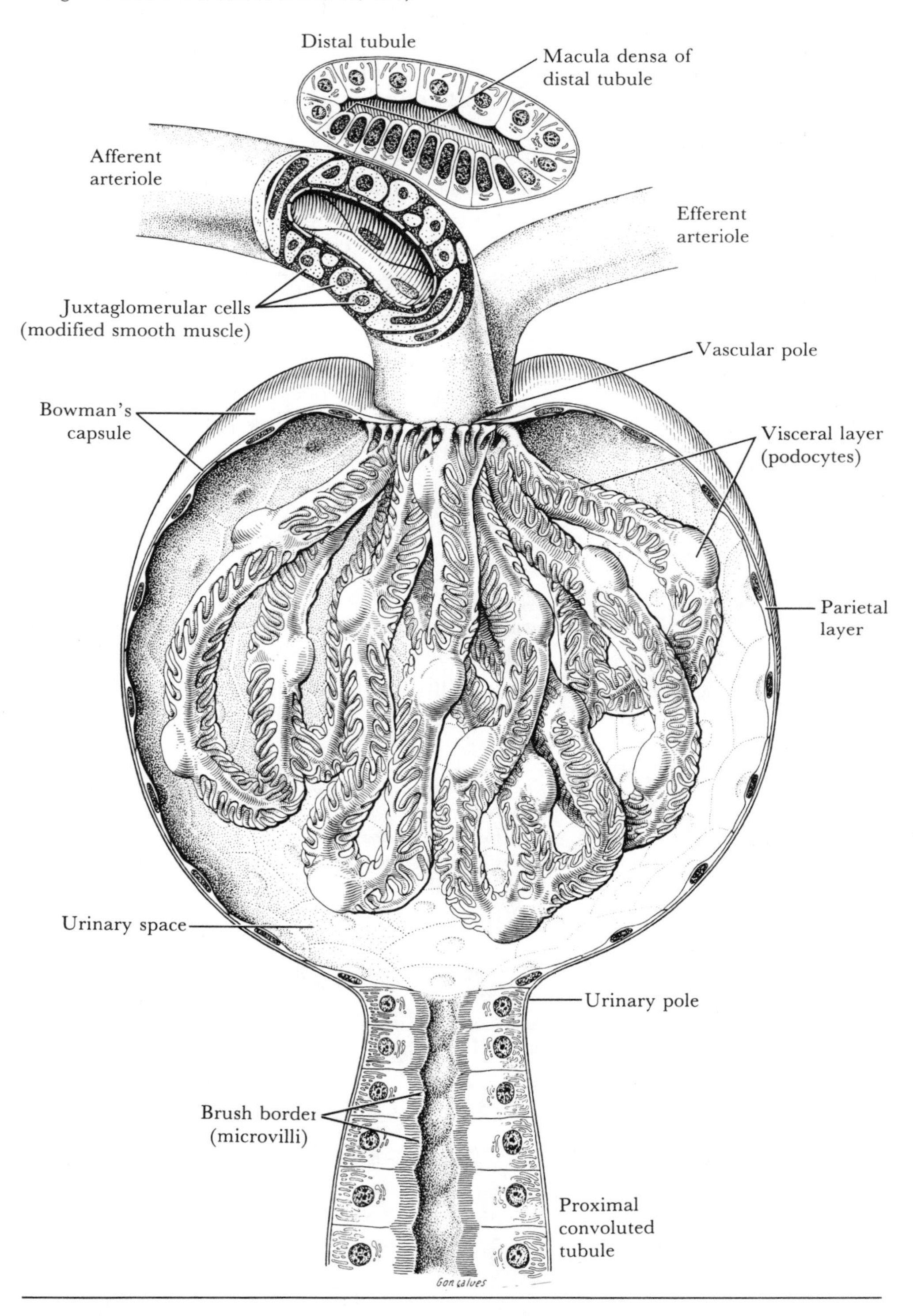

or 1200 ml of whole blood per minute. Of all the organs only the liver receives a greater proportion of the cardiac output (roughly 30%). Because only the plasma (not the red cells) is available for filtration, 660 ml of plasma per minute (the renal plasma flow) is available for modification during any given moment (55% of 1200). Roughly 120 ml per minute of the 660 ml per minute of plasma flow is filtered into Bowman's space as ultrafiltrate. From there it passes into the renal tubule system. Only about 1 percent of the filtered load of fluid eventually passes through the tubules and collecting system to emerge as urine. The average urine output or flow per minute ($\dot{V}$) is 1.2 ml per minute, i.e., 1 percent of the 120 ml per minute filtered.

DETERMINANTS OF THE GFR. The four determinants of the magnitude of the glomerular filtration rate are:

1. Ultrafiltration coefficient or constant, K_f, which is simply the capillary permeability per meter squared multiplied by the surface area available for filtration.
2. Net oncotic pressure, $\bar{\pi} = (\pi_c - \pi_t)$, in the glomerular capillary, in which π_c = glomerular capillary oncotic pressure and π_t = the oncotic pressure in Bowman's space. Because there is normally no protein in Bowman's space, $\pi_t = 0$ and, therefore, $\bar{\pi} = \pi_c$.
3. Net hydraulic pressure driving filtrate from the capillaries into Bowman's space, $\bar{P} = (P_c - P_t)$, in which P_c = the hydraulic pressure in the capillary and P_t = the hydraulic pressure in Bowman's space.
4. Initial or afferent glomerular capillary plasma flow, Q_A.

GLOMERULAR AND SYSTEMIC CAPILLARIES. Glomerular capillaries differ from systemic capillaries in several important respects. In a systemic capillary bed, such as those of the skeletal muscles, the hydraulic pressure driving filtration falls steadily as filtrate is forced out of the capillaries (Fig. 5-9). When the net

Fig. 5-8. *Electron micrograph of a portion of the wall of a glomerular capillary, showing pores in the extremely attenuated endothelium. On the outer surface of the basal (basement) lamina are the foot processes of the podocytes, with the narrow filtration slits between them. The portion on the left indicates the distribution of charges in the glomerular capillary wall. (From H. Valtin,* Renal Function: Mechanisms Preserving Fluid and Solute Balance in Health *[2nd ed.]. Boston: Little, Brown, 1983. P. 47.)*

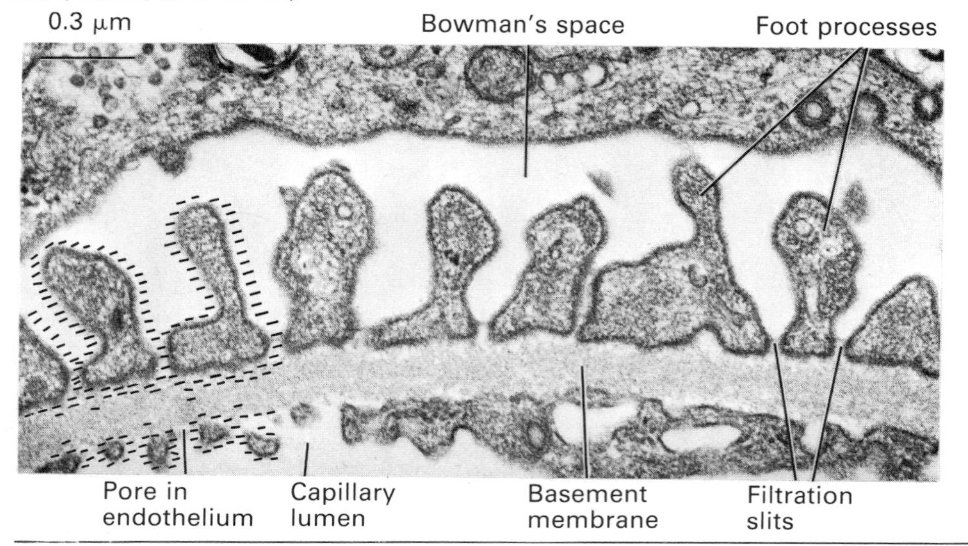

hydraulic pressure just equals the net oncotic pressure, which keeps fluid in the capillaries, filtration pressure equilibrium is reached and no more fluid can leave the capillaries. As the blood reaches the venous side of the capillary bed, there is a decrease in the net hydraulic pressure and reabsorption of fluid begins (Fig. 5-9). Almost all the fluid filtered out on the arterial side of the capillaries is reabsorbed on the venous side. The remainder is returned to the venous circulation by the lymphatics, as are the proteins that manage to leave the capillaries.

The situation in the glomerular capillaries is quite different. The mean hydraulic pressure driving filtration remains nearly constant along the length of the capillary (Fig. 5-10). Filtration is slowed as blood reaches the end of the capillary bed because the capillary oncotic pressure rises to oppose the hydraulic pressure, thus decreasing the net pressure driving filtration. The increase in capillary oncotic pressure is the result of the rapid loss of ultrafiltrate and consequent increase in protein concentration.

There is a large hydraulic pressure drop at the level of the efferent arteriole while the oncotic pressure remains nearly constant because no further filtration can occur. Thus, when the blood reaches the peritubular capillaries, the oncotic

Fig. 5-9. *Starling forces involved in skeletal capillaries. As shown, ultrafiltration pressure decreases mainly because hydrostatic pressure decreases. Hydrostatic pressure and plasma oncotic pressure are plotted against distance along the skeletal capillary. In this example, tissue hydrostatic pressure and tissue oncotic pressure fortuitously cancel. Initially, net hydrostatic pressure exceeds net oncotic pressure driving fluid out of the capillaries. At some point along the capillaries the net hydrostatic pressure equals the net oncotic pressure and filtration pressure equilibrium (FPE) occurs with cessation of net fluid movement. Subsequently, net oncotic pressure exceeds net hydrostatic pressure and reabsorption begins. Any fluid not reabsorbed by the end of the permeable capillary bed is returned to the circulation by the lymphatic system. (Adapted with permission from a lecture given by Channing R. Robertson, Ph.D., Stanford Medical School, 1984.)*

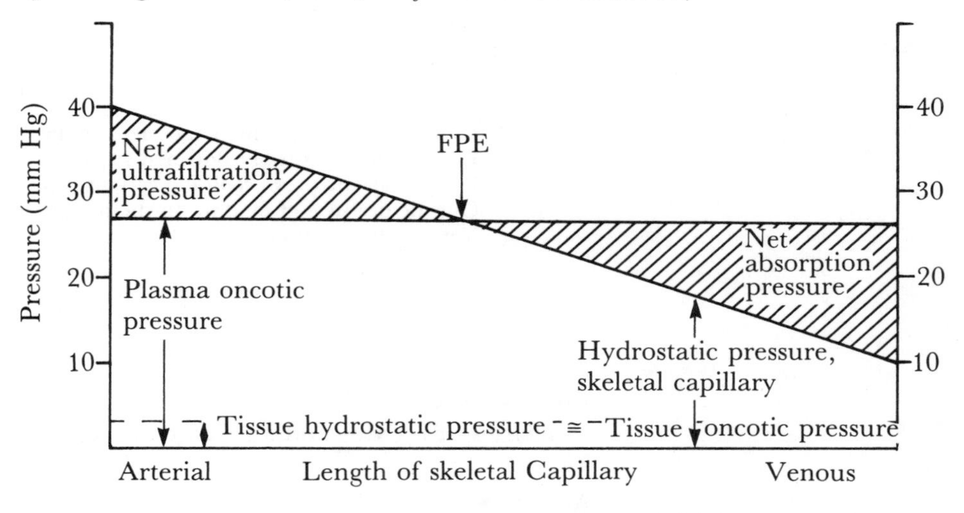

Balance of Mean Values	Initial Capillary Values (mm Hg)	End Capillary Values (mm Hg)
Hydrostatic pressure in skeletal capillary	40	10
Tissue hydrostatic pressure	3	3
Plasma oncotic pressure	27	27
Tissue oncotic pressure	3	3
Net ultrafiltration pressure	13	–17

Fig. 5-10. *Starling forces involved in glomerular ultrafiltration (specific values vary with species). As shown, ultrafiltration pressure declines in glomerular capillaries, mainly because plasma oncotic pressure rises. This contrasts with extrarenal capillaries, in which the decline in ultrafiltration pressure is due mainly to a decrease in intracapillary hydrostatic pressure (see Fig. 5-9). It is not yet known at what point in the capillary the sum of the hydrostatic pressure in Bowman's space and of plasma oncotic pressure exactly balances the hydrostatic pressure in the glomerular capillary. In some species such as dog or man, this point, filtration pressure equilibrium (FPE), may not be reached in the glomerular capillary. In those species in which it is attained, the pattern for the rise in plasma oncotic pressure as a function of capillary length is not known precisely, and it can vary. For example, an increase in the rate at which plasma enters the glomerular capillary will lead to a change in the pattern from curve A to curve B, and, consequently, to a rise in the mean net ultrafiltration pressure.*

Hydrostatic pressure drops rapidly at the level of transition between the efferent arterioles and the vasa recta. Net oncotic pressure decreases much more slowly and, therefore, exceeds hydrostatic pressure in the vasa recta. This leads to net reabsorption of fluid. (From H. Valtin. Renal Function: Mechanisms Preserving Fluid and Solute Balance in Health *[2nd ed.]. Boston: Little, Brown, 1983. P. 44.)*

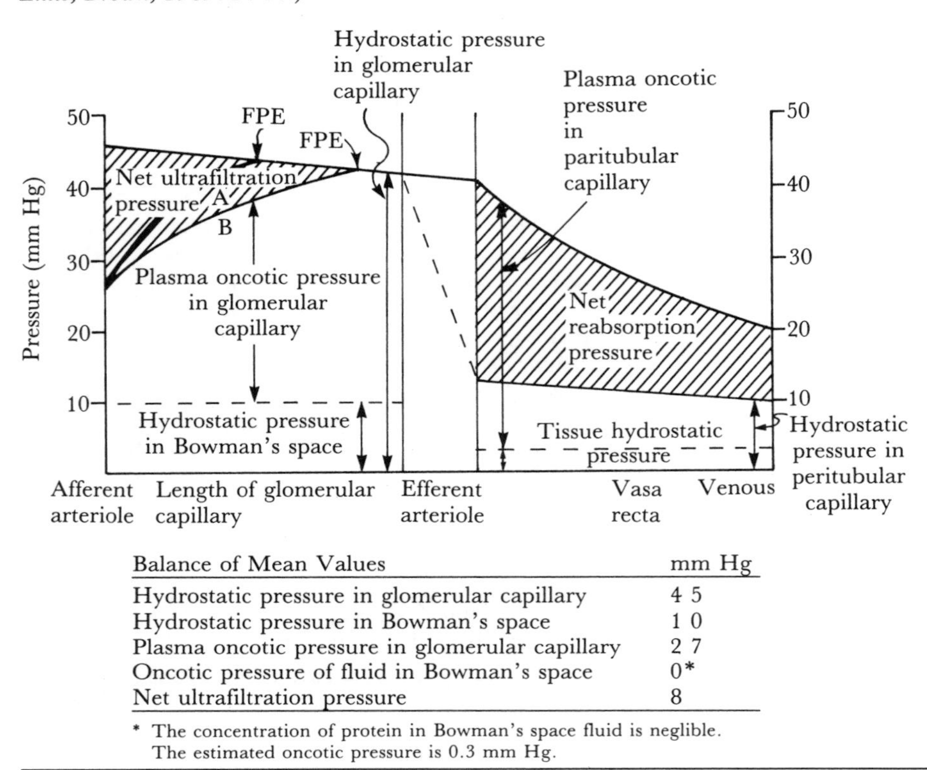

Balance of Mean Values	mm Hg
Hydrostatic pressure in glomerular capillary	45
Hydrostatic pressure in Bowman's space	10
Plasma oncotic pressure in glomerular capillary	27
Oncotic pressure of fluid in Bowman's space	0*
Net ultrafiltration pressure	8

* The concentration of protein in Bowman's space fluid is neglible. The estimated oncotic pressure is 0.3 mm Hg.

pressure exceeds the hydraulic pressure and fluid is reabsorbed into the capillaries of the descending vasa recta or peritubular plexi.

AUTOREGULATION. The kidneys are able to regulate the GFR under a wide range of conditions. This phenomenon is called autoregulation because there are intrinsic mechanisms in the kidney responsible for maintaining the nearly constant GFR. This is necessary because even small changes in the GFR can vastly alter the filtered load of solute and water.

Over a range of arterial pressure from 80 to 200 mm Hg both GFR and RBF remain quite constant (Fig. 5-11).

The kidneys are able to autoregulate glomerular capillary hydrostatic pressure, flow, and, indeed, filtration because there are precapillary sphincter muscles in both the afferent and efferent arterioles. Increasing or decreasing the size of the afferent arteriole allows more or less of the systemic pressure to be felt

Fig. 5-11. *Autoregulation in the kidney. GFR = glomerular filtration rate; RPF = renal plasma flow. (From R. E. Shipley, and R. S. Study. Changes in renal blood flow, extraction of inulin, glomerular filtration rate, tissue pressure and urine flow with acute alterations of renal artery blood pressure.* Am. J. Physiol. *167:676, 1951.)*

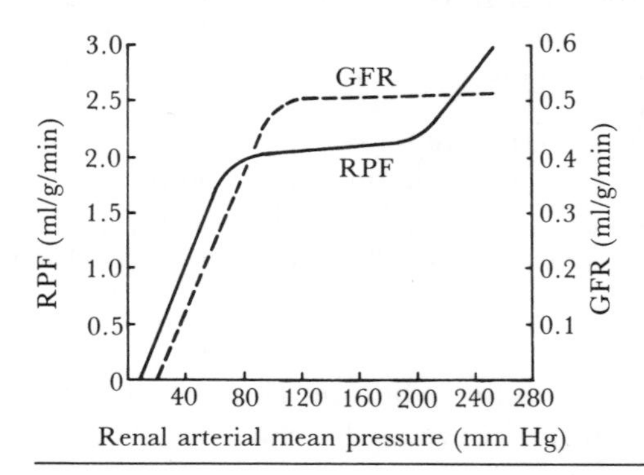

by the glomerulus, while changes in the diameter of the efferent arteriole can alter the filtration rate as well as help govern flow (Fig. 5-12).

CALCULATION OF RENAL CLEARANCE, GLOMERULAR FILTRATION RATE, AND RENAL BLOOD FLOW. Renal clearance is defined as that volume of plasma from which all of a given substance is removed per unit time in one pass through the kidneys (the units are, therefore, milliliters plasma per unit time). If a substance is neither secreted nor reabsorbed, then the clearance is equal to the GFR, and all of the plasma filtered has been cleared of the substance (e.g., inulin) (Table 5-7). The clearance would then be due to filtration of the substance into the glomerulus followed by its passage through the uriniferous tubule without further alteration (in the case of inulin). Clearance can, however, be the result of any possible combination of filtration, reabsorption, and secretion. This relation is examplified by the following equation:

Quantity excreted = Quantity filtered + Quantity secreted − Quantity reabsorbed

If all of the substance is reabsorbed (e.g., glucose) then the clearance would be zero (none of the plasma is cleared of the substance), but the GFR would still be

Fig. 5-12. *Changes in renal blood flow (RBF) and glomerular filtration rate (GFR) that occur when resistance is altered in either the afferent or the efferent arterioles, provided that renal perfusion does not change. (Reproduced with permission from H. Valtin.* Renal Function: Mechanisms Preserving Fluid and Solute Balance in Health *[2nd ed.]. Boston: Little, Brown, 1983. P. 105.)*

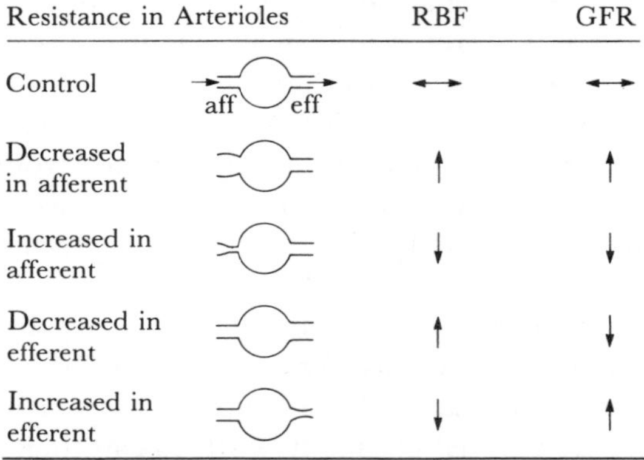

Resistance in Arterioles		RBF	GFR
Control	aff eff	↔	↔
Decreased in afferent		↑	↑
Increased in afferent		↓	↓
Decreased in efferent		↑	↓
Increased in efferent		↓	↑

Table 5-7. *Characteristics of a Substance Suitable for Measuring the GFR by Determining Its Clearance*

Freely filtered (i.e., not bound to protein in plasma or sieved in the process of ultrafiltration)
Not reabsorbed or secreted by tubules
Not metabolized
Not stored in kidney
Not toxic
Has no effect on filtration rate
Preferably easy to measure in plasma and urine

120 ml per minute as before. If all of the substance in the plasma is secreted, then the clearance would be equal to the renal plasma flow because all of the plasma volume passing through the kidneys would be cleared of the substance (e.g., para-aminohippurate [PAH]). In other words, the combined rates of filtration and secretion of the substance would equal the plasma flow rate.

Clearance can be calculated by the following equation:

$$C_x = \frac{(U_x)\ (\dot{V})}{(P_x)} = \frac{\text{Amount excreted/minute}}{\text{Plasma concentration of x}} \tag{5-2}$$

in which C_x = clearance of x, U_x = urine concentration of x, $\dot{V}$ = urine flow in ml per minute, and P_x = plasma concentration of x.

Clearance could be defined simply as the amount of substance excreted per minute; however, this value would not in itself be an adequate indication of how efficiently or in what manner the kidneys were excreting the substance. Use of the equation for clearance creates an index of renal function that can be usefully compared to the calculated GFR.

If $C_x > \text{GFR}$, then there must be *net* secretion.
If $C_x < \text{GFR}$, then there must be *net* reabsorption.
If $C_x = \text{GFR}$, then probably there is neither net reabsorption nor net secretion (unless the two fortuitously canceled).

The renal clearance of inulin is often used to estimate GFR, because it is neither secreted nor reabsorbed, and its elimination into the urine is, therefore, entirely due to filtration and independent of the plasma inulin concentration. Inulin is not bound to plasma proteins, its volume of distribution includes all of the ECF, and it is not metabolized. The renal clearance of creatinine is often used clinically in place of inulin because, unlike inulin, it is a normal component of the blood. Renal clearance, however, does not give as accurate an estimate, because it is secreted to a small extent in the proximal tubule (Fig. 5-13), and because in disease states the extent of its secretion is unpredictable.

At low concentrations of PAH, approximately 91 percent of the PAH entering the kidneys in the renal arteries is removed from the plasma by a combination of filtration into Bowman's space and active secretion into the proximal tubules. Excretion of PAH is thus the result of filtration plus net secretion. At plasma concentrations below the level at which the secretory transport maximum of PAH is reached, the clearance of PAH is used to estimate the renal plasma flow (the effective renal plasma flow, [ERPF]). Given a urine flow of 1 ml per minute, and urine and plasma PAH concentrations of 6 mg per ml and 0.01 mg per ml, respectively, renal plasma flow (RPF) can then be calculated using the formula

Fig. 5-13. *Changes in the fraction of the filtered amount of substances remaining in the tubular fluid along the length of the nephron* (U_{Inu}/P_{nu})*. (From L. P. Sullivan, and J. J. Grantham.* Physiology of the Kidney *[2nd ed.]. Philadelphia: Lea and Febiger, 1982. P. 93.)*

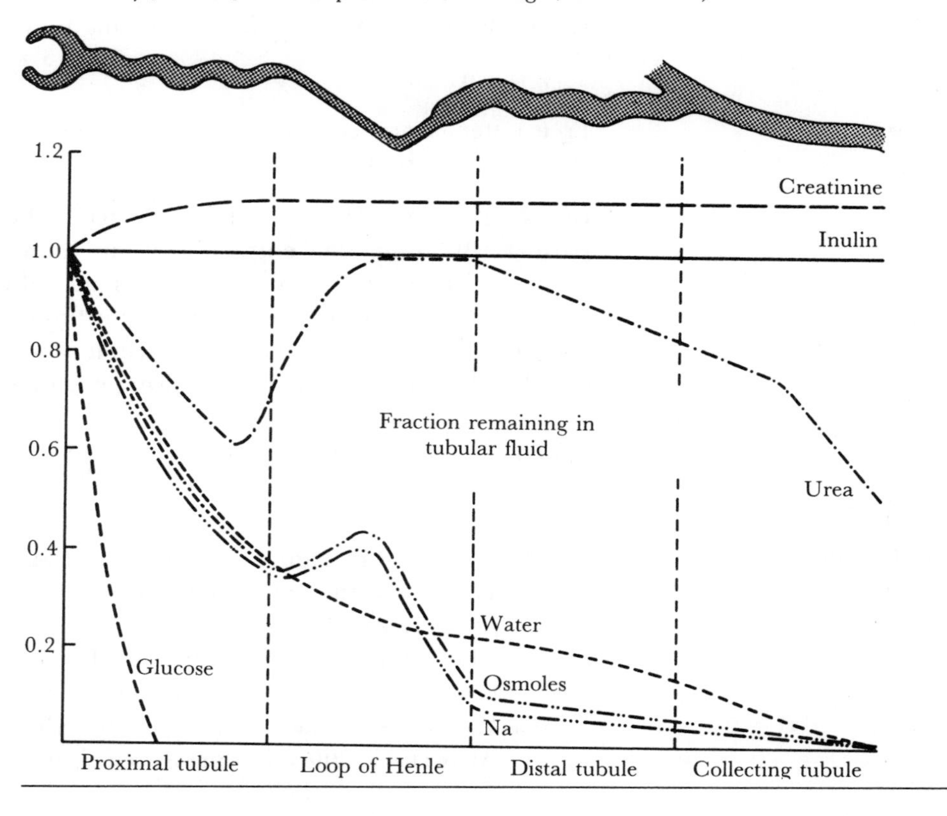

$$0.91 \text{ RPF} = \text{Clearance of PAH} = \frac{[U]\dot{V}}{[P]}$$

$$= \frac{(6 \text{ mg/ml})\ (1 \text{ ml/min})}{(0.01 \text{ mg/ml})} = 600 \text{ ml/min} \qquad (5\text{-}3)$$

$$\text{RPF} = \frac{600 \text{ ml/min}}{0.91} = 659 \text{ ml/min}$$

FREE WATER CLEARANCE. The kidneys may be remarkably miserly in their handling of water when such conservation is necessary, eliminating in the urine only that amount of water necessary to achieve excretion of waste products. Renal concentrating mechanisms can achieve urine concentrations of 1200 mosm/liter. Conversely, in a state of water excess, urine can be excreted in copious amounts with a solute concentration of less than 100 mosm/liter.

Free water is water without solute. Free water clearance (C_{H_2O}) is a calculation of the amount of distilled water that must be added to or removed from the urine to render the urine isosmotic with plasma. The concept of free water clearance is an attempt to quantify the states of water diuresis and conservation. It is calculated by subtracting the clearance of osmoles (C_{osm}) from the minute urine flow ($\dot{V}$).

$$C_{H_20} = \dot{V} - \frac{U_{osm}\dot{V}}{P_{osm}} \qquad (5\text{-}4)$$

If the C_{H_20} is greater than 0, then water is being excreted in excess of solute (diuresis). If the C_{H_20} equals 0, then the urine is being excreted isosmotically. If the C_{H_2O} is less than 0, then solute is being excreted in excess of water (antidiuresis).

Negative free water clearance (T_{H_2O}) is the free water clearance with the sign reversed for convenience when discussing water conserving states.

$$T_{H_2O} = -(C_{H_2O}) \quad (5\text{-}5)$$

Free water clearance is not a true clearance, although it does contain a true clearance as a part of its definition.

QUANTITATION OF WATER REABSORPTION. The ratio of tubular fluid (TF) to plasma (P) inulin concentrations is an indication of the degree of water reabsorption that has occurred. For instance, TF/P inulin is normally 3 for the proximal tubule, which means that the tubular fluid has 3 times the inulin concentration of the plasma. The fraction of water reabsorbed along the tubule can then be calculated by the following equation:

$$\text{Fraction of water reabsorbed} = 1 - \frac{1}{TF_{Inu}/P_{Inu}} \quad (5\text{-}6)$$

$$= 2/3$$

QUESTIONS

23. A. As shown in Figure 5-10, many oncotic pressure profiles are possible. How could the determinants of GFR be altered to change from curve 1 to curve 2?
    B. The glomerular capillaries filter far more fluid than do other capillary beds. Why are they able to do this?
    C. Why does filtration stop when the efferent arteriole is reached?
    Answer: 1. Filtration pressure equilibrium (FPE) is achieved.
    2. K_f becomes zero.
    3. $\bar{\pi} = \bar{p}$.
    4. Q_A nears zero.
24. A. Renal blood flow decreases with . . .
    B. GFR increases with . . .
    Answer: 1. Constriction of the efferent arteriole
    2. Constriction of the afferent arteriole
    3. Both
    4. Neither
25. Calculate GFR if U_{Inu} = 30 mg/ml, V = 1.1 ml/min, and P_{Inu} = 0.25 mg/ml.
26. Why is PAH used to calculate effective renal plasma flow? What naturally occurring substance could be used instead.
27. In the patient whose PAH clearance (600 ml/min) was previously illustrated, what was the total amount of PAH actually excreted and what was the total amount attributable to filtration, to reabsorption, and to secretion? (Note: $\dot{V}$ = 1 ml/min, U_{Inu} = 12 mg/ml, P_{Inu} = 0.10 mg/ml).
28. A 24-hour-urine collection yields 1296 ml of urine with 14 mg per milliliter of PAH. The patient is receiving PAH intravenously at a constant rate to maintain a PAH venous plasma concentration of 0.02 mg per milliter.
    A. Calculate the ERPF
    B. Estimate actual RPF
    C. Calculate the RBF given that the patient's hematocrit is 42 percent
    D. Calculate the rate of PAH excretion
29. A patient's total serum osmolality is 291 mosm per kilogram water prior to being placed on a nothing-by-mouth (NPO) diet. Would the patient's serum osmolality be increased, decreased, or remain the same 24 hours later if in the interim his clearance of water was

A. >0
B. <0
C. =0

Is the patient's urine isosmotic, hyperosmotic, or hyposmotic in A, B, and C?

30. Given a constant GFR and an increase in the urine-plasma inulin then
    A. Urine flow has decreased
    B. Free water clearance has decreased
    C. Inulin clearance has decreased
    D. Free water clearance has increased

PHYSIOLOGY OF THE NEPHRON AND COLLECTING SYSTEM: BEYOND THE GLOMERULUS

REABSORPTION AND SECRETION IN THE RENAL TUBULES. Renal tubular reabsorption and secretion can occur as a result of active transport, passive diffusion (either simple or facilitated), or secondary active transport. In the renal tubules reabsorption and secretion of many substrates are coupled to the active transport of Na^+ via specific membrane carriers, glucose and amino acids being of particular importance. The rate of secretion or reabsorption by membrane carriers is directly proportional to the concentration of the substrate and to the affinity of the carrier for the substrate. Also characteristic of such transport systems is that each has a maximum rate (Tm) at which it can transport solute above which the transport mechanism is said to be saturated (i.e., increases in solute delivery do not lead to any increase in transport because all the carriers are occupied). Transport maximums for the various solutes vary tremendously, from quite low to so high that their measurement is not practical. In some cases the transport may be bidirectional. The composition of the urine is determined by the balance that normally exists between filtration, secretion, and reabsorption (Table 5-8).

THE PROXIMAL TUBULE. The proximal tubule is the main reabsorptive region of the nephron (Figs. 5-13,14). All reabsorption that occurs there is isosmotic (Fig. 5-15). The proximal tubule, therefore, does not affect the overall concentration of the TF, but it does alter the composition of the TF. The proximal tubule has two subdivisions, the proximal convoluted tubule and the proximal straight tubule.

The proximal tubule is characterized by a large reabsorptive surface formed by thousands of microvilli projecting from the luminal surface of the proximal tubular cells and by extensive lateral intercellular channels which are important in the isosmotic reabsorption of NaCl and water (Fig. 5-16). The proximal tubule is the only region of the tubule in which the enzyme carbonic anhydrase may be found in the luminal membrane. Within the luminal membrane is a relatively weak H^+ ion secretory pump which is unable to acidify significantly the urine (i.e., does not contribute to H^+ excretion) but serves instead to facilitate the reabsorption of $NaHCO_3$ with carbonic anhydrase (Fig. 5-17, see Chapter 7). The basolateral membrane of the proximal tubule is well-developed and contains the active Na^+-K^+ exchange pumps (Fig. 5-18). Numerous mitochondria line the basolateral membrane to provide energy for the active Na^+-K^+ pumps. Cell-to-cell junctions in the proximal tubule are permeable to water and electrolytes to a greater degree than those of other parts of the tubule. This results in significant back-flux of fluid and solute which may be important in the regulation of solute and water reabsorption (see Fig. 5-16).

The functions of the proximal tubule are as follows:

Table 5-8. *Renal Handling of Various Substances*

Primarily filtered
Inulin
Urea[a]
Creatinine[b]
Primarily secreted[c]
Hydrogen ions
Organic acids (e.g., PAH, penicillins)
Organic bases (e.g., choline)
Filtration and reabsorption
Water
Sodium
Chloride
Bicarbonate
Calcium
Magnesium
Phosphate
Glucose
Amino acids
Filtration, reabsorption, and secretion
K^{+}[d]
Uric acid
Filtration and renal tubular catabolism
Low-molecular-weight proteins (e.g., insulin, glucagon, PTH and ADH)

[a] Approximately half of the filtered load may be passively reabsorbed into the medullary interstitium.
[b] Approximately 20 percent is secreted in humans.
[c] Some passive reabsorption is also possible.
[d] Virtually all the potassium filtered is reabsorbed somewhere in the tubule. Net excretion is by passive secretion in the distal nephron.

1. Reabsorption of two-thirds of the filtered Na^+ and water predominantly by active transport of Na^+ at the basolateral membrane with water following passively
2. Passive reabsorption of chloride and other electrolytes along their electrochemical gradient
3. Reabsorption of all the filtered glucose and amino acids by cotransport with Na^+
4. Preferential reabsorption of $NaHCO_3$ (carbonic anhydrase dependent)
5. Organic acid secretion (e.g., diuretics, salicylates, antibiotics, *p*-aminohippurate) by an anionic pump located in the basolateral membrane
6. Organic base secretion (e.g., procainamide, choline) by a cationic pump located in the basolateral membrane
7. Vitamin reabsorption
8. Pinocytosis and breakdown of any protein remaining in the TF
9. Reabsorption of organic anions by cotransport with Na^+ (e.g., phosphate, sulfate, and lactate)
10. Active K^+ reabsorption
11. Secretion of ammonia

The TF leaving this segment of the nephron is thus characterized by

1. A TF osmolality equal to that of plasma
2. Flow equals one-third of the GFR
3. Absence of glucose, amino acids, and proteins
4. A somewhat higher concentration of chloride. Because normally almost all of the HCO_3^- is reabsorbed in the proximal tubule, some anion must remain behind in greater concentration to preserve electroneutrality. Therefore,

Fig. 5-14. *Reabsorption of various solutes in the proximal tubule in relation to the potential difference (PD) along the tubule. TF/P = tubular fluid to plasma concentration ratio. (From W. F. Ganong.* Review of Medical Physiology *[12th ed.]. P. 573. © 1985 by Lange Medical Publications, Los Altos, CA.)*

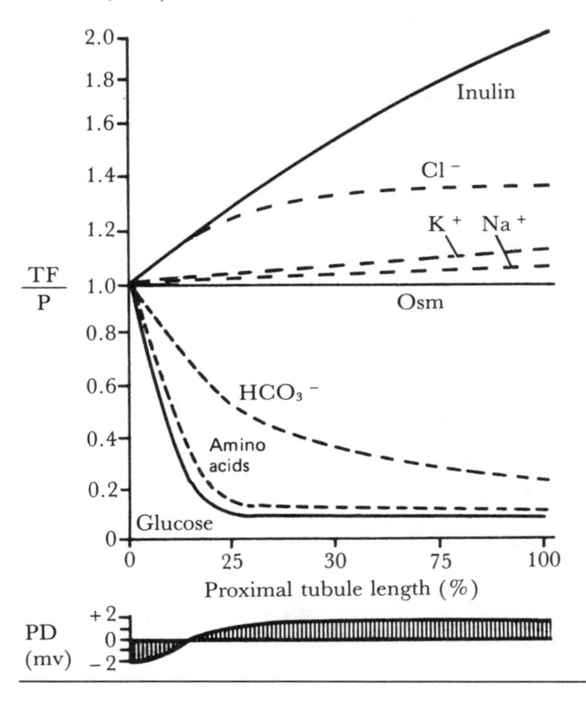

Fig. 5-15. *Changes in osmolality of the tubular fluid as it passes through the tubular system (From A. C. Guyton.* Textbook of Medical Physiology *[7th ed.]. Philadelphia: Saunders, 1986. P. 417.)*

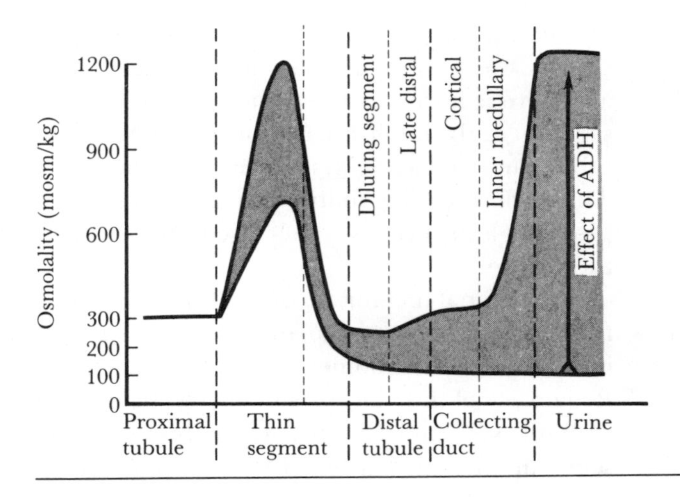

even though there is net reabsorption of chloride, its relative concentration nonetheless increases (see Fig. 5-14).

THE LOOP OF HENLE. Anatomically, the loop of Henle consists of the pars recta of the proximal tubule, the thin descending limb, the thin ascending limb (present only in long-looped nephrons), and the thick ascending limb. Because it is an anatomical and not a physiological entity, it is not surprising that its boundaries overlap those of other portions of the nephron. The thick ascending limb, for instance, may be considered a part of the distal tubule; its function is discussed separately. The physiological significance of the loop of

Fig. 5-16. *Diagrammatic summary of the movement of Na^+ from the tubular lumen to the renal capillaries in locations where it is actively transported. A. Situation in hydropenic state. B. Situation when capillary uptake is reduced by increasing capillary pressure or decreasing plasma protein concentration. This causes widening of the intracellular junctions with backflow of solute and fluid into the tubule. Heavy solid arrow indicates active transport; dashed arrows indicate passive movement; hatched arrow indicates movement into renal capillaries as a function of the Starling forces across these capillaries. TJ = tight junction; IC = lateral intercellular space; BI = basilar infoldings. (Reproduced with permission from P. F. Mercer, D. A. Maddox, and B. M. Brenner. Current concepts of sodium chloride and water transport by the mammalian nephron.* West. J. Med. *120:33, 1974.)*

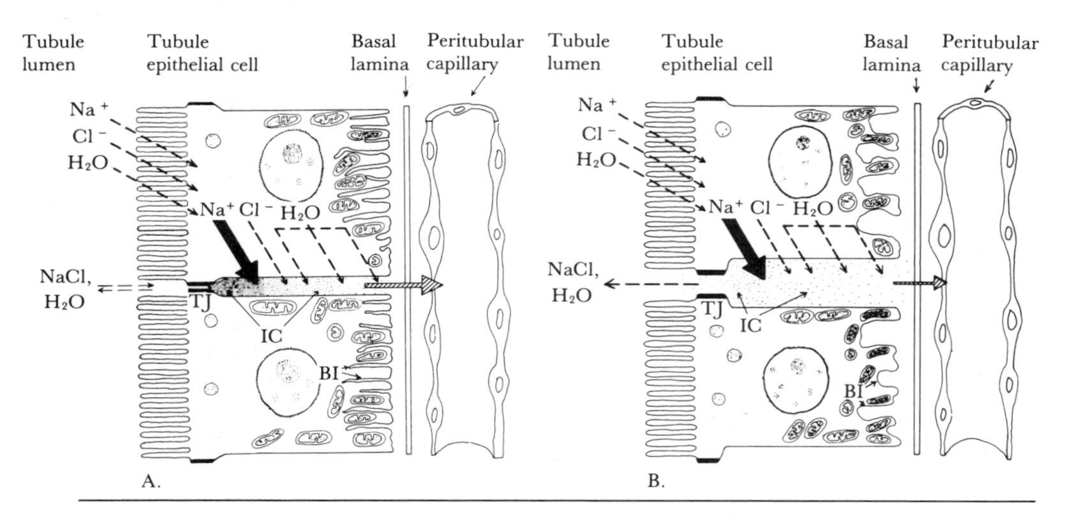

Henle is discussed in conjunction with the renal concentrating and diluting mechanism.

THE THICK ASCENDING LIMB OF THE LOOP OF HENLE. The thick ascending limb is the diluting segment of the nephron, because solute is actively pumped out of the thick ascending limb leaving behind dilute tubular fluid. In the absence of antidiuretic hormone the urine is dilute as well, because the fluid diluted in this segment of the nephron, by removal of NaCl, is diluted even further in the distal tubule and collecting system in which additional active Na^+ pumping occurs. The thick ascending limb is extremely impermeable to water (which makes sense physiologically because otherwise water would be osmotically obligated to follow the solute).

The luminal membrane of the thick ascending limb has a carrier that requires four sites to be occupied in order to carry solute from the lumen to the cell interior. The stoichiometry of the carrier is thought to be $1Na^+ : 1K^+ : 2Cl^-$. Regardless of the exact stoichiometry, the important point is that active pumping of both Na^+ and Cl^- occurs in this segment of the nephron. It is the only segment in which active Cl^- pumping has been demonstrated. A positive intraluminal potential difference provides the driving force for the reabsorption of all cations but particularly divalent cations (Mg^{2+} and Ca^{2+}) from the thick ascending limb. Medullary portions of the thick ascending limb contribute to the medullary hypertonicity by pumping solute into the interstitium and, therefore, contributing to the dilution of urine as well as to its concentration under the appropriate circumstances.

DISTAL CONVOLUTED TUBULE. The distal convoluted tubule is also impermeable to water. Sodium is actively reabsorbed in this segment with Cl^- following passively. The reabsorption of Na^+ and the active secretion of H^+ and K^+ are under the influence of the mineralocorticoid hormone aldosterone. Both the distal convoluted tubule and collecting duct are subject to the actions of aldosterone. The excretion of K^+ is dependent primarily upon the amount secreted in the distal nephron in response to aldosterone. The distal tubule has an extremely powerful H^+ secretory pump able to acidify urine to a pH of 4.5 (almost 3 pH units below the plasma pH).

Except for very distal portions, the distal convoluted tubule does not appear to respond to antidiuretic hormone.

THE COLLECTING SYSTEM. The major function of the collecting system is the reabsorption of water in response to antidiuretic hormone (ADH). In the presence of ADH the collecting system becomes very permeable to water and urea. A very concentrated urine is then formed as water flows along the osmotic gradient out of the tubule and into the interstitium. In the absence of ADH the collecting system is impermeable to water and dilute tubular fluid from the thick ascending limb and the distal tubule is excreted into the minor calyces. The collecting system, like the distal tubule, is subject to mineralocorticoid action with active reabsorption of Na^+ and secretion of K^+ and H^+.

QUESTIONS

31. List the following substances in order of decreasing clearance by the kidney:
    A. PAH
    B. Creatinine
    C. Glucose
    D. K^+
    E. Na^+
    F. Inulin

THE RENIN-ANGIOTENSIN SYSTEM

The juxtaglomerular apparatus (JGA) is the site of synthesis and release of the hormone renin (Fig. 5-19). The juxtaglomerular apparatus consists of four components:

1. Modified smooth muscle cells in the walls of the afferent arterioles
2. Modified smooth muscle cells in the walls of the efferent arterioles
3. Extraglomerular mesangium
4. The macula densa cells in a short region of the wall of the distal tubule

The JGA is located where the distal tubule of the nephron passes between the afferent and efferent arterioles of that same nephron. The JGA is, therefore, immediately adjacent to its own glomerulus.

Renin is synthesized and secreted by the modified smooth muscle cells in the walls of the arterioles when a decrease in ECF volume is sensed by the kidney. A decreased NaCl load reaching the macula densa cells in the TF or a decreased perfusion pressure detected by baroreceptors in the afferent arterioles results in increased renin secretion and release. Other causes of renin release include increased β-adrenergic stimulation, decreased serum potassium, and a decrease in the arterial pressure detected by baroreceptors in the major blood vessels (affecting vasomotor centers in the central nervous system (CNS) to increase sympathetic discharge).

These causes of renin release are interrelated. A decrease in the ECF volume, the perfusion pressure at the afferent arteriole, or the blood flow through the kidneys all cause a decrease in the GFR (Fig. 5-20). This in turn results in decreased flow in the tubule and more time for reabsorbing NaCl. The macula densa then senses the decreased load of NaCl and, in some unknown fashion,

Fig. 5-17. *A. Reabsorption of filtered HCO_3^- via H^+ secretion. The secreted H^+ reacts with HCO_3^- in the tubular fluid to form H_2CO_3, which then dissociates into carbon dioxide and water, i.e., a HCO_3^- ion is lost from the tubular fluid. Because the H^+ secretion process adds a HCO_3^- ion to the peritubular fluid and peritubular capillaries, however, the net result is HCO_3^- reabsorption. The proximal tubule is illustrated here, since approximately 90 percent of the filtered HCO_3^- is reabsorbed in the proximal tubule. Similar reactions occur in more distal portions of the nephron (except for the absence of carbonic anhydrase in the luminal surface) and are responsible for the reabsorption of most of the remaining HCO_3^- in the tubular fluid. An active Na^+-K^+ exchange pump in the basolateral membrane (not shown) actively pumps Na^+ from the cells into the peritubular fluid. This supplies the gradient for Na^+ reabsorption from the tubular fluid. (Reproduced with permission from J. B. West.* Best and Taylor's: Physiological Basis of Medical Practice *[11th ed.]. P. 522. © 1985 The Williams & Wilkins Co., Baltimore.)*
B. Generation of new HCO_3^- via H^+ secretion. The HCO_3^- generated in the H^+ secretion process represents a new HCO_3^- if the secreted H^+ ion reacts with NH_3 synthesized by the epithelial cells to form NH_4^+, or if the H^+ reacts with HPO_4^{-2} and other titratable acids in the

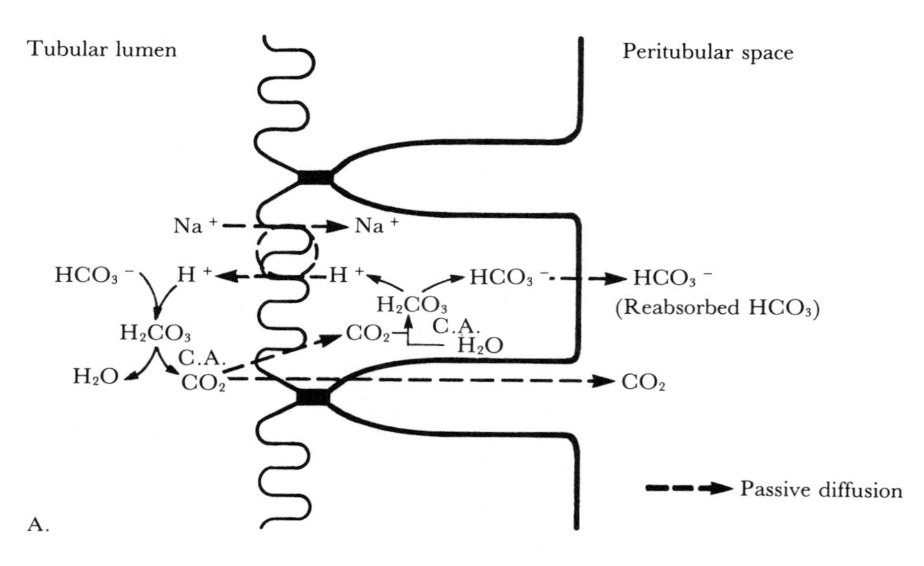

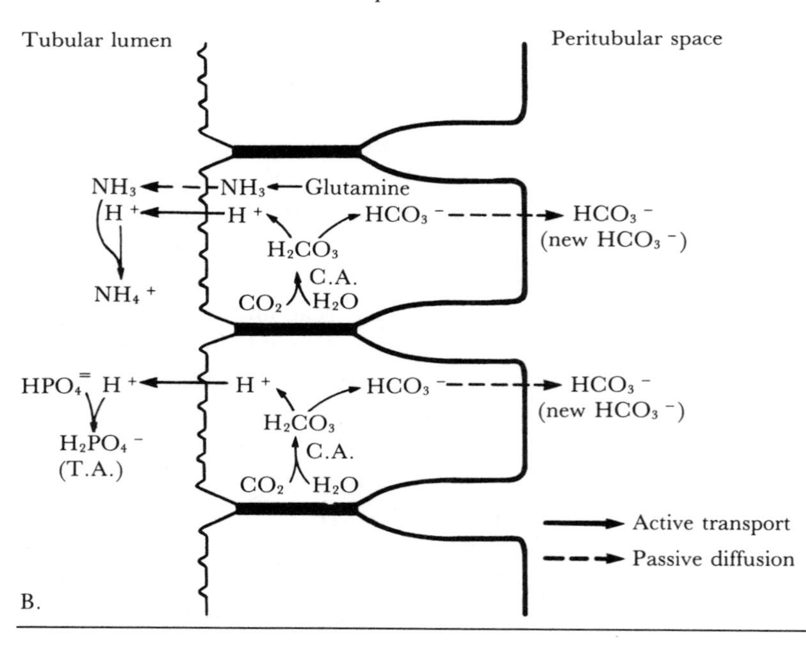

Figure 5-17 (Continued)

in the tubular fluid to form $H_2PO_4^-$ and to acidify the remaining titratable buffers. Although the distal nephron is illustrated here, similar reactions can occur in the proximal tubule (except that proximal tubular H^+ secretion is primarily coupled to Na^+ reabsorption and is gradient limited). Most of the H^+ secreted in the proximal tubule reacts with HCO_3^- in the tubular fluid and, therefore, accomplishes the reabsorption of filtered HCO_3^- rather than the generation of new HCO_3^-. The distal tubular H^+ secretory pump can create a 1000-fold H^+ concentration gradient acidifying the urine. (From J. B. West. Best and Taylor's: Physiological Basis of Medical Practice *[11th ed.]. P. 524. © 1985 The Williams & Wilkins Co., Baltimore.)*

Fig. 5-18.

Interaction of the transport of Na^+ and other solutes across early proximal tubule. (From B. M. Burg. Renal Handling of Sodium, Chloride, Water, Amino Acid and Glucose. In B. M. Brenner, and F. C. Rector The Kidney *(2nd ed.). Philadelphia: Saunders, 1981. P.335.*

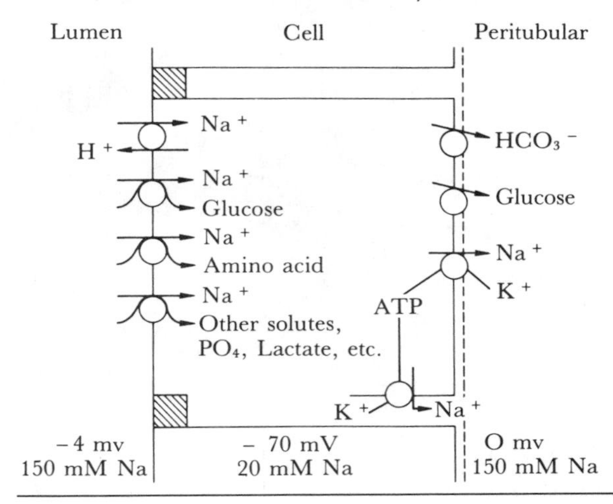

this activates renin release to correct the situation. Beta-adrenergic stimulation by its vasoconstrictor effect on the renal vasculature also results in a decreased GFR and has a direct effect on the release of renin. An acute increase in ECF volume, however, results in opposite actions (Fig. 5-21).

Renin causes circulating angiotensinogen (a decapeptide) to be cleaved resulting in angiotensin I (an octapeptide) (Fig. 5-22). Angiotensin I is then converted to angiotensin II, the most potent vasoconstrictor known, by a converting enzyme located primarily in the lung vasculature.

Angiotensin II has the following actions:

1. Potent vasoconstrictor effect on both arterial and venous beds.
2. Increases the synthesis and release of aldosterone.
3. Increases the release of ADH centrally.
4. Increases blood pressure by a central mechanism.
5. Increases thirst by a central mechanism.
6. Feedback inhibition of the release of renin.
7. Angiotensin II constricts both afferent and efferent arterioles but also increases levels of prostaglandins. The prostaglandins decrease the relative amount of afferent constriction and thereby act to maintain the GFR despite high levels of circulating angiotensin II.

QUESTIONS

32. Which of the following do not increase renin release?
   A. Converting enzyme inhibitor
   B. Standing up
   C. Almost everything (shock, fright, etc.)

Fig. 5-19. *Schematic diagram of the juxtaglomerular apparatus. (From A. W. Ham, and D. H. Cormack.* Histology *[8th ed.]. Philadelphia: Lippincott, 1979. P. 761 and from A. C. Guyton.* Textbook of Medical Physiology *[7th ed.]. Philadelphia: Saunders, 1986. P. 411.)*

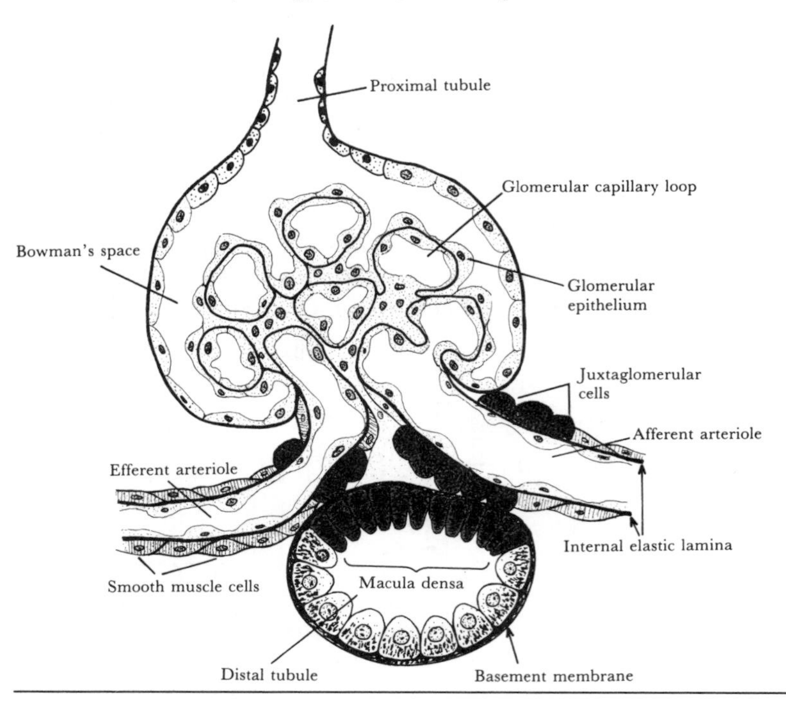

D. SIADH
E. Conn's syndrome
F. Hemorrhage
G. Low-sodium diet
H. Ureteral constriction
I. High blood pressure
J. Hypoglycemia
K. Epinephrine
L. Potassium depletion
M. Decreased Na^+ or K^+ in TF reaching the macula densa
N. Low central venous pressure
O. Low aortic or carotid pressure

33. Angiotensin II acts to elevate the filtration fraction (GFR/glomerular plasma flow) by
    A. Increasing renal arteriolar resistance
    B. Stimulating mesangial cell contraction
    C. A selective or disproportionate elevation of efferent relative to afferent resistance
    D. None of the above
    E. All of the above

ALDOSTERONE

The major stimulus for the release of aldosterone is increasing ECF levels of K^+. Adrenocorticotrophic hormone (ACTH) acts in a permissive role only, making possible the synthesis of aldosterone by its trophic effect on the adrenal cortex but not directly controlling the degree of production. Angiotensin II acts as a major stimulus for the synthesis and release of aldosterone as does one of its metabolites, angiotensin III. Angiotensin III is not known to have any other

Fig. 5-20. *Schematic view of afferent and efferent mechanisms in the response to acute volume depletion. These act to reduce the excretion of salt and water. In mild volume depletion, the pathways indicated by the heavy lines are probably the primary ones activated.* π_b = *the oncotic pressure of the plasma.* P_c = *hydrostatic pressure in the capillary. (Reproduced with permission from L. P. Sullivan, and J. J. Grantham.* Physiology of the Kidney *[2nd ed.]. Philadelphia: Lea and Febiger, 1982. P.173.)*

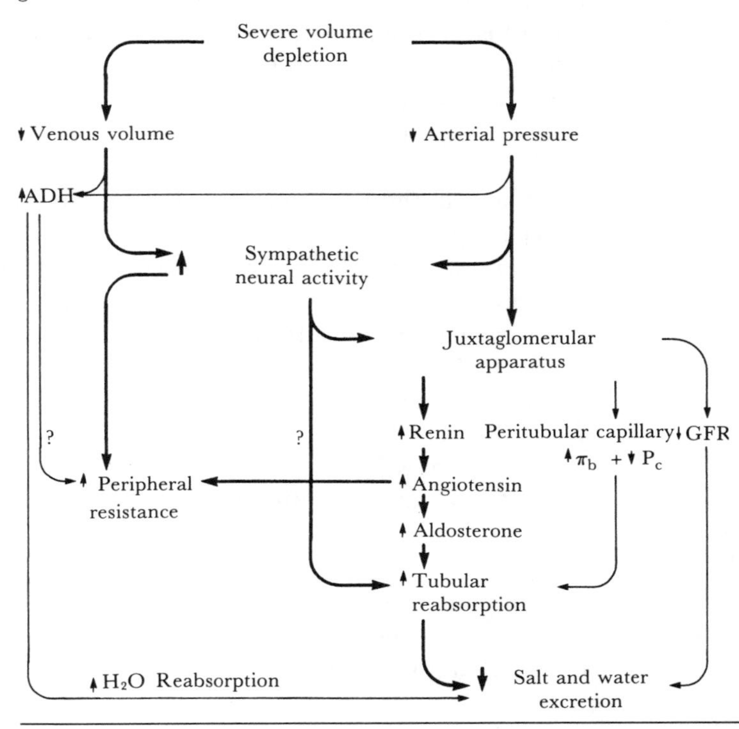

important physiological actions. The fourth and final regulatory mechanism known to control the secretion of aldosterone is the quantity of body sodium, the exact mechanism of which is unknown. Aldosterone is synthesized in the zona glomerulosa of the adrenal cortex and high levels of aldosterone have a negative feedback effect on the release of renin (Fig. 5-22).

Aldosterone acts on the distal tubule and collecting ducts to cause the secretion of K^+, and to a lesser extent H^+, in exchange for Na^+. Aldosterone, therefore, leads to the conservation of Na^+ and the excretion of K^+ and H^+. High aldosterone levels lead to the reabsorption of virtually all the Na^+ remaining in the TF. Loss of Na^+ in the urine may decrease to only a few meq/liter/day. Potassium loss in the urine, on the other hand, increases severalfold. The retention of Na^+ results in the rise of the Na^+ concentration by, at most, a few meq per liter because water is also retained. If the levels of aldosterone are excessive, however, total body sodium and water increases by up to 20 percent as the retained sodium osmotically obligates water to follow it into the body. The phenomenon known as escape from aldosteronism occurs when aldosterone has been high for more than a few days. Sodium and water reabsorption by the proximal tubule decreases as fluid levels in the body rise. Aldosterone continues to act on the distal and collecting tubules in an attempt to retain Na^+ and excrete K^+ and H^+ ions. However, the increased Na^+ load resulting from decreased reabsorption by the proximal tubule exceeds the distal and collecting tubules' ability to reabsorb Na^+ resulting in the return of Na^+ excretion toward normal levels. Potassium excretion continues at high levels. Sodium and water retention of up to 10 percent above normal and moderate to severe hypertension are frequent sequelae.

In Conn's syndrome an autonomously secreting tumor releases excessive amounts of aldosterone. As might be predicted, Conn's syndrome results in hypertension, hypokalemia, hypervolemia, and low plasma renin activity. When the serum K^+ levels drop to half of normal, muscle weakness or paralysis which can result in death secondary to hyperpolarization of nerve and muscle cells occurs. The increased exchange of H^+ for Na^+ may lead to a mild alkalosis.

Total lack of aldosterone causes plasma Na^+ concentrations to fall by 5 to 8 percent initially as Na^+ is lost in the urine. The ECF becomes volume depleted as Na^+ levels fall. If left untreated, progressive hyperkalemia eventually leads to cardiac arrhythmias and death.

QUESTIONS

34. Where in the nephron does aldosterone act to modulate Na^+ absorption?
35. Defects in the regulation of aldosterone may result in an increase or decrease in the ECF volume by up to 25 percent, yet serum Na^+ concentrations vary by only a few percent. Why?
36. Adrenocortical insufficiency causes an increase, decrease, or no change in
    A. ECF
    B. Total body Na^+
    C. Excretion of Na^+
    D. Serum K^+
    E. Which of these effects is a life-threatening emergency?

RENAL HANDLING OF SODIUM, POTASSIUM, AND CHLORIDE

Sodium, chloride, and potassium are freely filtered in the glomerulus. The proportions of these three ions reabsorbed in various parts of the uriniferous tubule are approximately:

Fig. 5-21. *Schematic view of afferent and efferent mechanisms in the response to severe volume expansion. These act to increase salt and water excretion. The heavy lines indicate the pathways that are probably activated in mild volume expansion.* π_b *= plasma oncotic pressure.* P_c *= hydrostatic pressure in the capillary. (From L. P. Sullivan, and J. J. Grantham,* Physiology of the Kidney *[2nd ed.]. Philadelphia: Lea and Febiger, 1982. P. 174.)*

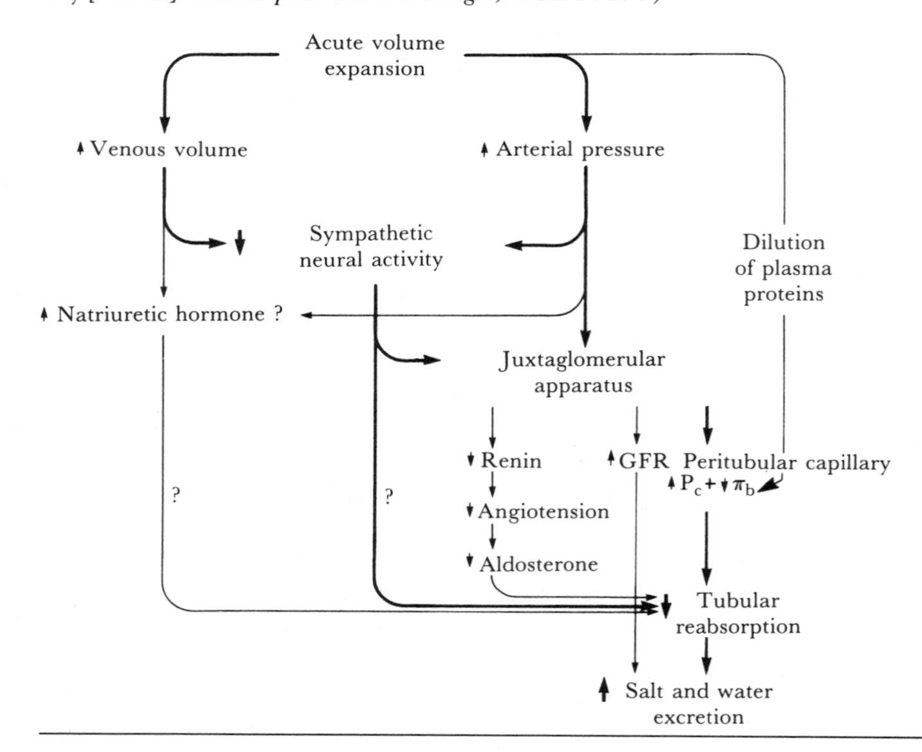

Fig. 5-22. *Schematic diagram of the renin-angiotensin-aldosterone system's response to a decrease in ECF volume or blood pressure, or an increase in sympathetic tone. Dashed lines indicate negative feedback.*

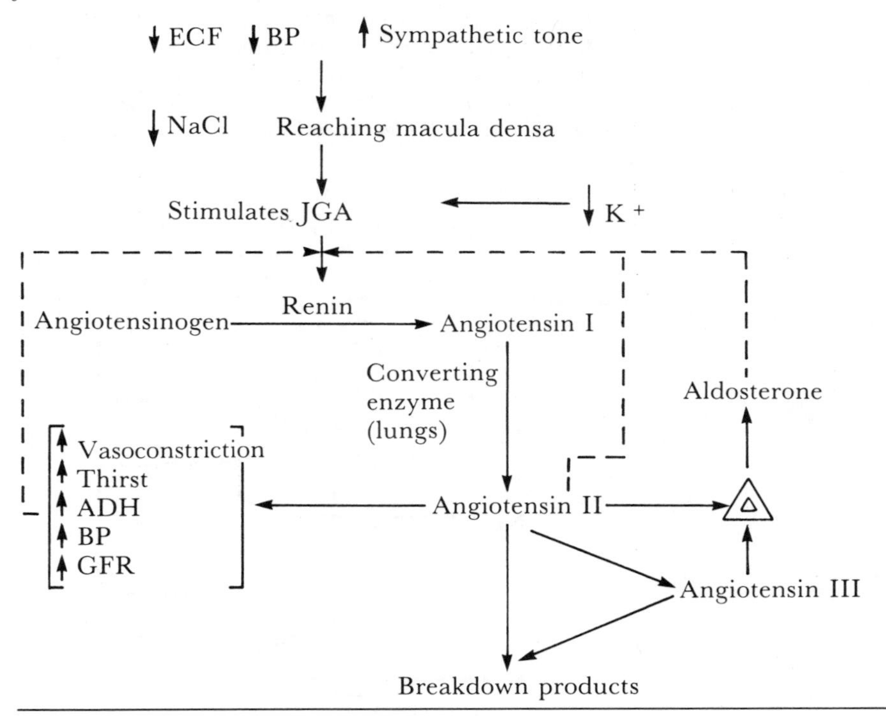

67 percent in the proximal tubule
25 percent in the loop of Henle
5 percent in the distal convoluted tubule
2 to 3 percent in the collecting tubule

While there is an average net reabsorption of greater than 99 percent of Na^+ and Cl^- and 80 to 90 percent of K^+, the mechanisms involved in each case are quite distinct.

The contributions of the distal convoluted tubule and the collecting tubules are particularly important despite the relatively small percentages reabsorbed there, because the final regulation of the composition of the urine takes place in these two segments. Thus, the fine control of excretion is governed by the most distal part of the nephron.

The primary active transport of Na^+ provides the driving force that is ultimately responsible for the reabsorption and secretion of many of the solutes transported by the kidneys. In the proximal tubules, roughly 70 percent of the Na^+ enters the proximal tubular cells passively by either facilitated or simple diffusion and is subsequently actively pumped out into the lateral intercellular space. The Na^+-K^+–ATPase exchange pumps located in the basolateral membrane provide the energy for the active transportation of Na^+ out of the cells in exchange for K^+ from the peritubular fluid (see Figs. 5-18, 5-23). The resulting low concentration of Na^+ in the cells creates a concentration gradient favoring the passive uptake of Na^+ from the tubular fluid. The transportation of Na^+ into the cells can be coupled to that of other solutes by either cotransport (glucose, amino acids, HCO_3^-, lactate, sulfate, and phosphate) or exchange (H^+). Because this transportation of solute is coupled to the transport of Na^+ rather than directly to metabolism, it is referred to as secondary active transport (see Fig. 5-18).

Fig. 5-23. *Mechanisms for Na^+ reabsorption in the proximal tubule. A. Unidirectional Na^+ transport (uniport). Sodium passively enters the cell along its electrochemical gradient and then is actively extruded across the basolateral surfaces by the Na^+-K^+ ATPase (illustrated here across the lateral surface). This reabsorption of Na^+ generates a lumen-negative transepithelial potential difference (PD) of about −2 mv because Cl^- accompanies Na^+ by diffusing across the "leaky" tight junctions. B. Na^+ -H^+ exchange (antiport). The passive diffusion of Na^+ into the cell is coupled (dashed circle) to the secretion of H^+ into the tubular fluid. The Na^+ is then actively extruded across the basolateral surfaces by the Na^+-K^+ ATPase, while the secreted H^+*

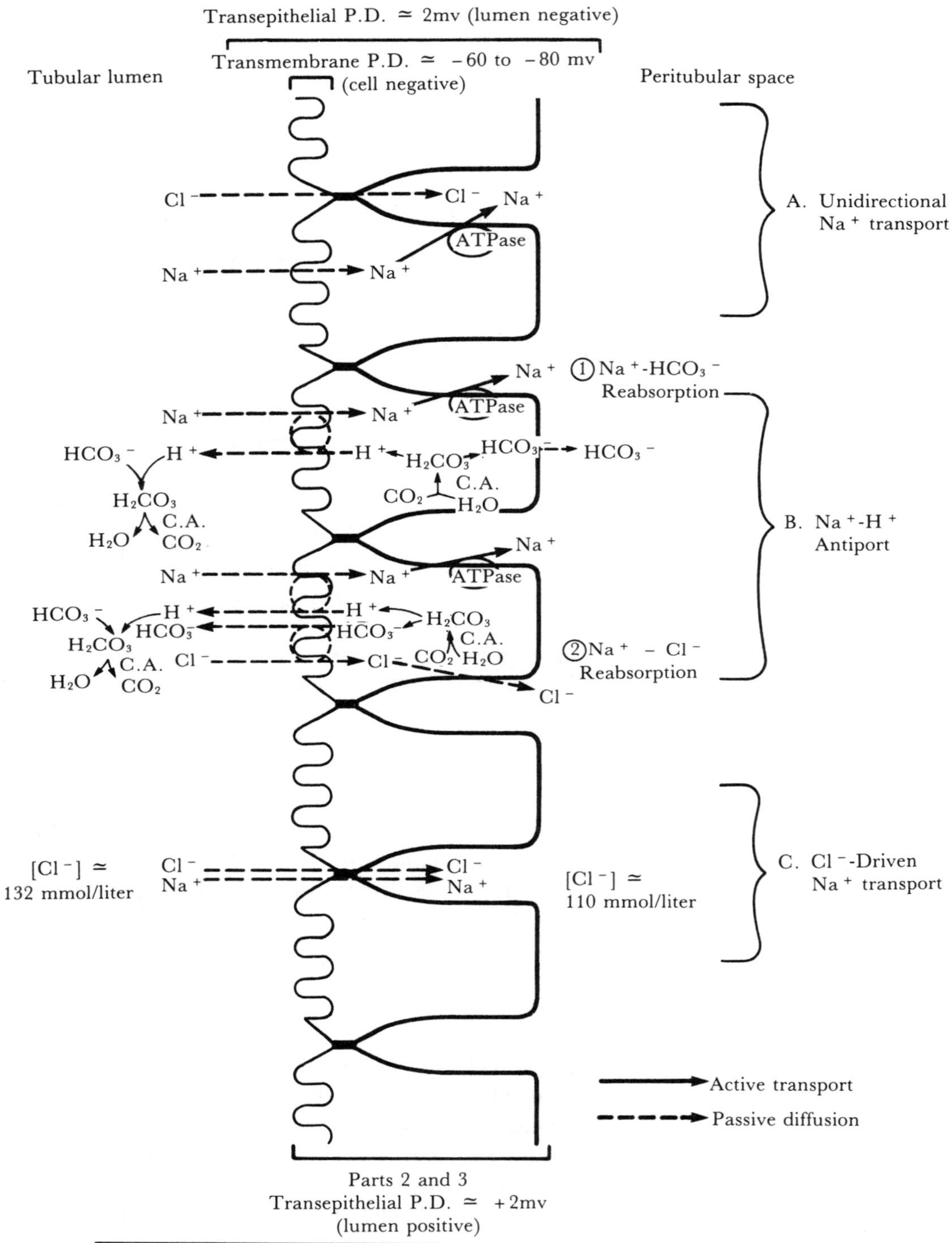

Figure 5-23 (Continued) *reacts with HCO_3^- in the tubular fluid to form carbonic acid, which then dissociates into carbon dioxide and water in a reaction catalyzed by carbonic anhydrase (CA) in the brush border. The HCO_3^- formed in the cell during the H^+ secretion process can diffuse across the basolateral surfaces, in which case the net result of the Na^+-H^+ antiport is Na^+-HCO_3^- reabsorption (B1), or diffuse across the brush border via a coupled exchange for Cl^- (dashed circle), in which case the net result of the Na^+-H^+ antiport is Na^+-Cl^- reabsorption (B2). C. Chloride-driven Na^+ transport. The Cl^- concentration in the tubular fluid increases to about 132 mmol per liter in parts 2 and 3, resulting in a concentration gradient for the diffusion of Cl^- across the "leaky" tight junctions into the peritubular fluid, in which the Cl^- concentration is about 110 mmol/liter. This reabsorption of Cl^- generates a lumen-positive transepithelial PD in parts 2 and 3, but only of about +2 mv, because Na^+ accompanies Cl^- across the "leaky" tight junctions. (Reproduced with permission from J. B. West.* Best and Taylor's: Physiological Basis of Medical Practice *[11th ed.]. P. 483. © 1985 The Williams & Wilkins Co., Baltimore.)*

Cotransport of Na^+ and glucose or amino acids results in a lumen negative potential difference in the most proximal portion of the proximal tubule because glucose and amino acids are electrically neutral (see Fig. 5-14). A positive intraluminal potential difference quickly supervenes, however, as a consequence of the preferential secretion of H^+ and reabsorption of bicarbonate in the early proximal tubule. Bicarbonate is less permeable to the proximal tubular membrane than is Cl^-. Because Cl^- is left behind while $NaHCO_3$ and water are reabsorbed, its concentration in the tubular fluid increases. As the Cl^- diffuses out of the tubular fluid into the cell along its concentration gradient, it creates a lumen positive transepithelial potential difference which results in additional Na^+ reabsorption. Unlike the balance of Na^+ reabsorption, the 30 percent of Na^+ reabsorption attributable to this lumen positive potential difference actually bypasses the proximal tubular cells and enters *between* cells (Fig. 5-23). The proximal tubule is so permeable to small ions that only a small potential difference is required to drive large amounts of Na^+ out of the tubule.

Of the Na^+ entering through the cell (70%), the following is a reasonable estimate of the contribution of each transport mechanism (Fig. 5-23):

1. Five percent by cotransport with glucose or amino acids using specific membrane carriers, or by cotransport with any one of several organic anions (sulfate, lactate, and phosphate)
2. Fifteen percent by a tightly coupled Na^+-H^+ exchange pump in the luminal membrane. Because the H^+ so secreted tritrates HCO_3^-, the net result is the reabsorption of $NaHCO_3$ without any net secretion of H^+
3. Fifty percent by simple diffusion with Cl^- (Cl^- is reabsorbed passively by the proximal tubular cells)

In the thick ascending limb of the loop of Henle both Na^+ and Cl^- are actively reabsorbed. In the distal convoluted and collecting tubules Na^+ is transported actively against both a large concentration gradient and a large electrical gradient in response to aldosterone. The Na^+ is reabsorbed in exchange for either H^+ or K^+. The collecting duct has a particularly powerful pump able to decrease the Na^+ concentration to almost zero. Chloride generally follows passively in the distal convoluted and collecting tubules, although in cases of severe Cl^- deprivation a Cl^- reabsorptive pump, normally masked by the quantitatively greater passive reabsorption of Cl^-, is demonstrable.

The magnitude of the net excretion of Na^+ and Cl^- depends, then, upon the extent to which they are reabsorbed following filtration.

Glomerular-tubular balance refers to the fact that under steady-state conditions a constant fraction of the filtered Na^+ is reabsorbed in the proximal tubule despite variations in GFR and the Na^+ load. This occurs in order to protect Na^+ balance from fluctuations in GFR because even a small change in GFR can result in large changes in the amount of filtered Na^+. Similar adjustments are made in more distal segments as well. The mechanism for this is unknown.

In contrast to Na^+ and Cl^-, the excretion of K^+ is dependent primarily upon the amount secreted in the distal portion of the nephron in response to aldosterone. Most of the K^+ filtered in the glomerulus is reabsorbed somewhere in the renal tubule. Furthermore, reabsorption of K^+ may occur anywhere in the tubule, and an active K^+ reabsorptive pump has been demonstrated in the luminal membrane of both the proximal convoluted tubule and the collecting ducts in addition to the Na^+-K^+ exchange pump in the basolateral membrane (Fig. 5-24). Under normal conditions, K^+ is reabsorbed in the proximal tubules and in the loop of Henle, and it is either secreted or reabsorbed in the distal portion of the nephron. Ninety percent or more of the filtered K^+ is reabsorbed in the proximal convoluted tubule and in the loop of Henle prior to reaching the distal tubule. In states of K^+ deficiency, the remaining 10 percent is reabsorbed in the distal tubule and collecting system. However, K^+ excretion is the norm, and it is the net secretion or lack of it in the distal portion of distal tubule and collecting system that determines the rate of K^+ excretion in the urine. Normally, an amount equal to 10 to 20 percent of the filtered load of K^+ is excreted. Net tubular excretion of K^+ is also possible, however, because the amount of K^+ excreted can range from 3 to 150 percent of the amount filtered. The absorption of K^+ is active while the secretion is thought to be passive.

An H^+-K^+ exchange pump in the basolateral membrane allows H^+ to be

Fig. 5-24. *Electrical potential gradient and transport of Na^+, Cl^-, and K^+ across a mammalian proximal tubular cell. The transepithelial electrical potential difference (PD) is directed lumen negative in the early proximal tubule, and it becomes lumen positive in the intermediate and late portions of the proximal tubule. (From H. Valtin.* Renal Function: Mechanisms Preserving Fluid and Solute Balance in Health *[2nd ed.]. Boston: Little, Brown, 1983. P. 253.)*

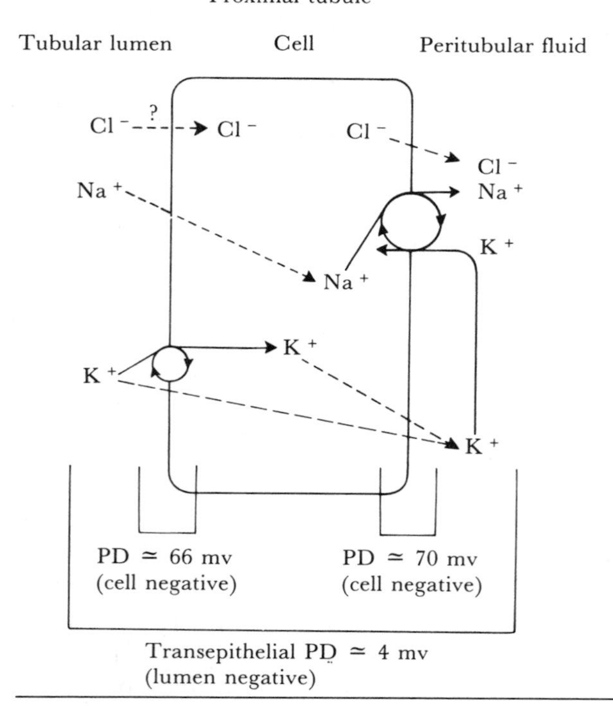

exchanged for K^+. Increasing plasma pH, for instance, causes H^+ ions to leave cells along its concentration gradient in exchange for extracellular K^+ (thus maintaining electroneutrality). Decreasing plasma pH causes K^+ to leave cells and H^+ to enter. Alkalosis, therefore, causes increased K^+ excretion by increasing the intracellular concentration of K^+ in the proximal tubular cells thereby increasing the gradient for K^+ secretion into the tubules. Unfortunately, acidosis also increases the excretion rate of K^+, presumably by some other mechanism.

Pharmacological doses of epinephrine and insulin (K^+ enters cells with glucose) drive K^+ into cells. High plasma pH, bicarbonate, and aldosterone also drive K^+ into cells. All of these, therefore, cause increased K^+ excretion by the kidneys by creating a greater gradient for K^+ secretion.

QUESTIONS

37. Active Na^+ transport must be
    A. Accompanied by movement of an anion in the same direction
    B. Accompanied by movement of a cation in the opposite direction
    C. A or B
    D. Neither A nor B
38. A. Can a patient with ascites ever be hyponatremic?
    B. Can such a patient ever be depleted of total body sodium?
39. Sodium ion is actively transported out of all segments of the nephron except two. Name them.
40. A person suddenly develops an insatiable appetite for potato chips and increases his dietary sodium intake by eightfold to 18 g per day. How much weight would this person gain the first day? If he continued to eat potato chips at the same rate, how much weight would be gained after 1 week? Would the weight gain continue indefinitely?
41. A. How would you treat acute hyperkalemia?
    B. Increases or decreases in plasma $[H^+]$ are generally accompanied by similar changes in ______.
42. Sodium ion is excreted by ______ while K^+ is excreted by ______. What would be the effect (increase, decrease, or no change) on urinary K^+ excretion of the following perturbations?
    A. Acute metabolic acidosis
    B. Low salt diet
    C. Acute metabolic alkalosis
    D. Diuretic that inhibits NaCl reabsorption
    E. Oliguria
    F. An aldosterone antagonist (e.g., spironolactone)
43. In which segments of the nephron has active Cl^- reabsorption been demonstrated?

RENAL CONCENTRATING AND DILUTING MECHANISMS

ANTIDIURETIC HORMONE. Antidiuretic hormone (ADH) or arginine vasopressin is an octapeptide synthesized in nerve cells in the hypothalamus. Release of ADH can occur in response to changes in both fluid volume and concentration and is controlled by:

1. Osmoreceptors responding to the effective osmolality of the ECF (Na^+ and its attendant anions Cl^- and HCO_3^-) but not to ineffective osmoles such as urea.
2. Receptors sensitive to changes in effective ECF and circulating blood volumes.
3. The action of circulating angiotensin II acting directly on the thirst center in the anterior hypothalamus.

4. A number of stimuli in a manner reminiscent of renin release. Nausea, vomiting, exercise, anxiety, fright, pain, nicotine, and anesthetic agents can increase the secretion of ADH.

Antidiuretic hormone increases the permeability of the entire collecting system to water. It also increases urea permeability of the medullary collecting ducts (see Fig. 5-25D). These responses are very rapid, occurring within minutes of the stimulation of ADH secretion.

When an individual is deprived of water, the osmotically obligated excretion of water continues until the plasma is rendered slightly hyperosmotic. This stimulates osmoreceptors in the hypothalamus and possibly others in the distribution of the internal carotids to initiate signals causing the secretion of ADH. The ADH increases the permeability of the entire collecting duct system to water. Water then flows out of the collecting system into the hypertonic medullary interstitium concentrating the urine up to a maximum (in humans) of 1200 mosm per liter.

In contrast, after drinking a large amount of water, the plasma is rendered hyposmotic, osmoreceptors call for decreased secretion of ADH, and the collecting system becomes impermeable to water once again. The dilute fluid entering the collecting system from the thick ascending limb and the distal convoluted tubule passes through the collecting system to be excreted as dilute urine (to a minimum of 50 mosm per liter).

THE COUNTERCURRENT MULTIPLIER: KUHN'S HYPOTHESIS. The countercurrent multiplier is really a countercurrent augmentor. Each successively deeper layer of the medullary interstitium augments the osmolality of the previous layer. This results in a continuously increasing osmolality or solute concentration parallel to the renal tubules which create it. By forming this osmotic gradient, the countercurrent multiplier enables the kidneys to excrete a concentrated urine. Furthermore, the same active process that creates the osmotic gradient simultaneously dilutes the TF and allows the kidneys to excrete a dilute urine (in the absence of ADH).

The term countercurrent refers to the flow in adjacent limbs of the U-shaped loop of Henle being in opposite directions (Fig. 5-25). The term multiplier is used because such countercurrent flow can establish large concentration gradients along the axis of the adjacent limbs. Concentration gradients in the transverse direction can remain relatively small and, therefore, within the capability of the renal tubules.

According to Kuhn there are three basic requirements for such a system to function:

1. Countercurrent flow
2. Differences in permeability between tubules carrying fluid in opposite directions
3. A source of energy

All three requirements are met by the kidneys: the loop of Henle provides the countercurrent flow, the descending limb is much more permeable to water than is the ascending limb, and active transport of sodium chloride from the ascending limb to the surrounding interstitium provides the energy.

The result of this arrangement is that active NaCl transport lowers the NaCl concentration (and osmolality) of fluid within the thick ascending limb and simultaneously raises the NaCl concentration in the interstitial fluid (Fig. 5-25A). Fluid that enters the descending limb still isosmotic with plasma and glomerular filtrate now becomes more concentrated as water is extracted across the limb's permeable wall by the higher osmolality of the interstitium (Fig. 5-25B). As this more concentrated fluid rounds the bend of the loop and begins its ascent, it delivers a higher NaCl concentration to the site of the active NaCl pump and the concentration of NaCl and hence the osmolality of the interstitium is raised still higher (Fig. 5-25B,C). Thus, the ability of the NaCl pump to establish a modest concentration difference across the wall of the ascending limb (perhaps 100 mM NaCl, equivalent to approximately 200 mosm per kilogram water) is augmented by countercurrent flow to achieve a large difference between the isosmotic fluid entering the descending limb and the hyperosmotic fluid at the tip of the papilla.

Fluid from the thick ascending limb enters the distal tubule in the cortex hyposmotic to the cortical interstitium. In the presence of ADH, which increases the water permeability of the entire collecting tubule, water is reabsorbed until osmotic equilibrium is achieved between the contents of the collecting tubule and the isosmotic cortical interstitium. As fluid enters the medullary portion of the collecting duct to be redirected for a second and final time through the medulla, more water is reabsorbed until equilibrium is established between luminal fluid and the hyperosmotic medullary interstitium. This results in the excretion of a concentrated urine. Additional active Na^+ reabsorption in the distal tubule and collecting system also contributes to the reabsorption of water.

Moreover, the same process that drives the multiplier also accounts for urinary dilution. In the absence of ADH the collecting tubule has a very low osmotic water permeability, so that fluid coming from the ascending limb remains dilute throughout the remainder of the distal and collecting tubule. Additional active Na^+ reabsorption in the distal tubule and collecting system now acts to dilute the tubular fluid.

Under normal circumstances, the regulation of urinary concentration is thus made elegantly simple; the degree to which urine is diluted or concentrated is controlled directly by the level of ADH in circulating plasma.

Only one major problem remained with this hypothesis. The thin ascending limb, while impermeable to water, does not actively pump out NaCl or any other substrate. Because the thick ascending limb is located in the outer medulla and the osmotic gradient continues to increase to reach a maximum in the inner medulla, some other process must be contributing to the renal concentrating mechanism, at least for the long-looped juxtaglomerular nephrons.

THE ROLE OF UREA. Urea is concentrated in the medullary interstitium. The active pumping of NaCl in the thick ascending limb provides the energy for the concentration of urea. The differential permeabilities of the tubules to water, solute, and urea provide the proper conditions. The resulting high concentrations of urea in the medullary interstitium can then account for the high osmolality of the inner medulla despite the absence of active solute transport there. The recycling of urea by the renal tubules accounts for the reabsorption of water by the descending thin limb. It also accounts for the "passive" reabsorption of NaCl by the ascending thin limb.

The concentration of urea in the renal medulla occurs in the following fashion (Fig. 5-25D). The thick ascending limb is impermeable to water and urea. Hence both urea and water remain behind in the lumen of the thick ascending limb as NaCl is actively pumped out. When, in the presence of ADH, the TF reaches the late distal tubule and the cortical and outer medullary collecting ducts, water is reabsorbed into the cortex and outer medulla. Because urea cannot penetrate, however, it remains behind in the TF. The concentration of urea in the TF, therefore, increases, and, consequently, a high concentration of urea reaches the inner medulla. The urea is concentrated as a result of the osmotic gradient

Fig. 5-25. *Countercurrent multiplier—Kuhn's hypothesis. A. The initial formation of the hypertonic medulla occurs as NaCl is actively pumped out of the thick ascending limb. The pump is gradient-limited to 200 mosm/liter (actual stoichiometry of the pump is $1Na^{+}: 1K^{+}: 2Cl^{-}$).*
B. Medullary tonicity is further augmented as the tubular fluid reaching the thin ascending limb becomes hyperosmotic.
C. A medullary tonicity of approximately 900 mosm per liter is achieved. To increase medullary osmolality further, using this method, requires longer-looped nephrons with longer thick ascending limbs; however, the long-looped juxtamedullary nephrons increase their length by incorporating thin descending and ascending limbs into their loops of Henle. No active pumping has ever been demonstrated in thin ascending limbs; however, medullary tonicity is increased by these nephrons to 1200 mosm per liter. This increased tonicity is attributed to the urea cycle.

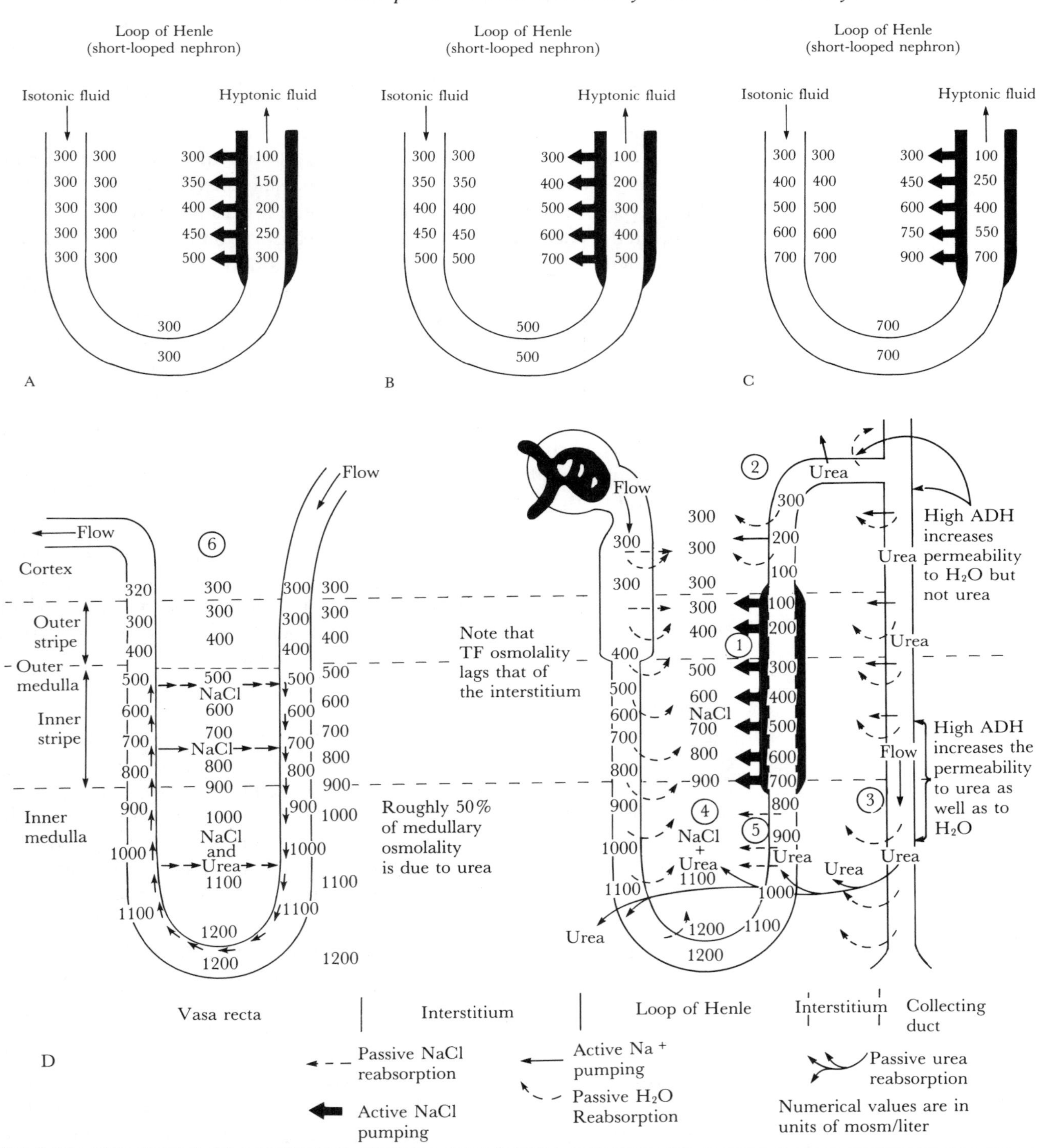

Figure 5-25 (Continued)

D. Recent modifications of the countercurrent hypothesis—the urea cycle. The thin ascending limb in the medulla, the thick ascending limb in the outer medulla, and the early distal tubule are impermeable to water.
(1) Active pumping of NaCl into the interstitium by the thick ascending limb renders the tubular fluid dilute and the outer medullary interstitium hypersomotic. This process provides the fuel for the urea cycle. Because the thick ascending limb is impermeable to both water and urea, urea remains behind in the diluted tubular fluid. (2) In the presence of ADH, the distal tubule and the cortical and outer medullary collecting ducts are permeable to water but not to urea. Because the tubular fluid leaving the thick ascending limb is hypotonic, water now leaves the distal tubule and cortical collecting ducts along its concentration gradient. This increases the concentration of urea that is left behind in the tubule. In the outer medulla, additional water is reabsorbed from the tubule due to the concentration gradient created by the thick ascending limb further increasing the tubular fluid urea concentration. (3) In the presence of ADH, the medullary collecting duct is permeable to both water and urea. The concentrated urea reaching the medulla now diffuses out of the tubule into the interstitium along its concentration gradient adding substantially to the tonicity of the medulla. Some urea reenters the thin ascending limb of the loop of Henle, which is somewhat permeable to urea. This medullary recycling of urea, in addition to the trapping of urea by countercurrent exchange in the vasa recta, causes urea to accumulate in large quantities in the medullary interstitium (large type) in the presence of ADH. Urea accounts for roughly 50 percent of the osmolality of the inner medulla in antidiuresis. (4) The thin descending limb is permeable to water but impermeable to NaCl and solute in general. The high concentration of urea in the interstitium osmotically extracts water from the thin descending limb thus concentrating NaCl in the descending limb fluid (i.e., the high concentration of NaCl in the tubular fluid is balanced osmotically by the high concentration of urea in the interstitium). (5) When the fluid rich in NaCl enters the NaCl-permeable but water-impermeable thin ascending limb, NaCl moves passively out of the tubule along its concentration gradient rendering the tubule fluid relatively hypoosmotic to the surrounding interstitium. The urea cycle can, therefore, account for both the reabsorption of water by the thin descending limb and the passive reabsorption of NaCl by the thin ascending limb. Note, however, that this process is dependent on the active pumping of solute by the thick ascending limb, and the configuration and differential permeabilities of the different parts of the nephron. (6) The countercurrent exchange system in the vasa recta (see also Fig. 5-26).

created by the thick ascending limb. The inner medullary collecting tubule is permeable to urea when ADH is present. Urea diffuses into the medullary interstitium along its concentration gradient and is trapped there, adding substantially to the tonicity of the medulla. Roughly 50 percent of the tonicity of the inner medulla is attributed to urea.

Because the descending thin limb is relatively impermeable to urea and NaCl, urea in the medullary interstitium acts to osmotically extract water from the thin descending limb, increasing the concentration of the NaCl in the luminal fluid above that in the interstitium (i.e., the high concentration of NaCl in the tubule is balanced osmotically by the high concentration of NaCl and urea in the interstitium). When the fluid now enters the water-impermeable but NaCl-permeable thin ascending limb, NaCl moves passively down its concentration gradient into the interstitium rendering the TF in the thin ascending limb hyposmotic to the surrounding hypertonic interstitium. Because the thin ascending limb is somewhat permeable to urea, some of the urea now diffuses along its concentration gradient into the thin ascending limb to begin the cycle again.

COUNTERCURRENT EXCHANGE AND THE VASA RECTA. The vasa recta perform two functions. They remove from the medullary interstitium fluid that has been reabsorbed from the thin descending limb and collecting tubules, and they minimize solute uptake from the medulla to preserve medullary hypertonicity. The vascular architecture is designed to facilitate transcapillary exchange of water and solute between the ascending and descending vasa recta, i.e., to maximize countercurrent exchange (see Figs. 5-25D, 5-26). This exchange occurs in such a fashion as to minimize both the amount of water retained in the medulla and the loss of solute out of the medulla.

In the descending vasa recta, the balance of the Starling forces is such that the net hydrostatic pressure favoring movement out of the capillaries and the opposing high oncotic pressure (resulting from filtration at the glomerulus) are nearly equal. Fluid nonetheless leaves the descending vasa recta because of the high interstitial osmolality due to NaCl and urea. Normally NaCl and urea do not exert an osmotic force across capillary membranes. However, due to the rapid flow in the capillaries and the high, but nonetheless finite, membrane permeability to these solutes, water is able to leave the capillaries faster than solute is able to enter (i.e., water is more permeable than solute, so the solute concentration in the capillary lags behind that in the interstitium). Thus there is a net egress of water from the descending vasa recta.

In the ascending vasa recta the solute concentration of the tubular fluid is now higher than that of the interstitium and rather than favoring egress of fluid now combines with oncotic pressure to favor influx of fluid. Meanwhile, the hydrostatic pressure has also decreased somewhat. The balance is such that, overall, more fluid is taken up by the ascending vasa recta than was lost by the descending vasa recta. This net uptake is ultimately attributable to the increased oncotic pressure resulting from loss of ultrafiltrate in the glomerulus. The net volume reabsorbed by the ascending vasa recta is equal to the volume of fluid taken up from the ascending thin limb and collecting system.

QUESTIONS

44. Does the countercurrent hypotheses require fluid entering the distal tubule to be hyposmotic at all times?
45. What happens to the renal concentrating mechanism because the medullary collecting duct is impermeable to urea in the absence of ADH?

Fig. 5-26. *The countercurrent exchange system in the vasa recta, (dotted arrows) including passive transport of solutes and passive movement of water (dashed arrows). The numbers refer to osmolalities (mosm/kg water) in the blood or interstitial fluid. (Adapted from R. W. Berliner, et al.* Am. J. Med. *24:730, 1958.)*

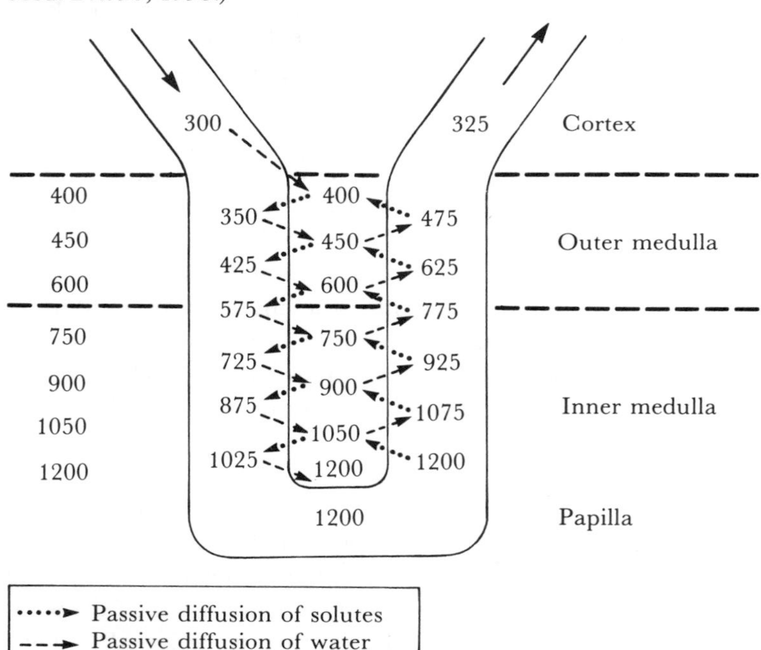

46. Why is urea a particularly good choice of substrate for use by the renal concentrating mechanism?
47. Why is the removal of NaCl in the thin ascending limb said to be passive?
48. Tubular fluid is removed from each of the labeled regions of the nephron in Fig. 5-27, first in the presence of ADH and then in its absence. State whether the fluid removed is hypertonic, isotonic, or hypotonic to plasma.

RENAL TUBULAR TRANSPORT: REABSORPTION AND SECRETION OF ORGANIC COMPOUNDS

Glucose, amino acids, and other solutes are reabsorbed in the early proximal tubule by cotransport with Na^+ (symport). The energy for this reabsorption is derived indirectly from the active Na^+-K^+ pump, and it is, therefore, called secondary active transport. They then diffuse out of the proximal tubular cells into the lateral intercellular space by either simple or facilitated (e.g., glucose) diffusion (see Figs. 5-16 and 5-18). There are five or more separate carriers that have been described for amino acid transport. These include carriers for neutral amino acids, dibasic amino acids, dicarboxylic amino acids, imino acids and glycine, and β-amino acids. Like glucose, all of these are symports.

Sodium transport interacts with that of glucose, amino acids, bicarbonate, citrate, and lactate. Furthermore, the carriers mediating the reabsorption of Na^+ and each of these solutes are much less efficient at transport of either Na^+ or the solute alone. Reabsorption of either solute or Na^+ is decreased if its symport is absent.

The transport maximum (Tm) for glucose is approximately 375 mg per minute in an adult male and somewhat less in an adult female. The renal threshold for spilling glucose into the urine should then be equal to the Tm divided by the GFR or 375 mg per minute divided by 125 ml per minute. The threshold would be surpassed, therefore, when the blood glucose exceeds 300 mg per deciliter. Measurements of the actual renal threshold show it to be only 200 mg per deciliter of arterial blood or 180 mg per deciliter of venous blood. This discrepancy from the theoretical value is called splay and is shown graphically in Figures 5-28 and 5-29. It may be explained by two observations. First, the Tm is not identical for all nephrons because it represents the average

Fig. 5-27. *Nephron (see question 48). (From P. J. Cannon. The kidney in heart failure.* N. Engl. J. Med. *296:26, 1977.)*

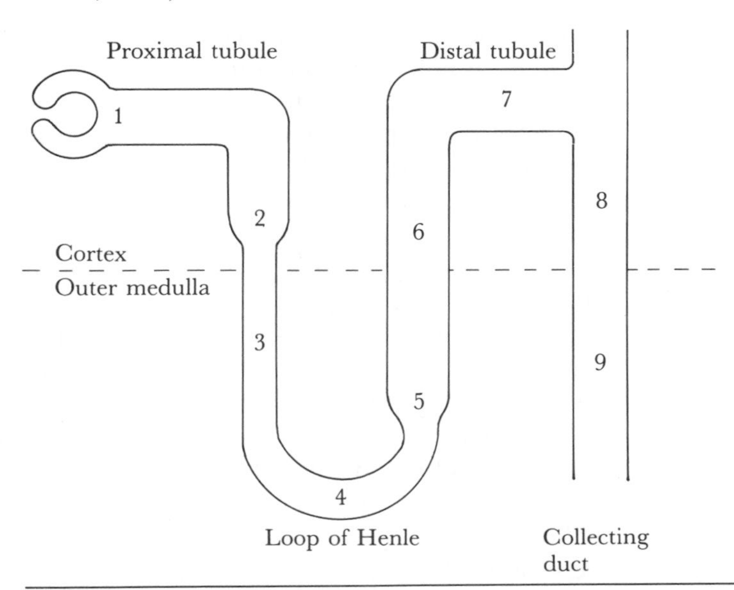

Fig. 5-28.

A. Relation between plasma levels (P) and excretion ($U\dot{V}$) of PAH and inulin.

B. Relation between plasma level (P) and excretion ($U\dot{V}$) of glucose and inulin.

C. Relation between plasma glucose level (P_G) and amount of glucose reabsorbed (T_G).

(From W. F. Ganong. Review of Medical Physiology *[11th ed.]. P. 573 © 1985 by Lange Medical Publications, Los Altos, CA.)*

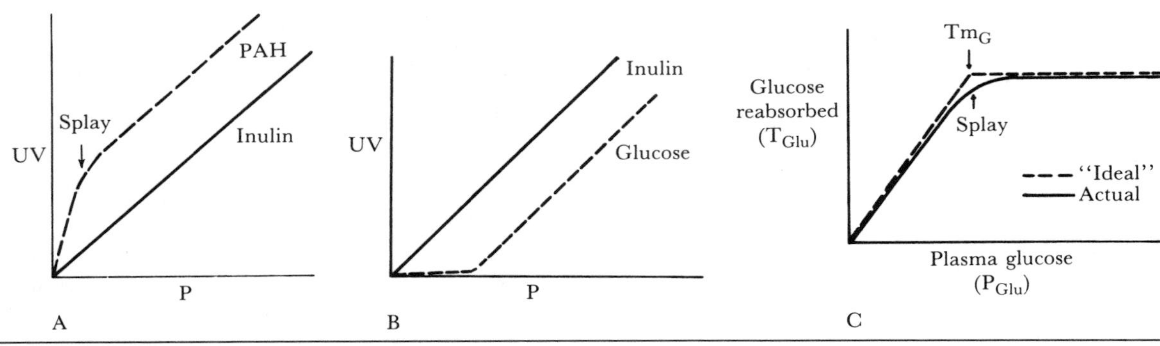

value for approximately 2 million nephrons. Consequently some nephrons may begin to spill glucose at values considerably above or below this value. Glucose is spilled into the urine as soon as the concentration of glucose in the plasma and, therefore, in the filtrate exceeds the ability of those nephrons least able to reabsorb glucose. The second explanation involves the fact that carriers, like enzymes, require supersaturating levels of substrate to operate maximally.

As the plasma glucose level exceeds the Tm for glucose, the clearance of glucose increases from zero to asymptotically approach the clearance of inulin (see Fig. 5-31). This is demonstrated by the following equations:

$$\text{Amount glucose excreted} = U_{PAH}\ \dot{V} = (C_{Inu})\ (P_{Glu}) - Tm \qquad (5\text{-}7)$$

Dividing both sides by P_{Glu} yields

$$C_{Glu} = C_{Inu} - Tm/P_{Glu} \qquad (5\text{-}8)$$

The ratio of the Tm of glucose to the plasma concentration of glucose approaches zero as the plasma glucose concentration increases and the amount of glucose filtered becomes many times greater than the amount reabsorbed.

There are two separate active mechanisms involved in secretion: an organic anion secretory pump and an organic cation secretory pump. The weak anions predictably compete with one another for secretion. Acetate and lactate facilitate this secretion while a number of substances inhibit it (including Kreb's cycle intermediates, probenicid, dinitrophenol, and mercurial diuretics).

Para-aminohippurate is excreted by a combination of filtration in the glomerulus and secretion in the proximal tubule. The filtered load of PAH, like that of all filterable solutes, is a linear function of the plasma level while PAH secretion increases only until its transport maximum for secretion (Tm_{PAH}) is reached (Fig. 5-30B). Thus, at low concentrations of PAH, the clearance of PAH is high (Fig. 5-30A). As the plasma concentration of PAH increases, however, C_{PAH} asymptotically approaches the clearance of inulin (i.e., the GFR) as the amount of PAH secreted becomes less significant.

$$\text{Amount PAH excreted} = U_{PAH}\dot{V} = (C_{Inu})\ (P_{PAH}) + Tm \qquad (5\text{-}9)$$

Fig. 5-29. *Renal handling of glucose as a function of increasing plasma glucose concentrations. The curve for reabsorption is known as the glucose titration curve because it determines the plasma concentration at which the carrier for glucose becomes saturated. Tm_{Glu} refers to the maximal amount of glucose that can be transported per unit time. The range of normal plasma glucose concentration (70–100 mg/100 ml) is spanned by the arrows on the abscissa; note that normally virtually all the filtered glucose is reabsorbed. The significance of the splay is explained in the text. In the clinical laboratory, plasma glucose concentrations are ordinarily expressed as milligrams per 100 ml of plasma; corresponding concentrations, in millimoles per liter, are given on the second abscissa. (From H. Valtin.* Renal Function Mechanisms Preserving Fluid and Solute Balance in Health *[2nd ed.]. Boston: Little, Brown, 1983. P. 72. Data from Smith, H. W.* Principles of Renal Physiology. *New York: Oxford University Press, 1956.)*

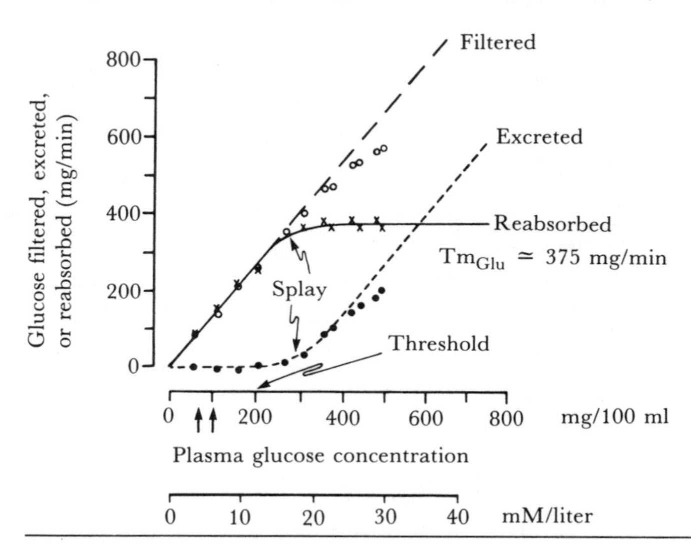

Dividing both sides by P_{PAH}

$$C_{PAH} = C_{Inu} + Tm/P_{PAH} \quad (5\text{-}10)$$

RENAL METABOLISM AND OXYGEN CONSUMPTION

Renal blood flow per gram tissue is very large, therefore, the arteriovenous oxygen difference (A-V O_2) is quite small; only 14 mls oxygen per liter of blood flow versus 60 mls per liter for skeletal muscle, 62 mls per liter for brain, and 140 mls per liter for heart. Renal oxygen consumption varies directly with the rate of active transport of sodium.

QUESTIONS

49. Which of the following regions of the kidney has the highest oxygen consumption?
    A. Cortex
    B. Medulla
    C. Columns of Bertin
    D. Perinephric fat

ANSWERS

1. A. Gucose, amino acids
   B. Na^+, K^+ (highly variable), Cl^-, HCO_3^-
   C. Urea, uric acid, H^+, phosphates, sulfates, creatinine
2. A. 400 ml/day × 1 day/1440 minutes = 0.278 ml/minute
   B. 4000 ml/day × 1 day/1440 minutes = 2.78 mls/minute
   C. 1200 mosm per liter urine in A 50 mosm per liter in B
3. All are true of renal failure.

Fig. 5-30.
A. Clearance of inulin, glucose, and PAH at various plasma levels of each substance in humans. (From W. F. Ganong. Review of Medical Physiology *[12th ed.]. P. 574 © 1985 by Lange Medical Publications, Los Altos, CA.) B. Rates of filtration, excretion, and secretion at increasing plasma concentrations of PAH. (Modified from Pitts, R. F.* Physiology of the Kidney and Body Fluids, *[3rd ed.]. Chicago: Year Book, 1974.)*

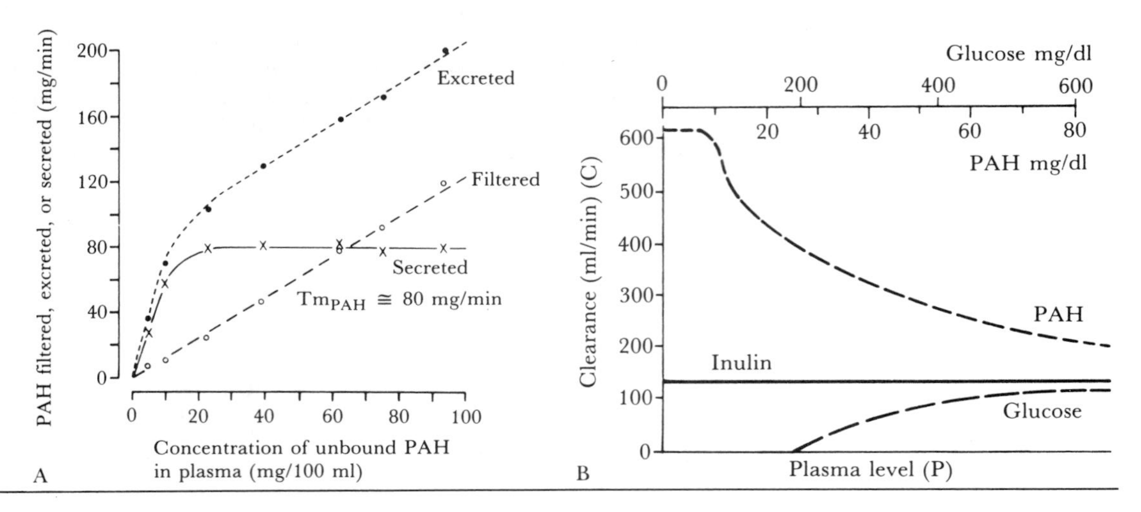

4. An ultrafiltrate of plasma is extruded into Bowman's space and flows from there directly into the proximal tubule, loop of Henle, distal tubule, and collecting tubules until it is dumped as urine into the calyceal system. The urine then flows through the minor and major calyces, into the renal pelvis, ureter, and bladder.
5. A. Fluid leaving in the efferent arterioles has an increased oncotic pressure secondary to the filtration that has occurred in the glomerulus (the first capillary bed). The hydrostatic pressure that had been driving filtration decreases rapidly as a consequence of the resistance of the efferent arterioles and of the extensive branching that occurs without any decrease in caliber of the lumens of the branches. The large cross-sectional area and low flow rates thus generated also favor fluid uptake. The second capillary beds, the vasa recta, and peritubular plexus are in close proximity to the tubules and have both a large reabsorptive surface area and high filtration coefficient. Thus all the Starling forces are primed and ready to reabsorb essential substrates (see Fig. 5-10).
   B. At the level of the afferent arterioles and again at the level of the efferent arterioles
6. E. Note: Hypoxia acts via increased sympathetic discharge.
7. C.
8. A. Volume of distribution = (1.0 − 0.005) / 0.024 = 41.5 liters
   B. Total body water
9. A. Volume of distribution = (12.5 − 1.5) / 0.588 = 18.7 liters
   B. Extracelluar fluid
   C. ICF = TBW − ECF = 41.5 − 18.7 = 22.8 liters
10. A. Volume of distribution = 10 − 0 / 0.0032 = 3100 ml
    B. Plasma volume
    C. Interstitial fluid volume = ECF − plasma = 18.7 − 3.1 = 15.6 liters
11. A. K^+ and Mg^{2+}; phosphate and HCO_3^- (Protein is third on a molar basis and second when measured in equivalents.)

B. Na^+ and K^+ (Ca^{2+} is third on a molar basis and tied for second when measured in equivalents); Cl^- and HCO_3^-.

12. A. Plasma osmolality = 2[Na^+] + glucose mg/100 ml/18 + BUN mg/100 ml/2.8 = 302 mosm/liter.

    B. A plasma osmolality of 380 suggests the presence of abnormally high concentration of some osmole other than glucose, sodium, or urea (e.g., alcohol).

13. Potential (millivolts) = −61 log 156/4 = −97.9 mv cell interior relative to the cell exterior.

14. Red cells constitute part of the ICF; therefore, the hematocrit is indicative of the ratio ICF : ECF.

    A. No change
    B. Decrease
    C. No change
    D. Decrease (lack of erythropoietin)
    E. No change

15. Both volume expansion and volume contraction may be accompanied by any one of the possibilities listed.

16. Total body exchangeable (i.e., accessible) Na^+, total body exchangeable K^+, and total exchangeable body water. Intracellular cation and intracellular water must be considered in the evaluation of changes in plasma sodium concentration. Therefore,

$$\text{Plasma } [Na^+] \cong \frac{Na^+_{TE} + K^+_{TE}}{H_2O}$$

in which the term TE ≅ total exchangeable and refers to that amount of the substance that is available for exchange over a reasonable equilibration period. Much of the calcium and phosphate in bone, for example, is not available for exchange.

17. [Na^+] × TBW ≅ Total body exchangeable cation *or*
    140 × (70)(0.60) = 5880 meq cation
    120 × (70)(0.60) = 5040 meq cation
    This method assumes that intracellular hypokalemia normally accompanies hyponatremia.

18. For a 60-kg nonobese female

    TBW = (0.5) (60 kg) = 30 liters

    ECF = (0.45) (30) = 13.5 liters

    ICF = (0.55) (30) = 1 = 16.5 liters

    A. Addition of 2 liters of urea, an ineffective osmole, would increase both compartments proportionately:

       TBW = 30 + 2 = 32 liters

       ECF = (0.45) (32) = 14.4 liters

       ICF = (0.55) (32) = 17.6 liters

    B. TBW = 30 + 1.5 = 31.5 liters

       0.15 M NaCl = 150 mM NaCl or 150 meq Na^+/liter

       NaCl is restricted to the ECF (i.e., it is an effective osmole). Using the total body exchangeable cation method and assuming a [Na^+] of 140 meq/liter

Cation ECF before	$= 140 \times 13.5$	$= 1890$ meq
Cation ICF before	$= 140 \times 16.5$	$= 2310$ meq
Total body cation before	$= 140 \times 30$	$= 4200$ meq

Cation ECF after	$= 1890 + (1.5)(150)$	$= 2115$ meq
Cation ICF after	$= 140 \times 16.5$	$= 2310$ meq
Total body cation after		$= 4425$ meq

[Cation] after = 4425 meq cation/31.5 liters = 140.5 meq/liter

ECF volume after = 2115 meq/140.5 meq/liter = 16.44 liters
ICF volume after = 2310 meq/140.5 meq/liter = 15.06 liters

Note that because the NaCl solution added was slightly hypertonic with respect to the patient's own fluids some water was drawn out of the ICF (approximately 60 mls).

C. Because urea is an ineffective osmole, the answer to this question is the same as that of B.

D. TBW = 30 + 3 = 33 liters

0.3 M NaCl = 130 mM NaCl or 300 meq Na^+/liter

Cation ECF before	$= 140 \times 13.5$	$= 1890$ meq
Cation ICF before	$= 140 \times 16.5$	$= 2310$ meq
Total body cation before	$= 140 \times 30.0$	$= 4200$ meq

Cation ECF after	$= 1890 + (3)(300)$	$= 2790$ meq
Cation ICF after	$= 140 \times 16.5$	$= 2310$ meq
Total body cation after		$= 5100$ meq

[Cation] after = 5100/33 = 154.5 meq/liter

ECF volume after = 2790/154.5 = 18.05 liters
ICF volume after = 2310/154.5 = 14.95 liters (i.e., the ICF has lost 1.55 liters to the ECF)

E. Same as A. Glucose is an ineffective osmole.

19. The new steady state following a loss of fluid is known as volume contraction and conversely a gain of fluid as volume expansion. The change in volume as well as the adjectives isosmotic, hyperosmotic, and hyposmotic, refer to the extracellular fluid in the new steady state. The answers are explained below and in Figure 5-31.
    A. 2. The ECF becomes volume contracted and hyperosmotic. Water moves out of the ICF in response to the osmotic gradient leading to hyperosmotic contraction of both compartments.
    B. 4. ECF is expanded without change in osmolality, hence ICF remains unchanged.
    C. 1. Loss of isosmotic fluid from the ECF causes the ECF volume to contract without changing the osmolality of the ECF. The ICF, therefore, is unaffected.
    D. 3. Decreased osmolality of the ECF causes water to shift into cells following the osmotic gradient. The ICF volume is expanded and both ICF and ECF are hyposmotic.
    E. 5. Osmolality of the ECF rises causing water to move from the ICF to the ECF until the two equilibrate at a higher osmolality. At equilibrium, therefore, the ECF is expanded, the ICF is contracted and both are hyperosmotic.

Fig. 5-31. *Darrow-Yannet diagrams of types of contraction and expansion of the body fluid compartments. The plasma and the interstitial fluid are depicted as a single extracellular compartment. In each diagram, the healthy state is drawn in solid lines and the new steady state in dashed lines. (From H. Valtin.* Renal Function: Mechanisms Preserving Fluid and Solute Balance in Health *[2nd ed.] Boston: Little, Brown, 1983. P. 272.)*

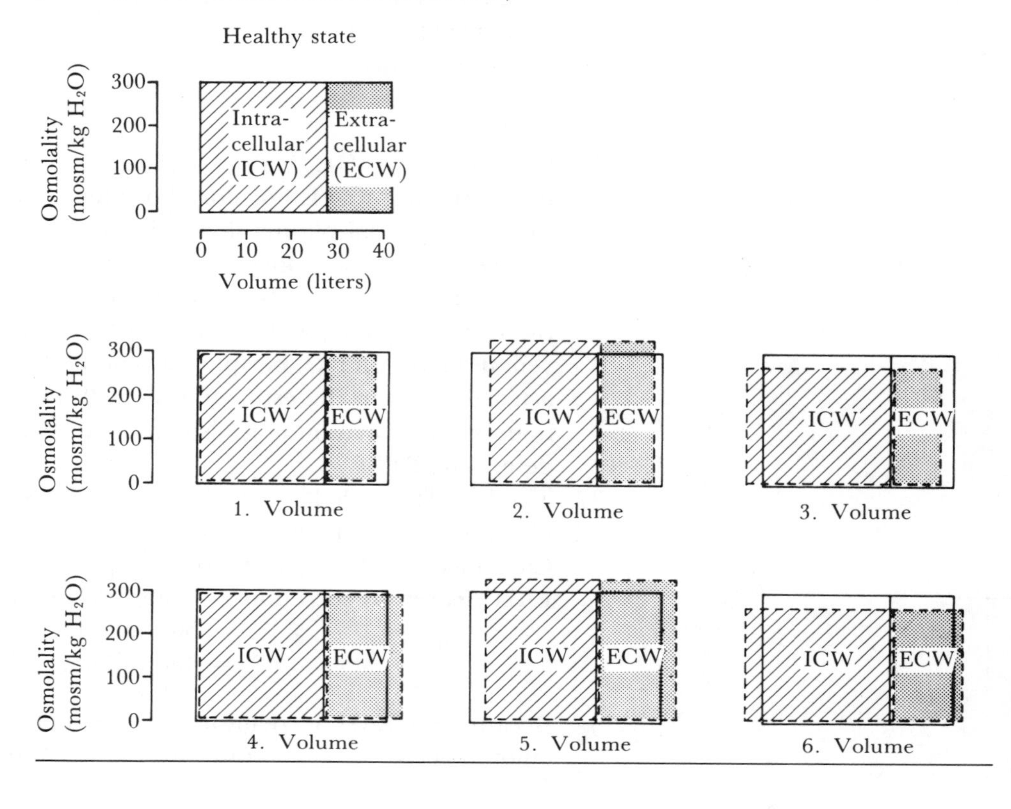

F. 6. The ECF volume is expanded while its osmolality is decreased. Water shifts into the ICF along its concentration gradient until both the ICF and ECF equilibrate at an expanded volume with hyposmotic tonicity.

20. Osmolality before × TBW before = Osmolality after × TBW after. For a 70-kg male

$$290 \times 142 = \text{osmolality after} \times (142 - 1.4)$$
$$\text{Osmolality after} = 300 \text{ mosm/liter}$$

Similarly,

$$[Na^+] \text{ after} = 145 \text{ meq/liter}$$

For a 60-kg female: 302 mosm/liter, 146 meq/liter

21. Yes, because they are large enough to remain segregated on one side of the semipermeable membrane.
22. His body water must be about 140/155th of what it was, hence body water loss (%) = (1.0 − 140/155) × 100 = 10%. Note: Up to 10 percent water deficit may be asymptomatic; greater than 10 percent may lead to CNS disturbances including disorientation, stupor, and coma; greater than 20 percent may be lethal.
23. A. Lower K_f, lower the mean hydraulic pressure, or increase Q_A.

    B. Due to their twofold greater surface area and 100-fold greater permeability, the glomerular capillaries are even more permeable to small molecules than other capillaries. The increased permeability may be due to the large endothelial fenestrae and slit pores. In addition, because the glomerular capillaries are interposed between two arterial circulations, the mean hydraulic pressure driving filtration is unusually high.

C. 2. Answers 1 and 3 both refer to FPE, and FPE may or may not be reached in humans. The efferent arteriole is impermeable; hence, filtration stops.

24. A. 3
    B. 1
25. $GFR = C_{Inu} = U_{Inu}\dot{V}/P_{Inu} = (30 \times 1.1) / 0.25 = 132$ ml/minute
26. Any substance that is not stored or metabolized by the kidney may be used to measure renal plasma flow using the Fick equation if the renal arterial and venous plasma concentrations of the substance can be measured. The PAH is preferred only because its high extraction ratio allows calculation of the RPF to be done without measuring the PAH concentration in the renal venous blood. In fact, peripheral venous blood may be used instead of renal arterial blood in the estimation of RPF making the procedure completely noninvasive. Creatinine is also secreted and could theoretically be used to determine RPF.
27. Total PAH cleared = (PAH cleared by secretion) + (PAH cleared by filtration) = 600 ml/min

    Clearance of PAH by filtration $= C_{Inu} = 12 \times 1 / 0.1 = 120$ ml/min

    PAH cleared by secretion = Total PAH cleared − cleared by filtration
    = 600 ml/min − 120 ml/min
    = 480 ml/min

    Because no PAH is reabsorbed, the total amount of PAH excreted = amount secreted + amount filtered = Clearance by secretion $\times P_{PAH} + C_{Inu} \times P_{PAH}$
    $= 480 \times 0.1 + 120 \times 0.1$
    = 6 mg/minute
28. A. $ERPF = (U x \dot{V}) / P = (14$ mg/ml $\times$ 1296ml/1440 minute)/0.02 mg/ml
    = 630 ml/minute

    B. Actual RPF = 630/0.91 = 692 ml/minute

    C. RBF = RPF / (1 − %HCT) = (692 ml/min) / 0.42 = 1193 ml/min

    D. PAH excreted $= U_{PAH} \times \dot{V} = 14$ mg/ml $\times$ 0.9 ml/minute
    = 12.6mg/minute
29. A. Increased and hyposmotic
    B. Decreased and hyperosmotic
    C. The same and isosmotic
30. B.
31. PAH >> creatinine > inulin > K^+ > Na^+ > glucose; note that K^+ clearance is very variable.
32. D, E, I
33. C.
34. Distal convoluted tubule and collecting tubule
35. Because Na^+ osmotically obligates water to be reabsorbed with it. Furthermore, should the $[Na^+]$ in the plasma rise, the resulting increase in plasma osmolality would elicit increased ADH secretion and thirst, returning the osmolality towards normal.
36. A. Decrease
    B. Decrease
    C. Increase

D. Increase
E. Increasing K^+

37. C. In the renal tubule, Na^+ reabsorption must be accompanied by either excretion of a cation (H^+ or K^+) or reabsorption of an anion (primarily Cl^-). Deficiency of any one of these three ions compels the other two to move excessively.
38. A. Yes, frequently.
B. No, essentially by definition, if there is sufficient ascites to be detectable clinically, total body sodium is elevated — ascites is "salt-water."
39. Thin descending limb and the thin ascending limb (and Bowman's capsule, of course).
40. Normally Na^+ intake equals Na^+ output. The body acutely responds to an excessive Na^+ load by excreting half the excess the first day, half the remaining excess salt water the following day, and so on until equilibrium is reestablished. The patient has increased his Na^+ intake from 2 to 18 g, i.e., he has bolused 16 g into his system. Roughly half of the excess would be excreted the first day. Because the molecular weight of sodium is 23, 16000 mg/23 equals approximately 700 meq that have been taken in. If 350 meq were excreted and the same amount retained, then assuming a plasma $[Na^+]$ of 140 in this patient, he would retain (350/140 =) 2.5 liters of fluid and, therefore, gain 2.5 kg weight. If the increased salt intake were to be continued, the kidney would continue to increase excretion by half the excess per day until input equals output. After 1 week equilibrium would be established but at the expense of 700 meq Na^+ retained (or roughly 5 liters of fluid).
41. A. Treatment with bicarbonate or with glucose and insulin would be expected to drive K^+ into cells. In addition, Ca^{2+} is used clinically to antagonize the effects of K^+.
B. Plasma $[K^+]$
42. Filtration. (Na^+ excretion is increased if reabsorption is decreased,)
Secretion in the distal nephron and collecting tubule
A. Increase
B. Increase
C. Increase
D. Increase
E. Decrease
F. Decrease
43. Thick ascending limb, distal convoluted tubule, collecting tubule.
44. Yes.
45. The osmolality and urea concentration both decrease until, at the papillary tip, the osmolality is only twice that of the plasma, while the urea concentration declines to very low levels. In the absence of ADH, diuresis occurs because the entire collecting system is impermeable to water. The urea entering the system is then excreted because it can no longer be reabsorbed in the inner medulla. The tonicity of the medullary interstitium is, therefore, decreased as its urea concentration is "washed-out." Dilute fluid leaving the thick ascending limb is, therefore, diluted even further in the remainder of the uriniferous tubule by active NaCl pumping.
46. Because urea is the major excretory product of metabolism, one whose excretion in the urine would otherwise osmotically obligate the excretion of large amounts of water were it not for its paradoxical use to concentrate the urine.

Fig. 5-32. *Summary of changes in urine osmolality in various parts of the nephron. The thickened wall of the ascending limb of the loop of Henle indicates relative impermeability of the tubular epithelium to water. In the present of vasopressin, the fluid in the collecting ducts becomes hypertonic, while in the absence of this hormone, the fluid remains hypotonic throughout the collecting duct. Aldosterone promotes reabsorption of Na^+ and secretion of H^+ and K^+ in the distal convoluted tubule. (From P. J. Cannon. The kidney in heart failure.* N. Engl. J. Med. *296:26, 1977.)*

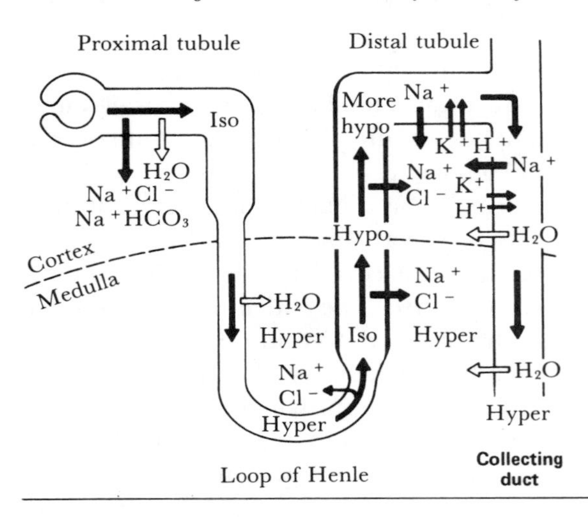

47. Because the energy used to concentrate urea in the medulla results from the active pumping of NaCl in the thick ascending limb and not from active pumping in the thin limb itself.

48.

ADH Present	*ADH Absent*
1. Isotonic	1. Isotonic
2. Isotonic	2. Isotonic
3. Hypertonic	3. Hypertonic
4. Hypertonic, >3	4. Hypertonic, >3
5. Isotonic	5. Isotonic
6. Hypotonic	6. Hypotonic
7–9. Increasingly hypertonic	7–9. Increasingly hypertonic

See Figure 5-32.

49. A. The cortex has the highest oxygen consumption because it is there that Na^+ transport is most active. The medulla has an oxygen consumption one-twentieth that of the cortex.

BIBLIOGRAPHY

Brenner, B. M., and Rector, F. C. (eds.). *The Kidney* (3rd ed.). Philadelphia: Saunders, 1986.

Ganong, W. F. *Review of Medical Physiology* (11th ed.). Los Altos, CA: Lange, 1983.

Guyton, A. C. *Textbook of Medical Physiology* (7th ed.). Philadelphia: Saunders, 1986.

Harvey, A. M., et al. (eds.). *The Principles and Practice of Medicine* (21st ed.). Norwalk, CT: Appleton-Century-Crofts, 1984.

Jamison, R., and Maffly, R. H. The Urinary Concentrating Mechanism. *N. Engl. J. Med* 295:1059, 1976.

Petersdorf, R. G., et al. (eds.) *Harrison's Principles of Internal Medicine* (10th ed.). New York: McGraw-Hill, 1983.

Stanford University School of Medicine Faculty, *Renal Physiology Syllabus* (lecture handout), 1984.

Valtin, H. *Renal Function: Mechanisms Preserving Fluid and Solute Balance in Health* (2nd ed.). Boston: Little, Brown, 1983.

West, J. B. *Best and Taylor's: Physiological Basis of Medical Practice* (11th ed.). Baltimore: Williams & Wilkins, 1985.

6 Respiratory Physiology

Charles H. Tadlock

STRUCTURE AND FUNCTION OF THE LUNG

The primary function of the lungs is gas exchange. Oxygen utilized by the tissues is replaced and carbon dioxide generated during metabolism is excreted. The gas exchange unit of the lung is the alveolus. Normal human lungs are composed of approximately 300 million alveoli. The pulmonary capillary bed contains roughy 150 ml of blood spread out over a surface of 70 m^2 (1 m^2/kg) creating a film 10 microns thick which surrounds the alveoli. At basal heart rates, the red cells spend approximately three-fourths of a second in the pulmonary capillaries, 2 to 3 times the amount of time needed to equilibrate with the alveolar gases. This results in a gas exchange system which is so efficient that over half the lung volume may be lost before any abnormality of gas exchange appears.

THE AIRWAYS. Air enters the oropharynx and is conducted to the lung alveoli through a series of branching tubes. The trachea, mainstem bronchi, and their major branches contain cartilage which acts to maintain a patent lumen during changes in intrathoracic pressure. More distally, the conduction system consists of smaller diameter membranous bronchioles. The first 16 divisions of the tracheobronchial tree comprise the conduction system. The final division of the conducting system is called the terminal bronchiole. These areas of the lung are supplied by the bronchial circulation and do not participate in gas exchange. The volume of gas contained within these tubes is, therefore, referred to as the anatomical dead space and normally equals 150 cc.

The gas exchange portion of the lungs consists of the respiratory bronchioles, the alveolar ducts, and the alveoli. The portion of lung distal to a terminal bronchiole forms an anatomical unit called the primary lobule or acinus.

Smooth muscle in the walls of the airways may alter regional airflow by modifying the resistance properties of the airways. Resistance to flow occurs primarily in the larger airways with only minor resistance encountered in the small airways (Fig. 6-1). Resistance to laminar flow is given by the equation

$$R = \frac{8nl}{\pi r^4} \qquad (6\text{-}1)$$

in which n is fluid viscosity, l is length, and r is the radius of the airway.

This equation would lead to the conclusion that resistance to air flow would be greatest in the smaller diameter airways; however, this did not prove to be the case. The total cross-sectional area of the smaller airways is so great that resistance actually decreases in the smaller airways of the lungs (compare Figs. 6-1 and 6-2).

BULK FLOW AND DIFFUSION. During respiration air moves by bulk flow from the atmosphere into the tracheobronchial tree to the level of the terminal bronchioles (the last nonexchanging branch of the tracheobronchial tree). At the level of the terminal bronchioles the total cross-sectional area of the airways becomes so great that bulk flow essentially ends. Diffusion of gases is responsible for gas transport from terminal bronchioles through the respiratory ducts, the alveolar ducts, the alveoli, and into the blood.

Fig. 6-1. *Location of the chief site of airway resistance. Note that the intermediate-sized bronchi contribute most of the resistance and that relatively little is located in the very small airways. (Redrawn from T. J. Pedley, R. C. Schroter, M. F. Ludlow. The prediction of pressure drop and variation of resistance within the human bronchial airways.* Respir. Physiol. *9: 387, 1970.)*

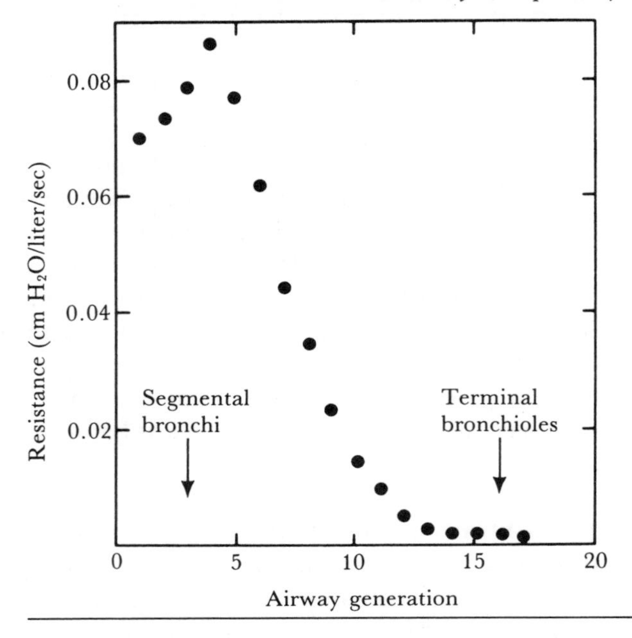

Fig. 6-2. *Diagram to show the extremely rapid increase in total cross-sectional area of the airways in the respiratory zone. As a result, the forward velocity of the gas during inspiration becomes very small in the region of the respiratory bronchioles, and gaseous diffusion becomes the chief mechanism of ventilation. (From J. B. West.* Respiratory Physiology: The Essentials *[3rd ed.]. P. 7. © 1985 The Williams & Wilkins Co., Baltimore.)*

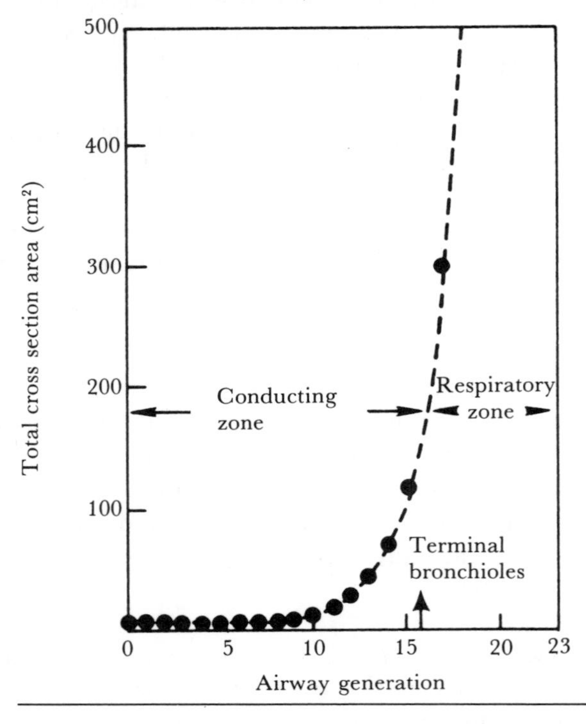

Fig. 6-3. *Ultrastructure of the respiratory membrane as shown in cross section. (From A. C. Guyton.* Textbook of Medical Physiology *[7th ed.]. Philadelphia: Saunders, 1986. P. 488.)*

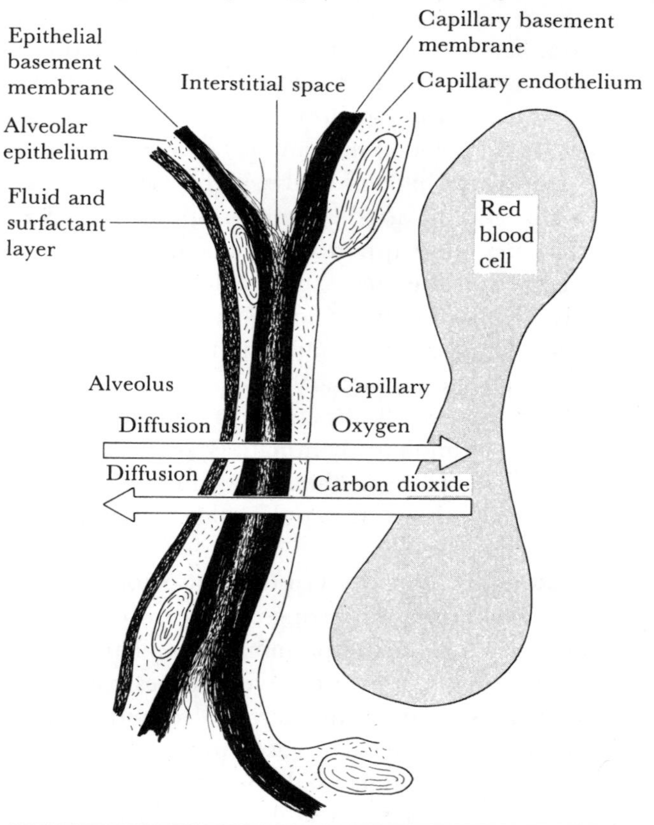

BLOOD-GAS BARRIER. As illustrated in Figure 6-3, the blood-gas barrier consists of

1. Layer of pulmonary surfactant
2. Alveolar epithelium
3. Epithelial basement membrane
4. Interstitium
5. Endothelial basement membrane
6. Capillary endothelium
7. Plasma
8. Red cell membrane
9. Intracellular fluid

Oxygen must transverse these structures in the order shown as it diffuses from the alveolar air sacs into the red cells to ultimately combine with hemoglobin. Carbon dioxide diffuses along its concentration gradient in the opposite direction.

PULMONARY CIRCULATION. The lungs are supplied by two independent arterial circulations: the pulmonary circulation and the bronchial circulation. The pulmonary arteries carry deoxygenated venous blood from the right ventricle to supply the gas-exchanging units of the lungs. The pulmonary arteries are larger and thinner than their systemic counterparts. The pulmonary circulation, therefore, is a high-compliance and low-resistance system. The pulmonary artery branches accompany branches of the bronchi. Small pulmonary arteries surrounded by alveoli respond to changes in alveolar gas composition. Alveolar hypoxia (i.e., a decrease in P_AO_2) causes pulmonary artery hypoxia and pulmonary vasoconstriction. This results in the shunting of blood from poorly ventilated areas of lung to better ventilated areas. This is in contrast to systemic circulations in which hypoxia causes vasodilatation.

In contrast, the bronchial circulation accounts for only 1 to 2 percent of pulmonary blood flow and originates from branches of the aorta. The bronchial arteries are high-pressure, high-resistance systemic vessels with thick muscular walls and a vasodilatatory response to hypoxia. The bronchial circulation supplies the supporting tissues of the lung, including the connective tissue, septa, and the large and small bronchi, with oxygenated arterial blood. After supplying the supporting tissue, the bronchial venous blood drains directly into the pulmonary veins diluting the well-oxygenated blood which has just passed through the pulmonary capillaries with poorly oxygenated venous blood. This results in a small physiological right-to-left shunt.

The pulmonary lymphatic system extends from the perivascular and peribronchial spaces in the supporting tissues of the lung to the hilus of the lung. From the hilus the pulmonary lymphatics drain primarily into the right lymphatic duct.

Roughly 10 percent of the body's total blood volume is present in the lungs at any given time. At resting cardiac outputs, an erythrocyte spends an average of 0.75 seconds in the pulmonary capillaries. During heavy exercise, the cardiac output may increase to several times basal levels. The erythrocyte transit time decreases to approximately 0.3 seconds, still leaving adequate time for equilibration with alveolar gases to occur. Further decreases in transit time are avoided as additional pulmonary capillaries are recruited to handle the increased flow and as those vessels already open are distended. Indeed, increases in cardiac output cause no increase in pulmonary artery pressure because recruitment of additional capillary beds and distention of the capillaries result in a decrease in pulmonary vascular resistance.

Gravity causes uneven distribution of blood in the lung as illustrated in Figure 6-4. The lung may be divided into three zones. In zone 1, the apex of the upright lung may have a pulmonary alveolar pressure that is greater than the pulmonary

Fig. 6-4. *Model to explain the uneven distribution of blood flow in the lung based on the pressures affecting the capillaries. A graph of blood flow versus height is appended on the right. (From J. B. West, C. T. Dollery, and A. Naimark. Destribution of blood flow in isolated lung; Relation to vascular and alveolar pressures,* J. Appl. Physiol. *19:713, 1964.)*

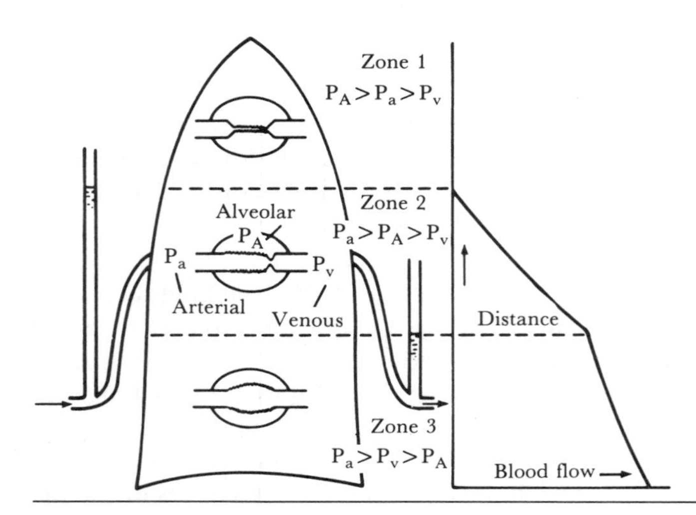

artery pressure causing collapse of the pulmonary arteries and capillaries, and cessation of blood flow (Q). This may occur only intermittently, for example during expiration. The result is ventilation-perfusion mismatch (V/Q mismatch). In zone 2, pulmonary artery pressure is greater than pulmonary alveolar pressure which, in turn, is greater than pulmonary venous pressure. Blood flow, therefore, proceeds normally and a normal V/Q ratio results.

In zone 3, the capillaries are engorged due to the increased pressure in the pulmonary capillaries resulting from the action of gravity on the column of blood in the lungs. Perfusion now exceeds ventilation and V/Q mismatch again results.

PULMONARY INNERVATION AND THE MUCOCILIARY BLANKET. The bronchi and trachea, especially the carina, are sensitive to irritation. This stimulates the cough reflex which is mediated by the vagus nerve and a regulatory center in the medulla. Initially, 2 to 3 liters of air are inspired. The epiglottis and vocal cords are then closed, and contraction of the expiratory muscles causes rapid development of up to 100 mm Hg of positive pressure in the lungs. The vocal cords and epiglottis open suddenly and air rushes out at 75 to 100 miles per hour. During this explosive expiration, the noncartilaginous portions of the bronchi and trachea collapse forming narrow slits, all of which serves to propel foreign objects from the respiratory passageways. The sneeze reflex is similar to the cough reflex except that the uvula is depressed shunting air flow through the nose. The sneeze reflex is mediated by the trigeminal nerve and the medullary regulatory center.

The pulmonary artery walls are richly innervated with branches of the sympathetic nervous system. No clear function has yet been elucidated for these.

Air is warmed, humidified, and filtered during its journey through the nasal passageways and the trachea. The numerous obstructions to air flow in the nasal passageways, the septa, turbinates, and hair, serve to increase the surface area available for equilibration of temperature and humidity and to remove large particles by turbulent precipitation. Ciliated epithelium propels the precipitated particles and the mucus secreted by the mucous membrane covering of the nose and oropharynx toward the pharynx where it is either swallowed or expectorated. Similarly, the "mucociliary blanket" of the tracheobronchial tree propels particles out of the lungs and into the oropharynx to meet the same fate.

Particles greater than 4 to 6 microns are removed by turbulent precipitation in the nose and tracheobronchial tree as previously described. Smaller particles, 1 to 5 microns in size, settle out by gravitational precipitation in the smaller bronchioles. Particles less than 1 micron in size diffuse against the walls of the alveoli and adhere to the alveolar fluid where they are ultimately removed by alveolar macrophages. Particles less than 0.5 microns in diameter may remain suspended in the alveolar air to be expired later.

QUESTIONS

1. A. Halving the diameter of a tube will cause the resistance to increase by what amount?
   B. What will happen if, at the same time, its length is halved?
2. Alveolar ventilation is acutely decreased to half of normal. What will happen to P_aCO_2?
3. In the absence of ventilation, diffusion may sustain life for a time because enough oxygen may diffuse into the lungs to supply the body's needs if the gradient driving diffusion is great enough (i.e., in the presence of 100% O_2). Why is this technique not utilized?
4. What are the four major cell types of the alveoli and what are their major functions?

5. Decreased PCO_2 in the alveoli of one region of lung will cause
   1. Bronchodilatation
   2. Bronchoconstriction
   3. Shunting of blood away from that area of lung
   4. Alteration in the resistance of the airways causing ventilation to be shunted to better perfused areas of lung

   Answer: A. 1,2, and 3
   B. 1 and 3
   C. 2 and 4
   D. 4
   E. None of the above
6. Describe the effect of the following perturbations on the pulmonary and systemic circulations, respectively.
   A. Increased PCO_2
   B. Decreased PO_2
   C. Decreased pH
7. The pressure gradient determining pulmonary blood flow in
   1. The apex of the lung (Zone 1)?
   2. Midlung (Zone 2)?
   3. The base of the lung (Zone 3)?

   Answer A. $P_A > Pa > Pv$
   B. $Pa > P_A > Pv$
   C. $Pa > Pv > P_A$

MECHANICS OF RESPIRATION

The lungs are encased within a double sac. The inner sac is the visceral pleura which is tightly adherent to the lung surface. The outer sac is tightly adherent to the chest wall and is called the parietal pleura. The pleural space is a potential space located between these two sacs. Normally it contains only a small amount of fluid that serves both to lubricate the pleura as they move relative to one another during breathing and to cement the two together much as a film of water causes two panes of glass to adhere.

A negative intrapleural pressure causes air to enter the lungs and results from

1. Downward movement of the diaphragm as it contracts lengthening the chest cavity
2. Elevation of the ribs by the inspiratory muscles causing the anterior-to-posterior diameter of the chest to increase

These two processes create a vacuum or bellows effect sucking air into the lungs. Because of the extremely high compliance of the lung and the chest wall, only a few millimeters of mercury of negative intrapleural pressure is sufficient to produce a normal inspiration.

During quiet breathing, relaxation of the diaphragm and chest wall and the elastic recoil properties of the lungs, chest wall, and abdominal structures cause the lung to be compressed resulting in expiration.

Heavy breathing requires active expiratory effort with the abdominal muscles contracting to force the abdominal contents against the bottom of the diaphragm.

QUESTION

8. The functional residual capacity or resting position of the lungs is determined by what two factors.

LUNG VOLUMES AND CAPACITIES

LUNG VOLUMES. Figure 6-5 illustrates the use of a spirometer to determine lung volumes and capacities. Note that a lung capacity is equal to two or more lung volumes. The following are the common lung volumes used in physiology:

1. Tidal volume (TV) is the volume expired or inspired with each breath at rest and is normally equal to about 500 cc.
2. Inspiratory reserve volume (IRV) is the additional volume of air that can be inspired on maximum forced inspiration at the end of a normal tidal inspiration (i.e., in addition to the TV). Normally, IRV is approximately 3000 cc.
3. Expiratory reserve volume (ERV) is the volume of air that can still be expired by forceful expiration after the end of a normal tidal expiration. Normally, ERV is approximately 1100 cc.
4. Residual volume (RV) is the volume of air that remains in the lungs after a maximal expiration. Note that the residual volume cannot be measured by spirometry but must be determined by helium dilution or plethysmography.

ALVEOLAR MINUTE VENTILATION VERSUS DEAD SPACE VENTILATION. Approximately 500 cc of air are inspired with each resting ventilation, the TV. At a respiratory rate of 15 breaths per minute, the minute volume (volume respired in 1 minute or $\dot{V}$ would be 7500 cc per minute. That portion of each TV that remains in the nonexchanging portions of the airways (the anatomic dead space) is referred to as the dead space ventilation or $\dot{V}_{Dan}$. The volume of the anatomic dead space ($\dot{V}_{Dan}$) is roughly 150 cc, and the amount of air actually reaching the alveolus with each breath, the alveolar gas volume (V_A), is, therefore, 350 cc. Alveolar minute ventilation ($\dot{V}_A$), therefore, equals 350 × 15 or 5250 cc per minute. The ratio of $\dot{V}_{Dan}$ to $\dot{V}_A$ is normally 0.2 to 0.35. Figure 6-6 illustrates typical lung volumes and flows.

Alveolar air is a mixture of freshly inspired air and the air remaining in the lungs between breaths, the functional residual capacity (roughly 3000 cc). The alveolar air remaining in the lungs between breaths serves to prevent the wide

Fig. 6-5. *Lung volumes and capacities. Note that the functional residual capacity and residual volume cannot be measured by spirometry alone. (From A. C. Guyton.* Textbook of Medical Physiology *[6th ed.]. Philadelphia: Saunders, 1981. P. 481.)*

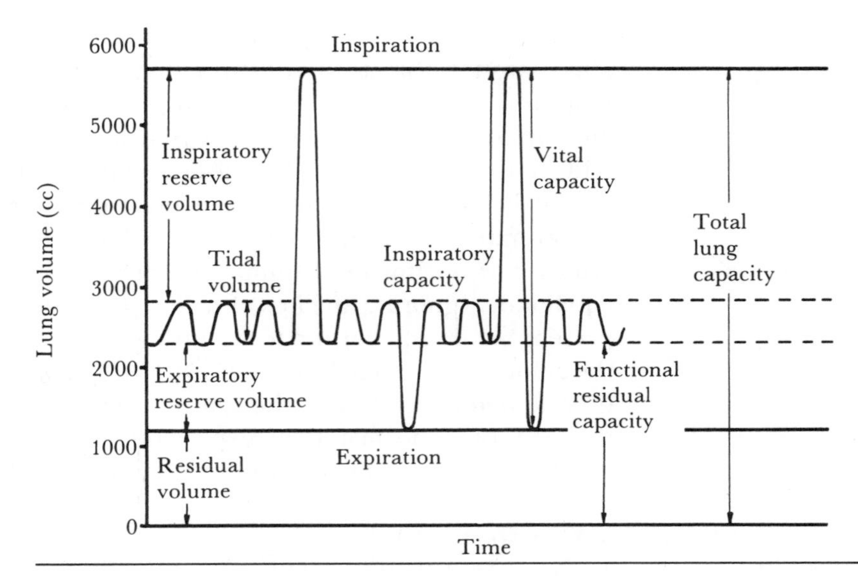

Fig. 6-6. *Diagram of a lung showing typical volumes and flows. There is considerable variation around these values. (Modified from J. B. West.* Ventilation/Blood Flow and Gas Exchange. *Oxford: Blackwell, 1977. P. 3.)*

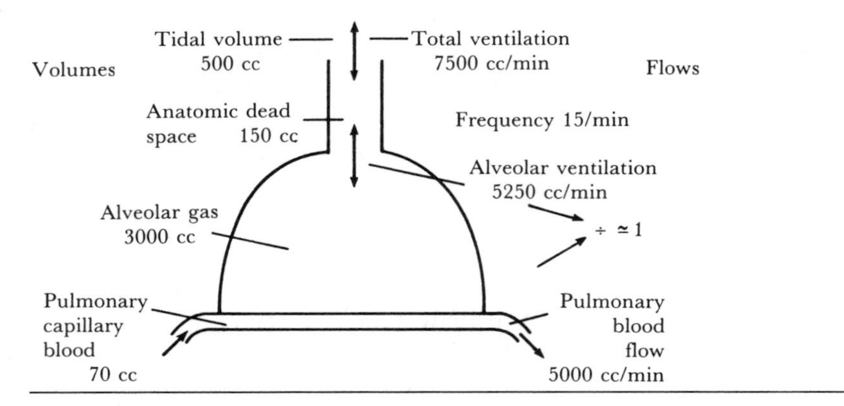

swings in alveolar and arterial gas composition that would occur if no air were present between breaths.

LUNG CAPACITIES. The following are the common lung capacities used in physiology.

1. Inspiratory capacity (IC) = IRV + TV; the inspiratory capacity is the maximal volume of air that can be inspired at the end of a normal tidal expiration.
2. Functional residual capacity (FRC) = ERV + RV; the functional residual capacity is the total volume of air remaining in the lungs at the end of a normal expiration. It may be calculated using either helium dilution or plethysmography.
3. Vital capacity (VC) = IRV + TV + ERV; vital capacity is the maximum volume of air that can be expelled from the lung following a maximal inspiration. It is equal to the inspiratory capacity plus the expiratory reserve volume.
4. Total lung capacity (TLC) = IRV + TV + ERV + RV = VC + RV; the total lung capacity is equal to the total volume of the lung following a maximal inspiration.

MEASUREMENT OF LUNG VOLUMES. Spirometry cannot be used to calculate the total lung volumes because the lung never entirely empties of air. A certain amount of air, the RV, necessarily remains. Thus RV and, therefore, FRC and TLC must be determined by utilizing some other technique.

The helium dilution technique utilizes a spirometer containing a known concentration of helium (Fig. 6-7). Helium is virtually insoluble in blood and after a few breaths the helium concentration in the spirometer and the lungs equilibrate. Ignoring any loss of helium from the system, the amount of helium present before equilibrium (concentration × volume) is $C_1 \times V_1$ and is equal to the amount after equilibration, $C_2 \times (V_1 + V_2)$. The following equation can be used to determine FRC if the subject begins breathing from the spirometer at the end of a normal tidal expiration and if measurement of the helium concentration at equilibrium is done at the end of a normal tidal expiration:

$$C_1V_1 = C_2\,(V_1 + V_2)$$
$$V_2 = (C_1\,V_1 / C_2) - V_1 = \text{FRC} \qquad (6\text{-}2)$$

Fig. 6-7. *Measurement of the functional residual capacity by the helium dilution method. (From J. B. West.* Respiratory Physiology: The Essentials *[3rd ed.]. 1985. P. 13. © 1985 The Williams & Wilkins Co., Baltimore.)*

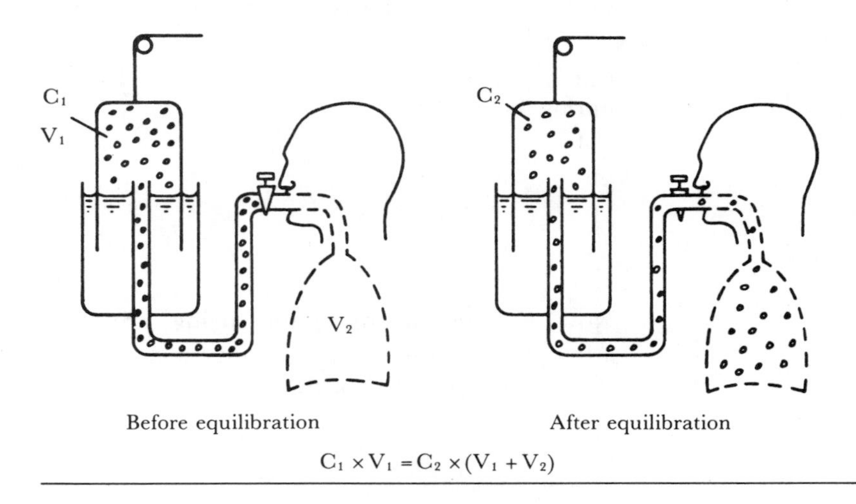

in which

C_1 is the initial concentration of helium in the spirometer.
C_2 is the concentration of helium after equilibration has occurred.
V_1 is the volume of the spirometer.
V_2 is the volume of the lungs at end-tidal expiration (i.e., the FRC).

The technique of total body plethysmography uses a large air-tight box like a telephone booth in which a man can sit (Fig. 6-8). The volume of the booth is known and the pressure within the booth can be monitored. A subject enters the booth and is instructed to make a respiratory effort at the end of a normal tidal expiration against a mouthpiece with an automatic shutter. As the subject tries to inhale, the automatic shutter closes. The subject expands the volume of his lungs as his intrathoracic pressure becomes increasingly negative in a fruitless attempt

Fig. 6-8. *Measurement of the functional residual capacity by whole body plethysmography. When the subject makes an inspiratory effort against a closed airway, he slightly increases the volume of his lung. Airway pressure decreases and box pressure increases. (From J. B. West.* Respiratory Physiology: The Essentials *[3rd. ed.]. P. 14. © 1985 The Williams & Wilkins Co., Baltimore.)*

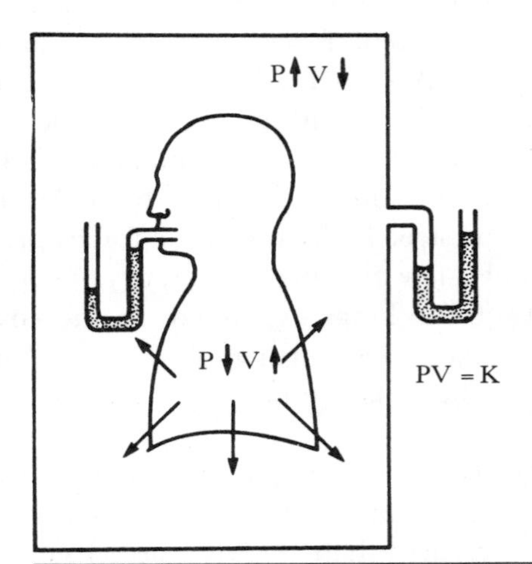

to move air against the closed mouthpiece. The pressure in the box, in contrast, rises as the thorax expands. Boyle's law states that pressure times volume is equal to a constant ($P \times V = C$). Thus, because the change in pressure can be measured, the change of volume of the lungs can be calculated as follows:

Change in lung volume = Change in box volume = Constant / change in box pressure during the inspiratory effort

or

$$\Delta V = C/\Delta P \qquad (6\text{-}3)$$

Now, by simultaneously measuring pressures at the mouth as the subject attempts a tidal inspiration against a closed mouthpiece, and by again applying Boyle's law, the FRC of the lung can be calculated.

$$P_1V = P_2\,(V + \Delta V)$$
$$V = P_2\,\Delta V\,/\,(P_1 - P_2) \qquad (6\text{-}4)$$

in which

P_1 is the mouth pressure at the beginning of inhalation.
P_2 is the maximal pressure developed.
V is the FRC.
ΔV is the change in volume of the box as calculated above.

SURFACE TENSION, ELASTIC RECOIL, AND LUNG COMPLIANCE. The lung's compliance characteristics may be described as an elastic component in parallel with an inelastic component. As long as the elastic component has some stretch left, the lungs are very compliant. Once the inelastic component is required to stretch, however, the lungs become much less compliant.

The elastic recoil properties of the lung tend to collapse it while the action of the chest wall acts to keep the lungs expanded. The elastic recoil properties result from the interaction of the stretched elastic elements of the structure of the lung and from the surface tension of the fluid lining the alveoli. The elastic fibers, analogous to a rubber band, are stretched and tend to regain their resting length. The fluid lining the alveoli tends to minimize its surface area (i.e., the fluid attempts to become a sphere) as a result of the molecular attraction between the water molecules. This also tends to collapse the lungs. Indeed, surface tension is such a potent force tending to collapse the lungs that surfactant (dipalmitoyl lecithin) is required to decrease the surface tension of the alveolar water to the extent that breathing becomes practical. At rest, roughly two-thirds of the tendency to collapse is attributable to surface tension and one-third to elastic forces. At high lung volumes, the inelastic supporting structures (e.g., collagen) must be stretched, and elastic forces predominate.

Normally only a few millimeters mercury of negative intrapleural pressure are required to keep the lungs expanded. In the absence of surfactant, however, lung expansion becomes much more difficult, and a much greater negative intrapleural pressure may be necessary to overcome the tendency of the alveoli to collapse.

Surfactant works by forming a monomolecular layer between the fluid lining the alveoli and the air in the alveoli. The surface tension of a water-air interface is many times greater than the surface tension of a surfactant-air interface.

Surface tension also tends to pull fluid into the alveoli from the alveolar wall, and this, too, is minimized by surfactant preventing pulmonary edema.

Several factors interact to stabilize the alveoli relative to one another. By the law of Laplace, the transalveolar pressure required to keep the alveoli expanded is proportional to the tension in the alveolar wall divided by the diameter of the alveolus. Assuming that the alveoli are perfect spheres

$$\text{Transalveolar pressure} = 2\ (\text{wall tension}) / \text{diameter} \qquad (6\text{-}5)$$

This equation implies that the transalveolar pressure required to keep the alveoli open should increase linearly as diameter decreases. If wall tension remained constant, this would inevitably lead to collapse of the alveoli.

Surface tension, however, is the major cause of wall tension. In the presence of surfactant, surface tension decreases as diameter decreases. This occurs because the concentration of surfactant increases as diameter decreases. Therefore, the need for greater distending pressures at lower alveolar diameters is negated by the decrease in wall tension associated with increasing concentrations of surfactant. In addition, the alveoli share walls in common and are thus interdependent, which helps maintain patency of the alveoli because each helps stabilize the other.

Compliance is defined as the change of volume per unit change of pressure. For normal lungs, this value is 0.200 liters per centimeter water.

The work of breathing refers to the energy expended to perform the following work:

1. Expand the elastic tissues of the chest wall and lungs (compliance work)
2. Overcome the viscosity of the inelastic structures of the chest wall and lungs (i.e., move everything out of the way and then move it back)
3. Move air against the resistance of the airways

The work of breathing normally accounts for about 2 to 3 percent of the body's total energy expenditures. At rest compliance work predominates while on exercise overcoming airway resistance is more important.

The pressure-volume curve of the lung is illustrated in Figure 6-9. Note that the inspiratory and expiratory curves are not the same and that compliance is

Fig. 6-9. *Pressure-volume curve of the lung showing the inspiratory work done overcoming elastic forces (area OAECDO) and viscous forces (hatched area ABCEA). Line AEC is the compliance line. The inflation and deflation curves are not the same. This is called hysteresis. (From J. B. West.* Respiratory Physiology: The Essentials *[3rd ed.]. P. 110. © 1985 The Williams & Wilkins Co., Baltimore.)*

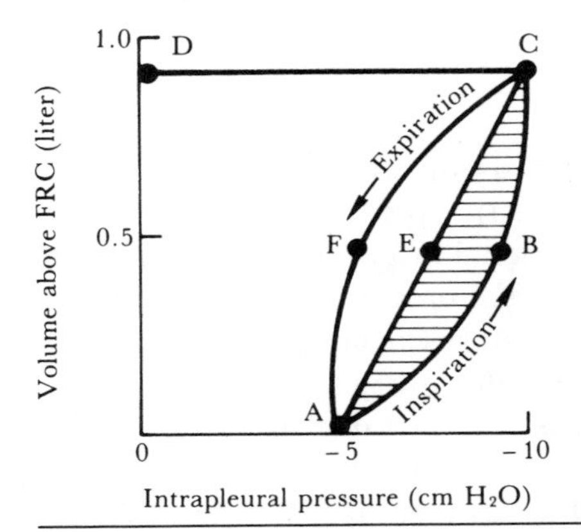

somewhat greater for expiration than for inspiration. This is called hysteresis and is due to the viscous properties of the lung.

QUESTIONS

9. Use of helium dilution and plethysmography with the same subject give consistently divergent results even when done consecutively and without any apparent change in the subject's condition. What could account for this?
10. Total lung volume = 2.4 liters; rate = 16/minute; change in intrapleural pressure = 3 cm H_2O; Tidal volume = 75 cc. The compliance of the lung would be
    A. 0.04 cm/ml
    B. 1.2 liter/minute
    C. 25 cm/ml
    D. 400 ml/cm
    E. 25 ml/cm

PULMONARY GAS EXCHANGE

ALVEOLAR GAS EXCHANGE. Inspired air moves from the environment to the level of the terminal bronchioles by bulk flow. From there, the air diffuses into the aveoli to mix with air already present there. Oxygen then diffuses along its concentration gradient from the alveoli into the pulmonary capillaries. Carbon dioxide diffuses along its concentration gradient in the opposite direction.

At sea level the barometric pressure is 760 mm Hg. Dry air is 79 percent nitrogen, 21 percent oxygen, 1 percent inert gases and 0.04 percent carbon dioxide. Dalton's law states that each gas contributes to the total pressure in direct proportion to its relative concentration. The partial pressure of any gas can, therefore, be calculated by multiplying its percentage composition by the ambient barometric pressure. The partial pressure of nitrogen at sea level is 600 mm Hg and that of oxygen is 160 mm Hg.

Important changes occur to the air as it traverses the conduction system to enter the gas exchanging regions of the lungs. The temperature and humidity of the gas approach body temperature and 100 percent humidity. The contribution of the partial pressure or vapor pressure of water must then be taken into consideration when calculating the proportion of gases in the alveoli. The vapor pressure of water at 37°C is 47 mm Hg. The water vapor expands the volume of air and thereby dilutes the gases present in the inspired air. The sum of the partial pressures of the other gases present is now 760 − 47 = 713 mm Hg. Furthermore, carbon dioxide is being excreted into the lungs causing further dilution of the inspired gases. The alveolar gas equation allows one to estimate the partial pressure of oxygen in the alveolus.

$$P_{A}O_2 = (BP - 47) \times F_{I}O_2 - (P_{A}CO_2 / RER) \qquad (6\text{-}6)$$

in which

$P_{A}O_2$ is the partial pressure of oxygen in the alveolus.
BP is the barometric pressure.
$F_{I}O_2$ is the fractional inspired oxygen concentration.
$P_{A}CO_2$ is the partial pressure of carbon dioxide in the alveolus.
RER is the respiratory exchange ratio or respiratory quotient and is equal to the ratio of carbon dioxide given off by the body to the oxygen consumed by the body. This is usually calculated as the minute carbon dioxide production divided by the minute oxygen consumption ($\dot{V}CO_2 / \dot{V}O_2$). The RER may change with alteration in diet and metabolism but is normally considered to

average 0.8. The respiratory exchange ratio for glucose and other carbohydrates is 1.0, for fat 0.7, and for protein 0.8.

The alveolar-arteriolar oxygen difference $P(A\text{-}a)O_2$ is simply the difference between the partial pressure of oxygen in the alveoli calculated by the alveolar gas equation and the arterial partial pressure of oxygen. Normally, this difference should not exceed 15–20 mm Hg. The arterial partial pressure of oxygen is age dependent and may be estimated by the equation

$$PaO_2 = 103 - 0.4 \text{ (age in years)} \qquad (6\text{-}7)$$

As oxygen diffuses from the alveoli into the capillaries the partial pressure of oxygen in the alveolar gas and that in the liquid equilibrate. As Henry's law states, the actual concentration of dissolved gas in a liquid is equal to the partial pressure of the gas in contact with the liquid, multiplied by the solubility coefficient of the gas in that particular liquid. The solubility coefficient of oxygen in blood is 0.003 ml O_2/mm Hg/liter blood while that of carbon dioxide is 0.03 mmoles CO_2/mm Hg/liter blood. Note that the two solubilities use different units!

DIFFUSION CAPACITY. The gradient for diffusion is the partial pressure of the gas in one area minus the partial pressure in a second area divided by the distance (T) over which diffusion must occur

$$\frac{P_1 - P_2}{T} = \Delta P/T \qquad (6\text{-}8)$$

Factors which influence diffusion include

1. Temperature (assumed to remain constant at 37°C and, therefore, ignored hereafter)
2. Molecular weight of the gas (MW)
3. Distance over which diffusion must occur (T)
4. Cross-sectional area available for diffusion to occur (A)
5. Solubility of the gas in the fluid (S)
6. The pressure gradient driving diffusion (ΔP)

The overall rate of diffusion at any given temperature can be estimated by the following relationship:

$$\text{Rate of diffusion or } \frac{\Delta P \times A \times S}{T \times MW} \qquad (6\text{-}9)$$

The diffusing capacity (or volume/mm Hg/minute/pair of lungs) is the volume of a gas that diffuses through the respiratory membrane each minute with a pressure difference of 1 mm Hg (Fig. 6-10). For normal lungs, the diffusing capacity for oxygen is 21 ml/mm Hg/minute at rest. It may be increased by recruitment and dilatation of pulmonary capillaries. The diffusing capacity for carbon dioxide is so high that it cannot be measured due to its lipid solubility. Carbon monoxide is less lipid soluble and, therefore, more likely to be diffusion limited. Because of its high affinity for hemoglobin, carbon monoxide is often used to calculate the diffusion capacity of the lungs in an attempt to determine whether a limitation to diffusion exists. Only the inspiratory carbon monoxide pressure need be measured because carbon monoxide binds to hemoglobin with

Fig. 6-10. *Diffusion through a tissue sheet. The amount of gas ($\dot{V}_{gas}$) transferred is proportional to the area (A), a diffusion constant (D), and the difference in partial pressure ($P_1 - P_2$) and is inversely proportional to the thickness (T). The constant is proportional to the gas solubility (Sol) but inversely proportional to the square root of its molecular weight (MW). (From J. B. West.* Respiratory Physiology: The Essentials *[3rd ed.]. 1985. P. 22. © 1985 The Williams & Wilkins Co., Baltimore.)*

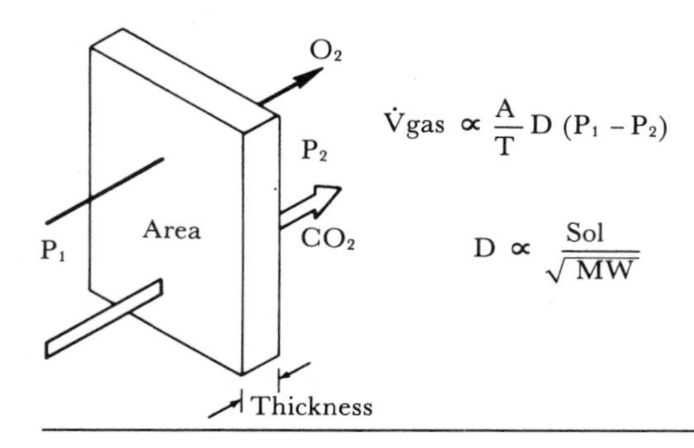

such great affinity that the partial pressure of carbon monoxide in the blood is almost zero.

VENTILATION-PERFUSION (MIS)MATCH. Ventilation is better in the more dependent portions of the lung. Perfusion is better in the more dependent portions of the lung. The ventilation-perfusion ratio, V/Q, is best at the apex of the lung. Due to the effect of gravity, perfusion is greatest at the bases. Ventilation is also greatest at the lung bases, in this case because of the action of the diaphragm pulling the alveoli open at the bases. The ratio of ventilation to perfusion, however, is greatest at the apex of the lungs because perfusion is more affected by gravity than ventilation is by the action of the diaphragm. Ideally, a ventilation to perfusion ratio of 1 should exist with ventilation being perfectly matched to perfusion. As Figure 6-11 illustrates, however, the normal relationship between minute alveolar ventilation ($\dot{V}A$) and minute perfusion ($\dot{Q}$) in the upright lung is such that ventilation is roughly 3 times greater than necessary at the apex and only 60 percent of that required at the bases. During exercise,

Fig. 6-11. *Distribution of ventilation and blood flow down the upright lung. Note that the ventilation-perfusion ratio decreases down the lung. (From J. B. West.* Ventilation/Blood Flow and Gas Exchange *[3rd ed.]. Oxford: Blackwell, 1977, P. 30.)*

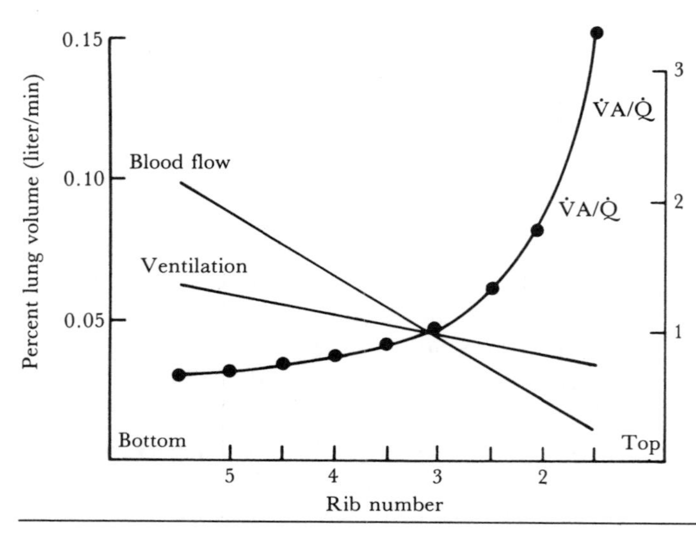

Table 6-1. *Oxygen Content of Whole Blood*

PO_2 (mm Hg)	% Saturation of Hb[a]	O_2 Content of Hb (ml/dl)[b]	Dissolved O_2 (ml/dl)
10	13.5	2.71	0.03
20	35	7.04	0.06
27	50	10.05	0.08
40	75	15.08	0.12
60	89	17.89	0.18
90	96.5	19.40	0.27
100	97.5	19.60	0.30
600	99+	19.90	1.80

[a] pH = 7.40, temperature = 38° C.
[b] Estimated using 1.34 ml O_2/g hemoglobin (Hb) and 15 g Hb/dl blood.

blood flow to the upper lung increases minimizing ventilation-perfusion mismatch.

Oxygen exchange is exquisitely sensitive to V/Q changes. Carbon dioxide excretion, while less sensitive to minor alterations in V/Q relationships (because increasing the rate and depth of ventilation can compensate to a large extent), is affected in extreme examples.

If the lung is perfused but not ventilated, then the V/Q ratio is zero and no gas exchange takes place. Similarly if the lung is ventilated but not perfused, then the ratio approaches infinity and no gas exchange takes place (Fig. 6-12). At one extreme a sampling of alveolar air resembles venous blood and at the other inspired air.

If ventilation and perfusion are ideally matched, then a ratio of one exists and the efficiency of gas transport is maximized. The normal response to either hypoxia or hypercarbia is to increase the rate and depth of ventilation, i.e., to hyperventilate. This works quite well for carbon dioxide because underventilation of one region of lung, perhaps secondary to obstruction of a bronchus, is readily compensated for by hyperventilation of other regions of the lung. The arterial carbon dioxide would then be the arithmetical average of the blood coming from the under- and overventilated regions of the lungs.

The oxygen content of blood, however, is only minimally affected by dissolved oxygen and is instead dependent on the amount of oxygen bound to hemoglobin (Fig. 6-13 and Table 6-1). Blood which has been exposed to a partial pressure of oxygen sufficient to saturate hemoglobin does not carry substantially more if either ventilation or FiO_2 is increased. Similarly, if blood with poorly oxygenated hemoglobin is mixed with well-oxygenated blood, the partial pressure of oxygen

Fig. 6-12. *The normal PO_2-PCO_2, $\dot{V}/\dot{Q}$ diagram. (From A. C. Guyton.* Textbook of Medical Physiology *[7th ed.]. Philadelphia: Saunders, 1986. P. 491.)*

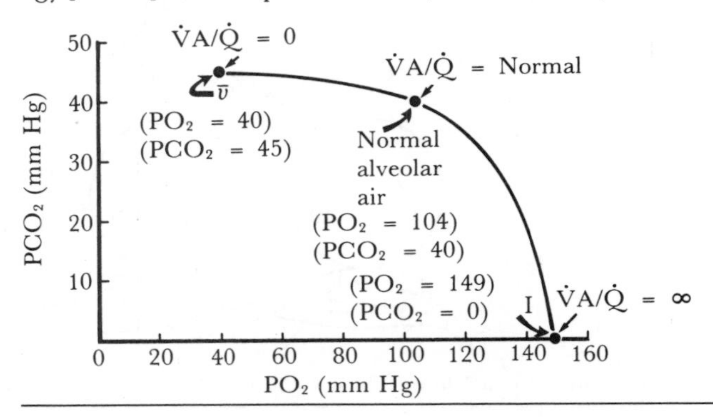

Fig. 6-13. *Oxygen dissociation curve (solid line) for pH 7.4, PCO_2 40 mm Hg, and 37°C. The total blood oxygen concentration is also shown for a hemoglobin concentration of 15 g/100 ml of blood. (From J. B. West.* Respiratory Physiology: The Essentials *[3rd ed.]. P. 69. © 1985 The Williams & Wilkins, Co., Baltimore.)*

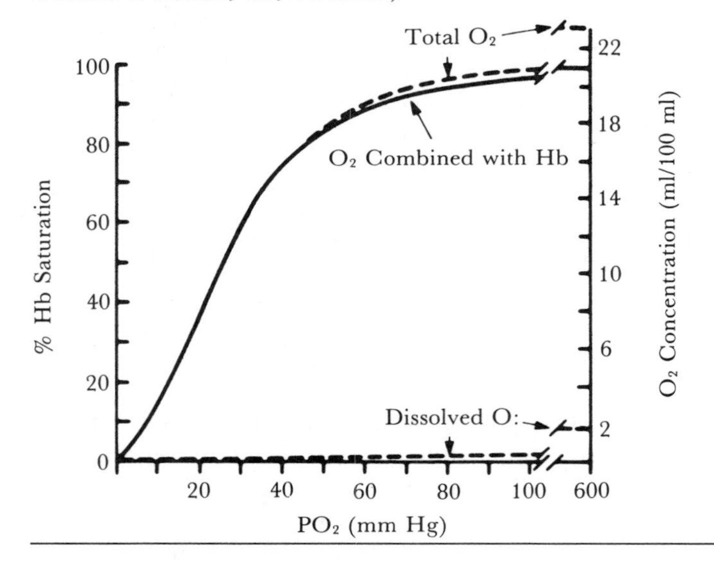

resulting is remarkably affected as oxygen-starved hemoglobin soaks up the small amount of oxygen dissolved in the well-oxygenated blood and then begins to pull oxygen out of the saturated hemoglobin until a new equilibrium is reached (Fig. 6-14).

Normally, the lungs are able to keep the V/Q ratio within reasonable limits. Underventilation of a region of lung causes blood to be shunted preferentially to other better ventilated regions of the lungs as the decrease in P_AO_2 and increase in PCO_2 cause reflex constriction of the pulmonary vessels. Similarly, if perfusion is compromised in a certain region of the lungs, a reflex bronchoconstriction results (see Question 5). Nonetheless, these changes only minimize and not eliminate V/Q mismatch.

Fig. 6-14. *Depression of arterial PO_2 by mismatching of ventilation and blood flow. The lung units with a high ventilation-perfusion ratio add relatively little oxygen to the blood compared with the decrement caused by alveoli with a low ventilation-perfusion ratio. $\bar{V}$ = mixed venous blood; a = arterial blood. (Modified from J. B. West.* Ventilation/Blood Flow and Gas Exchange *[3rd ed.]. Oxford: Blackwell, 1977. P. 59.)*

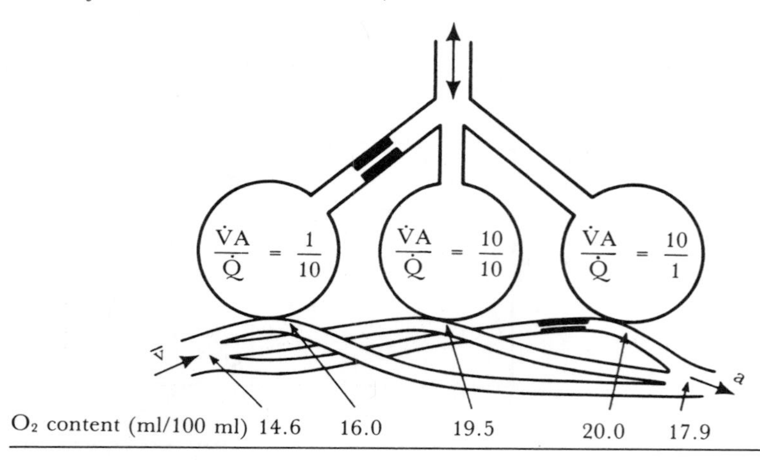

PHYSIOLOGICAL SHUNTS AND PHYSIOLOGICAL DEAD SPACE. A physiological shunt occurs when the ratio of ventilation to perfusion is abnormally low. It is that fraction of the blood passing through the pulmonary vasculature that does not become fully oxygenated, assuming that the hypoxemia results from blood passing through unventilated alveoli. The bronchial vessels constitute a normal component of physiological shunt accounting for roughly 2 percent of the cardiac output. The physiological shunt may be calculated using the following equation:

$$\frac{Qsp}{Q_T} = \frac{CiO_2 - CaO_2}{CiO_2 - CvO_2} \qquad (6\text{-}10)$$

in which

Qsp is the physiological shunt blood flow.
Q_T is the total blood flow through the lungs.
CiO_2 is the total oxygen content of blood draining from an ideal alveoli (calculated using the alveolar gas equation to determine P_AO_2 and then using the calculated P_AO_2 to determine the total oxygen content of the blood, i.e., the sum of the amount bound to hemoglobin and the amount free in solution at that PaO_2).
CaO_2 is the arterial oxygen content.
CvO_2 is the venous oxygen content.

The physiological shunt equation does not differentiate between the three types of shunts: intrapulmonary shunts (e.g., pulmonary artery to pulmonary vein), anatomical shunts (e.g., right to left shunting of blood in the heart), and physiological shunting due to inadequate ventilation relative to perfusion.

Physiological dead space occurs when the V/Q ratio is greater than normal. It, therefore, refers to wasted ventilation much as the physiological shunt refers to wasted perfusion. It is equal to the dead space ventilation (including unperfused alveoli) plus the amount of alveolar ventilation that exceeds that required to adequately supply the blood flowing through the lungs. The Bohr equation may be used to estimate the physiological (or total) dead space. In normal adults, the anatomical dead space comprises most if not all of the physiological dead space. The total ventilation (V_T) is equal to alveolar ventilation (V_A) plus dead space ventilation (V_D). A tidal breath of 500 ml normally consists of 350 ml V_A and 150 ml V_D.

The Bohr equation calculates the ratio of dead space to total ventilation by attributing all of the lowering of the expiratory partial pressure of carbon dioxide (P_ECO_2) relative to arterial partial pressure of carbon dioxide ($PaCO_2$), to unperfused alveoli, and to the anatomical dead space.

$$V_D/V_T = \frac{P_ACO_2 - P_ECO_2}{P_ACO_2} = \frac{PaCO_2 - P_ECO_2}{PaCO_2} \qquad (6\text{-}11)$$

The reasoning is as follows. If no gas exchange is occurring in a portion of lung because it is not being perfused adequately, then that portion does not add carbon dioxide to the alveolar air. Because carbon dioxide diffuses so readily across the respiratory membrane, its partial pressure in the expiratory air should equal that in the pulmonary capillary blood, and any discrepancy must be due to air that has not been exposed to pulmonary capillary blood. Arterial blood is used in the equation because pulmonary capillary blood is difficult to obtain.

11. Calculate the partial pressure of oxygen in the alveolus of a normal lung at sea level. $PaCO_2 = 40$ mm Hg.
12. What is the effect on compliance of removing one lung?
13. At standard temperature and pressure, 1 mole of a gas occupies 22.3 liters volume. If the solubility of carbon dioxide in plasma is 0.03 mmoles/mm Hg/liter plasma, calculate its solubility in ml/mm Hg/liter plasma. How does this compare with the solubility of oxygen in ml/mm Hg/liter plasma?
14. If a patient's $\dot{V}CO_2/\dot{V}O_2$ is measured at
    (1) 0.98
    (2) 0.81
    (3) 0.69
    (4) 1.1
    The patient is
    A. On a fat-burner diet (the all-cholesterol diet)
    B. Is a vegetarian
    C. On an average American diet
    D. On dextrose IV
15. A normal subject has a PaO_2 of 95 and a P_ACO_2 of 40. Calculate the $P(A\text{-}a)O_2$ if the subject is metabolizing
    A. Carbohydrate
    B. Protein
    C. Fat
16. In the following examples, is lung oxygen exchange normal or abnormal? What could account for the abnormality if one is present?
    A. $P_AO_2 = 90$, $P_ACO_2 = 8$
    B. $P_AO_2 = 50$, $P_ACO_2 = 80$
17. What is the normal A-a gradient for carbon dioxide?
18. The respiratory membrane averages some 70 m^2 and approximately 0.5 micron thick. Name its components.
19. Which of the following statements is correct:
    A. Low concentrations of carbon monoxide are used to calculate the diffusing capacity of the lung by the single-breath method.
    B. Carbon monoxide has a higher affinity for hemoglobin than does oxygen.
    C. Only the partial pressure of carbon monoxide in the blood must be measured to calculate the diffusion gradient between alveolar air and the blood.
    D. In carbon monoxide poisoning the victim's PaO_2 remains normal as long as he continues to breathe.
    E. Carbon monoxide diffuses even more rapidly than oxygen and carbon dioxide.
20. Calculate the diffusing capacity of the lung for carbon monoxide given the following data:
    Patient is breathing 0.001% carbon monoxide
    Carbon monoxide uptake by the blood is 28 ml per minute
    Barometric pressure is 700 mm Hg (dry gas)
21. Tuberculosis develops in areas of high oxygen tension, i.e., where the ratio of V/Q is maximal. Where do bats develop tuberculosis?
22. One liter of arterial blood with a PaO_2 of 27 mm Hg is mixed with 1 liter of arterial blood with a PaO_2 of 100 mm Hg.
    A. What is the total oxygen content of the first liter and of the second?
    B. What is the percent hemoglobin saturation of the first liter and of the second liter? (Hint: refer to the oxygen-hemoglobin dissociation curve [see Fig. 6-13] and to Table 6-1).

C. What is the total oxygen content and percentage saturation of hemoglobin of a liter of the mixture? What would be the PaO_2?
D. The $PaCO_2$ of the first liter is 20, that of the second liter is 40. What would be the $PaCO_2$ of the mixture?

23. Abnormally low V/Q ratios cause abnormally high $P(A\text{-}a)O_2$ by increasing the physiological shunt fraction. Abnormally high V/Q also causes high $P(A\text{-}a)O_2$. Why?
24. What is an anatomical shunt? Give an example.
25. A. Calculate the dead space volume (V_D) given the following data: total ventilation = tidal volume = 500 ml
    P_ECO_2 = 30 mm Hg
    $PaCO_2$ = 45 mm Hg
    B. What is the normal value for the ratio of V_D/V_T?

BLOOD GAS TRANSPORT AND TISSUE GAS EXCHANGE

CARBON DIOXIDE TRANSPORT AND EXCHANGE. Carbon dioxide transport and exchange are covered in detail in Chapter 7. Briefly, then, $PaCO_2$ is directly proportional to the effective ventilation with the lungs excreting approximately 10,000 meq of carbonic acid per day. The carbon dioxide content of the blood is carried in three forms:

1. As dissolved gas
2. As bicarbonate
3. Combined with proteins, particularly hemoglobin

Carbon dioxide is extremely lipid soluble, and it, therefore, equilibrates rapidly across cell membranes. Carbon dioxide diffuses 20 times as rapidly as does oxygen.

OXYGEN TRANSPORT AND EXCHANGE. Oxygen is transported in the blood in two forms—as dissolved gas and bound to hemoglobin. Oxygen is less soluble in blood than carbon dioxide. The solubility of oxygen in blood is equal to 0.003 ml oxygen/mm Hg/dl blood. Assuming a PaO_2 of 100 mm Hg, only 0.30 ml of oxygen as the dissolved gas is carried per deciliter of blood.

In a person with a normal hematocrit, the hemoglobin concentration in blood is 15 g per deciliter of blood. Because hemoglobin can carry 1.34 ml oxygen per gram hemoglobin, 20.1 ml oxygen can theoretically be carried bound to hemoglobin. At a PaO_2 of 100 mm Hg, hemoglobin is over 97 percent saturated (see Table 6-1). Thus, increasing the partial pressure of oxygen above this level can add only a small fraction of additional oxygen. Similarly, the solubility of oxygen in blood is so low that little benefit is gained by increasing the amount dissolved in blood. The actual amount of oxygen carried by hemoglobin must be calculated to equal

$$134 \text{ ml } O_2/\text{g Hb} \times \text{g Hb}/100 \text{ ml blood} \times \text{percent } O_2 \text{ saturation}/100\text{ml blood}$$

The total oxygen content of blood is, therefore, equal to the amount bound to hemoglobin plus that dissolved as gas. At a PaO_2 of 100 mm Hg, 19.7 ml oxygen are bound to hemoglobin and 0.3 ml is dissolved in the blood.

HEMOGLOBIN. Hemoglobin is composed of two alpha polypeptide chains and two beta polypeptide chains, each of which contains a separate heme group. A total of four oxygen molecules can, therefore, be carried by a single hemoglobin molecule.

Binding of the initial oxygen molecule to hemoglobin results in a change in configuration of hemoglobin. This change increases the affinity of hemoglobin for oxygen and is an example of positive cooperativity. Such enhanced binding yields the sigmoidal curve characteristic of such an interaction (see Fig. 6-11). The $P_{50}O_2$ is defined as the PO_2 at which hemoglobin is 50 percent saturated. At pH 7.4, the $P_{50}O_2$ is 27 mm Hg.

The oxyhemoglobin dissociation curve is shifted to the right by acidosis, hypercapnia, elevated temperature, or increased levels of 2,3-diphosphoglycerate (2,3-DPG). The rightward shift caused by increased levels of carbon dioxide is referred to as the Bohr effect (Fig. 6-15).

The hemoglobin dissociation curve is shifted to the left by alkalosis, hypocapnia, hypothermia, and decreased levels of 2,3,-DPG.

At a PO_2 greater than 90 mm Hg, hemoglobin is saturated irrespective of any shift in the hemoglobin dissociation curve. Similarly at an extremely low PO_2, shifts in the curve have little effect. The rightward shift that occurs in the tissues as the pH decreases and carbon dioxide levels increase improves tissue oxygen unloading. The leftward shift that occurs in the pulmonary capillaries does not adversely affect oxygen uptake because oxygen levels are still sufficiently high to saturate hemoglobin.

Deoxyhemoglobin takes up oxygen and releases H^+ and carbon dioxide as it traverses the pulmonary capillaries. The H^+ released inside the erythrocyte combines with bicarbonate to form carbonic acid which is then dehydrated to form additional carbon dioxide and water. Carbon dioxide diffuses out of the red cells into the alveoli and is excreted by the lungs. This results in the conversion of venous blood with a pH of 7.2, PCO_2 of 45 mm Hg, and PO_2 of 40 mm Hg to arterial blood with a pH of 7.4, PCO_2 of 40 mm Hg, and PO_2 of 97 mm Hg.

Fig. 6-15. *Shift of the oxygen dissociation curve by pH, PCO_2, temperature, and 2,3-diphosphoglycerate (DPG). A shift to right indicates increased hydrogen ion, increased carbon dioxide, increased temperature, and/or increased 2,3-DPG. (From J. B. West.* Respiratory Physiology: The Essentials *[3rd ed.]. P. 71. © 1985 The Williams & Wilkins, Co., Baltimore.)*

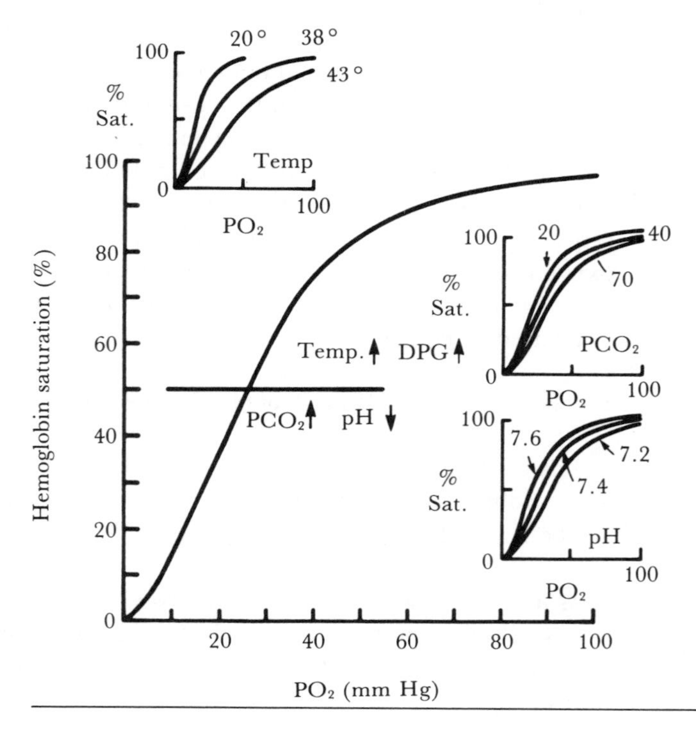

In the tissue capillaries the arterial blood enters an area of low PO_2, high PCO_2, and low pH, all of which act to facilitate oxygen unloading. Deoxyhemoglobin then combines with H^+ (by protonation of the imidazole group of histidine) and with carbon dioxide (by formation of carbaminohemoglobin). The low pH and high PCO_2 facilitate oxygen unloading by causing a rightward shift in the hemoglobin dissociation curve. Thus hemoglobin acts both to supply the tissues with oxygen and to buffer the excretory acid load of the tissues.

QUESTIONS

26. If minute oxygen consumption ($\dot{V}O_2$) at rest is equal to 200 ml per minute, what would the resting cardiac output have to be to provide that amount of oxygen to the periphery when
    A. Hemoglobin is not present (assume that the tissues can extract 100 percent of the dissolved oxygen).
    B. The concentration of hemoglobin is 15 g per deciliter, PaO_2 is equal to 90 mm Hg, and the $P\bar{v}O_2$ is equal to 40 mm Hg.
27. If the oxygen consumption doubles and blood flow remains constant, which of the following is true?
    A. $P(a\text{-}v)O_2$ is doubled.
    B. $C(a\text{-}v)O_2$ is doubled.
    C. $P(a\text{-}v)O_2$ is halved.
    D. $C(a\text{-}v)O_2$ is halved.
    E. Cardiac output doubles.
28. True or false.
    A. Hemoglobin binds oxygen with greater affinity than it binds carbon monoxide.
    B. 2,3-DPG causes hemoglobin to bind oxygen less tightly.
    C. A change in pH from 7.2 to 7.4 causes hemoglobin to bind oxygen more avidly.
    D. Increasing temperature from 37°C to 38°C causes hemoglobin to release additional oxygen (desaturate).
    E. Hemoglobin saturation varies directly with PO_2.
29. Diffusing capacity for oxygen = 21 ml/minute/mm Hg
    Arterial PO_2 = 100 mm Hg
    Minute oxygen consumption = 210 ml/minute

Calculate the $P(A\text{-}a)O_2$. The normal $P(A\text{-}a)O_2$ is 15 to 20 mm Hg and is age dependent.

REGULATION OF RESPIRATION

Regulation of respiration includes coordinated activity on the part of the central regulatory centers, the central and peripheral chemoreceptors, lung receptors, and the respiratory muscles.

CENTRAL REGULATORY CENTERS. Respiration is a reflex activity in humans. The rate, depth, and rhythm of inspiration and expiration are directly controlled by the inspiratory and expiratory centers in the medulla and pons. The cortex can override these centers for voluntary control on a temporary basis, as in voluntary breath holding or hyperventilation.

Three major respiratory centers are located in the brainstem.

1. The medullary respiratory center contains inspiratory and expiratory cell groups which predominantly effect the respiratory rhythm.
2. The apneustic center is located in the lower pons. Sectioning the brain of an

experimental animal just above the site of the apneustic center results in inspiratory gasps or apneuses interrupted by transient expiratory efforts.

3. The pneumotaxic center is located in the upper pons and acts to inhibit inspiration and regulate the respiratory rate.

In order to function properly, the central regulatory centers are dependent on central and peripheral sensory receptors for information regarding the effectiveness of the respiratory efforts and dependent on the neuromuscular effector organs to actually implement changes in respiratory effort.

CENTRAL CHEMORECEPTORS. The central chemoreceptors are located on the ventral surface of the medulla. These receptors respond to changes in $[H^+]$ and are involved in the minute-by-minute control of ventilation. An increase in the concentration of H^+ in the brain extracellular fluid (ECF) stimulates ventilation while a decrease inhibits it. Carbon dioxide levels affect the central chemoreceptors only to the extent that H^+ ion concentration is affected.

Cerebrospinal fluid (CSF) composition is the most important determinant governing the H^+ and carbon dioxide concentrations seen by the central receptors. Local metabolism and adequacy of perfusion can also affect the composition of the ECF bathing the central receptors.

The composition of the CSF, in turn, is governed by the blood-brain barrier and by active pumping of ions by the choroid plexi. The blood-brain barrier separates the blood and CSF compartments. This barrier is relatively impermeable to H^+ and HCO_3^- although carbon dioxide readily diffuses across the barrier. When blood PCO_2 rises, carbon dioxide diffuses into the CSF from the cerebral blood vessels and liberates H^+ which stimulates the central chemoreceptors and ultimately results in hyperventilation.

Normally, the pH of CSF is 7.32 and is much more labile than blood pH for a given change in PCO_2. The CSF has less protein and, therefore, less buffering capacity. Because the choroid plexus contains active HCO_3^- secretory pumps, chronic elevations in PCO_2 may be compensated for by increased pumping of HCO_3^- into the CSF. Furthermore, this pumping occurs more rapidly and more completely than does renal compensation.

PERIPHERAL CHEMORECEPTORS. The carotid body chemoreceptors located at the bifurcation of the common carotids and the aortic bodies located above and below the aortic arch constitute the peripheral chemoreceptors.

The carotid body chemoreceptors respond to decreases in arterial PO_2 and pH, and to increases in PCO_2 by increasing the rate and depth of respiration. The carotid bodies have the greatest blood flow per unit tissue of any tissue in the body and are, therefore, able to maintain a very low $C(a\text{-}v)O_2$ despite a high metabolic rate. Because of the very small $C(a\text{-}v)O_2$ difference, the carotid receptors respond to arterial PO_2 rather than venous PO_2.

The carotid body chemoreceptors are responsible for all increases in ventilation in response to arterial hypoxia. Responses to decreases in PaO_2 are nonlinear, and little response of any kind is seen until the PaO_2 is less than 100 mm Hg (Figs. 6-16 and 6-17). Complete loss of hypoxic drive has been shown with bilateral carotid body resection.

The aortic body chemoreceptors respond to increased levels of carbon dioxide by increasing the rate and depth of respiration. In man the carotid bodies not the aortic bodies respond to a fall in arterial pH. The response occurs regardless of the cause of the acid-base disturbance, whether respiratory or metabolic.

Only 20 percent or less of the ventilatory response to increased carbon dioxide is attributed to peripheral chemoreceptors. They may be important, however, in

Fig. 6-16. *Approximate effects on alveolar ventilation of changing the concentrations of carbon dioxide, hydrogen ions, and oxygen in the arterial blood when only one of the humoral factors is changed at a time and the other two are maintained at absolutely normal levels. (From A. C. Guyton.* Textbook of Medical Physiology *[7th ed.]. Philadelphia: Saunders, 1986. P. 510.)*

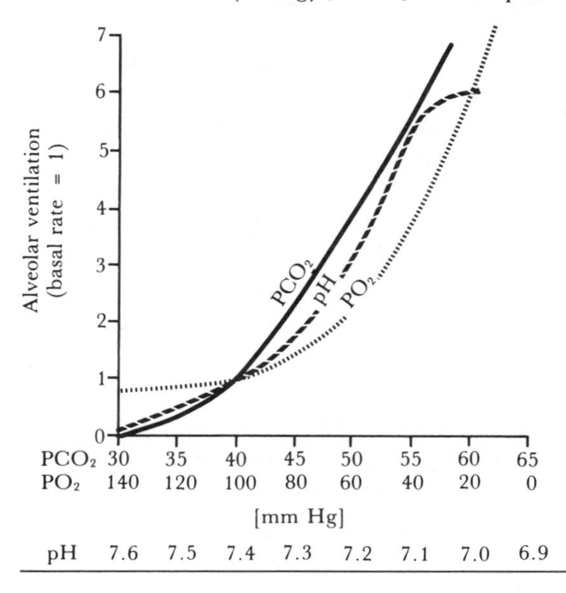

Fig. 6-17. *Hypoxic response curves. Note that when the PCO_2 is 36 mm Hg, almost no increase in ventilation occurs until the PO_2 is reduced to about 50 mm Hg. (Modified from H. H. Loeschke, and K. H. Gertz.* Arch. Ges. Physiol., *267:460, 1958.)*

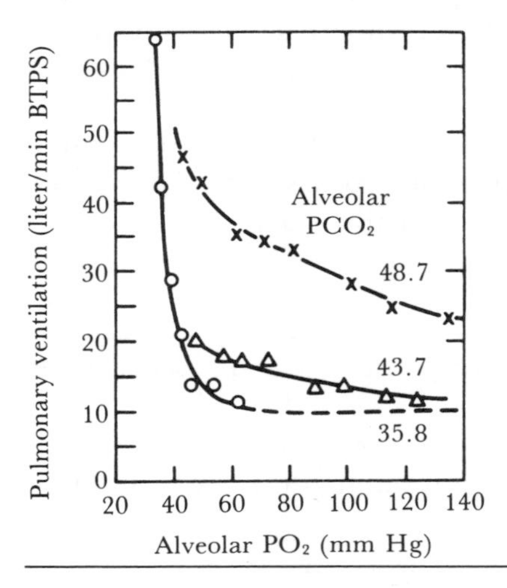

matching ventilatory responses to abrupt changes in PCO_2, and they do account for the entire ventilatory response to hypoxia.

Ventilatory responses to changes in PO_2, PCO_2, and pH may be potentiated or abrogated by changes in the other two components of ventilatory drive.

PULMONARY RECEPTORS. There are three types of pulmonary receptors described. Afferent impulses from all three receptor types travel to the CNS in the vagus nerve.

1. Pulmonary stretch receptors are located in the smooth muscle and respond to distension of the lung with sustained firing. Stimulation of these receptors results in slowing of the respiratory rate due to an increase in expiration time.

This is known as the Hering-Breur reflex. The normal inspiration to expiration ratio (I : E) is 1 : 2.

2. Irritant receptors are believed to lie interposed between airway epithelial cells and are stimulated by noxious gases, cigarette smoke, dusts, and cold air. Reflex effects include bronchoconstriction and hyperpnea.
3. J receptors refer to juxtacapillary receptors. Stimulation results in rapid shallow breathing with more intense stimulation resulting in apnea. Engorgement of pulmonary capillaries and increases in the interstitial fluid volume of the alveolar walls activate these receptors.

VENTILATORY RESPONSES TO ALTERATIONS IN PCO_2, PO_2, AND PH. Under normal conditions, PCO_2 is the most important variable in the regulation of ventilation. $PaCO_2$ is normally maintained to within a few mm Hg of normal value.

Figure 6-18 illustrates the ventilatory response to rising concentrations of inspiratory CO_2 (with alveolar PO_2 held constant). Note that alveolar end tidal PO_2 and PCO_2 are generally accepted to reflect arterial levels because the end tidal air is presumably alveolar air that has equilibrated with arterial blood.

Altering alveolar PO_2 while maintaining alveolar PCO_2 at 36 mm Hg results in the ventilatory response curve shown in Figure 6-17. This illustrates that alveolar PO_2 can be reduced to approximately 50 mm Hg prior to any appreciable increase in ventilation. Increasing the PCO_2 increases the ventilation at any given PO_2; however, when the $PaCO_2$ is increased, any reduction in PO_2 below 100 mm Hg causes some stimulation of ventilation in contrast to the situation when PCO_2 is normal.

The role of hypoxia in the day-to-day control of ventilation is, therefore, fairly small. However, in some patients with pulmonary disease, the hypoxic drive becomes very important (as indeed it does in normal people when ascending to high altitudes).

The cause of increased ventilation with exercise is unknown. The PO_2 increases slightly with exercise, PCO_2 decreases, and pH is generally unchanged until lactate accumulation occurs.

ACCLIMATIZATION TO HIGH ALTITUDE

The following responses occur on ascent to high altitude:

1. Acutely, alveolar ventilation increases by up to 65 percent (the response is limited by the resulting decrease in PCO_2 and increase in pH).

Fig. 6-18. *Ventilatory response to different concentrations of inspired CO_2. Note that airway obstruction reduces the response. (From J. B. West.* Respiratory Physiology: The Essentials *[3rd ed.]. P. 122. © 1985 The Williams & Wilkins Co., Baltimore.)*

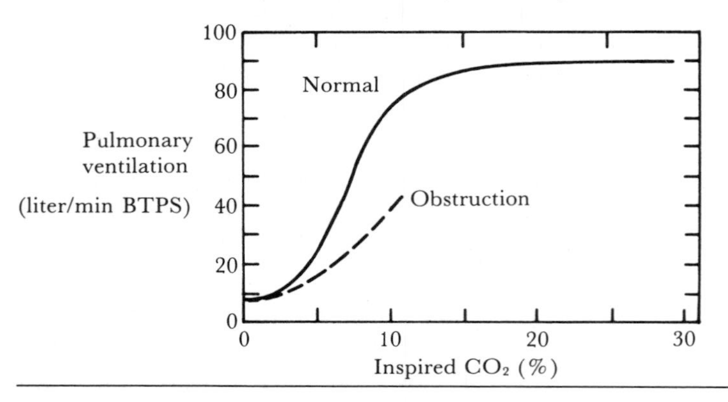

2. Chronically, alveolar ventilation may increase by 3- to 7-fold as the body apparently develops tolerance for the resulting alkalosis and decreased PCO_2.
3. Hematocrit increases from 40-to-45 to 60-to-65 percent (polycythemia).
4. Hemoglobin concentration increases by 20 to 30 percent.
5. Circulatory blood volume increases by up to 50 to 90 percent.
6. Levels of 2,3-DPG increase causing decreased affinity of hemoglobin for oxygen thus enhancing tissue oxygen unloading.
7. Diffusing capacity for oxygen increases in a manner similar to that seen in exercise. This effect may be secondary to the increased capillary blood volume, to increased number and size of pulmonary capillaries, or to increased pulmonary artery pressure.
8. Increased tissue vascularity occurs.
9. Acclimatization at the cellullar level, at least in animals born at a high altitude, may also occur with increased numbers of mitochondria and increased levels of certain cellular oxidative enzymes.

QUESTIONS

30. A patient with longstanding carbon dioxide retention and chronic pulmonary obstructive disease (COPD) has an elevated $PaCO_2$, acidic arterial pH, and an abnormally low ventilatory rate for his blood gases yet the CSF pH is normal. How can this be?
31. Dyspnea is frequently associated with left heart failure and interstitial lung disease. It is associated with stimulation of:
    A. Pulmonary stretch receptors
    B. J receptors
    C. Irritant receptors
    D. All of the above
32. The Hering-Breur reflex is associated with which of the above receptors.

GENERAL AND CELLULAR NONRESPIRATORY FUNCTIONS

The lungs have metabolic, immunological, and filtration functions. The lungs filter out small blood clots (small pulmonary emboli) preventing the serious complications of systemic embolization. Bronchial secretions contain secretory immunoglobulins (IgA) and other substances which serve to resist infection. Pulmonary alveolar macrophages are actively phagocytic removing bacteria and small particles inhaled by the lungs. They also perform the regular functions of macrophages (e.g., attraction of polymorphonuclear leukocytes and release of vasoactive and chemotactic substances).

The metabolic and endocrine functions of the lung include production of surfactant, conversion of angiotensin I to angiotensin II, and synthesis of prostaglandins, histamine, kallikrein and a number of other substances. The lungs contain a fibrinolytic system that breaks up clots in the pulmonary veins. Metabolic breakdown of certain substances also occurs in the lungs, the significance of which is unknown.

QUESTIONS

33. Which of the following would be expected to produce acutely a larger increase in the rate and depth of ventilation. Why?
    A. Infusion of HCl intravenously
    B. Breathing carbon dioxide
34. A patient on 100% O_2 has a PaO_2 of 100 mm Hg. Which of the following is probably true.
    A. Patient has totally normal lungs.
    B. Pulmonary fibrosis is present.
    C. Right-to-Left shunting is present.

D. Hemoglobin Kansas (an abnormal hemoglobin with increased affinity for oxygen) is present.

35. Ascent to high altitude may result in an increase (decrease) in the pH of the blood secondary to hyper (hypo) ventilation. This results in
    A. Respiratory alkalosis
    B. Respiratory acidosis
36. The acute hyperventilatory response to altitude is followed by normalization of pH as acclimatization occurs. The acclimatization occurs because
    A. The kidneys respond to the change in pH by retaining bicarbonate.
    B. The kidneys respond to the change in pH by increasing excretion of bicarbonate.
    C. Enhanced erythropoiesis results in polycythemia and increased buffering as the hemoglobin levels rise.
    D. Levels of 2,3-DPG increase.
37. What happens to an acclimatized person when he returns to sea level from a mountain top?

ANSWERS

1. A. 16
   B. Resistance will increase by 8-fold rather than 16.
2. Double.
3. Because carbon dioxide excretion would cease and death secondary to carbon dioxide retention would occur.
4. The four major cell types in the lung and their major functions are
   (1) Type I pneumatocytes are epithelial cells lining the alveoli.
   (2) Type II pneumatocytes are also alveolar epithelial cells. They produce surfactant and contain osmiophilic granules.
   (3) Endothelial cells line the capillaries in the lung.
   (4) Alveolar macrophages phagocytize dust and other particles which reach the alveolar walls.
5. C. Alveolar hypocapnia occurring secondary to decreased perfusion causes ventilation to be shunted to better perfused areas of the lung.
6. A. Vasoconstrict, vasodilate
   B. Vasoconstrict, vasodilate
   C. Vasoconstrict, vasodilate
7. (1) A
   (2) B
   (3) C
8. The tendency of the chest wall to expand and the tendency of the lungs to collapse.
9. If the lungs are entirely normal the results should be quite similar. However, if any appreciable region of lung is not ventilated, i.e., does not communicate with the airways, then the helium dilution method would underestimate the volume of the lung because helium must be able to diffuse into all regions of the lung and equilibrate for the estimate to be accurate. The whole body plethysmography technique does not require that all spaces in the lung communicate. Plethysmography would, therefore, be preferred in a patient with emphysema, for instance.
10. C
11. $(760 - 47)0.21 - (40 / 1) = 109.73$
12. Compliance is halved.
13. One mole occupies 22.3 liters at standard temperature and pressure (STP). Because the solubility of carbon dioxide in plasma is 0.03 mmoles CO_2/mm Hg/liter plasma.

(22.3 liters/mole) (1000 ml/liter) (0.03 mmoles CO_2/mm Hg/liter plasma) (1 mole/1000 mmoles) = 0.67 ml CO_2/mm Hg/liter plasma

Because the solubility of oxygen in plasma is 0.003 ml/mm Hg/liter plasma, carbon dioxide is roughly 300 times as soluble as oxygen in plasma.

14. (1) B
    (2) C
    (3) A
    (4) D
15. A. $P_AO_2 = F_IO_2\ (BP - 47) - (P_ACO_2/RER)$
    $= 0.21\ (760 - 47) - (P_ACO_2/RER)$
    $= 150 - (P_ACO_2/R)$
    For carbohydrate with a RER of 1.0
    $P_AO_2 = 150 - 40/1 = 110$ mm Hg
    $P(A\text{-}a)O_2 = 110 - 95 = 15$ mm Hg
    B. For protein with RER = 0.8
    $PAO_2 = 150 - P_ACO_2/0.8 = 150 - 40/1 = 110$ mm Hg
    $P(A\text{-}a)O_2 = 100 - 95 = 5$ mm Hg
    C. For fat with RER = 0.7
    $PAO_2 = 150 - P_ACO_2/0.7 = 150 - 57 = 95$ mm Hg
    $P(A\text{-}a)O_2 = 95 - 95 = 0$ mm Hg
16. A. Abnormal with a $P(A\text{-}a)O_2$ of 50. A decrease in ventilation secondary to a decrease in ventilatory drive could result in this picture (e.g., a nervous disorder leading to a decreased rate of breathing).
    B. Normal with a $P(A\text{-}a)O_2$ of 0. The range of normalcy increases with advancing age from 0 to 25.
17. 0
18. The respiratory membrane is composed of surfactant, alveolar epithelium, epithelial basement membrane, interstitial space, capillary basement membrane, and capillary endothelium (see Fig. 6-3).
19. A, B, D, and E
20. $P_ACO = (700 - 47) \times 0.001 \cong 0.7$ mm Hg (ignoring any dilution by carbon dioxide; how does this affect the answer?)
    $PaCO = 0$ mm Hg
    Therefore, the gradient is 0.7 mm Hg and the diffusing capacity is 28/0.7 = 40 ml/mm Hg/minute.
21. In humans the ratio of ventilation to perfusion is greatest in the upper lobes of the lungs. Bats hang upside down, so the situation is reversed: TB develops in the base of the lungs in bats.
22. A. Oxygen content of the first liter = (15 g Hb/dl plasma) (1.34 ml O_2/g Hb) (0.5) + (27 mm Hg) (0.003 ml O_2/mm Hg/dl blood) = 10.13 ml O_2/dl. A similar calculation for the second liter yields a value of 19.9 ml O_2/dl.
    B. PaO_2 = 50% and 97.5%, respectively.
    C. $$\frac{10.13 \text{ ml } O_2/\text{dl} + 19.9 \text{ ml } O_2/\text{dl}}{2} = 14.87 \text{ ml } O_2/\text{dl}$$
    Looking at Table 6-1 it is evident that such an oxygen content would correspond to a hemoglobin saturation of a little less than 75 percent or a PaO_2 of approximately 40 mm Hg. However, the exact saturation can be calculated as

$$(1.34)\ (15)\ (x) + (x)\ (0.003) = 14.87$$
$$20.103x = 14.87$$
$$x = 0.739, \text{ i.e., } 73.9 \text{ percent saturation}$$

Note that the arithmetical mean of the PaO_2 of the two liters would be equal to (100 + 27) / 2 or 63.5 mm Hg, a far cry, indeed, from the actual value of somewhat less than 40 mm Hg.

D. (20 + 40) / 2 = 30. For carbon dioxide, taking the mean works!

23. Obviously the alveolar oxygen is not able to equilibrate with the blood for some reason other than physiological shunting. If inadequate ventilation relative to perfusion is not the problem, then perhaps too much ventilation relative to perfusion is. If ventilation is being wasted on areas of lung that cannot exchange oxygen, then one would also expect P_AO_2 to exceed that of the arterial blood (i.e., physiological dead space is being ventilated).

24. The term anatomical shunt refers to shunting of blood that occurs in the heart. Atrial and ventricular septal defects may cause anatomical shunting. Other examples of anatomical shunts include pulmonary artery to left atrial shunting and coronary venous blood dumping into the left ventricle via the thebesian veins.

25. A. Remember, $PaCO_2 = P_ACO_2$

$V_D/\dot{V}_T = (P_ACO_2 - P_ECO_2) / P_ACO_2$

$= 45 - 30/30 = 0.5$

Because $\dot{V}_T = 500$ ml, $V_D = 250$ ml

B. 0.2 to 0.35

26. A. In the absence of hemoglobin the oxygen content of blood is equal to the amount dissolved in blood, i.e., 0.003 ml O_2/mm Hg/dl blood. Assuming a PaO_2 of 100 mm Hg and using the Fick equation

Cardiac output = minute oxygen consumption/$C(a\text{-}v)O_2$

= 200 ml/minute 0.3 ml O_2/mm Hg/dl blood

= 666.67 deciliters blood/minute

= 66.67 liter/minute

B. At a PaO_2 of 90 mm Hg, hemoglobin is roughly 97 percent saturated; hence,

Total arterial oxygen content = (15) (0.97) (1.34) + (0.003) (90) = 19.77 ml O_2/dl blood

Total venous oxygen content = (15) (0.75) (1.34) + (40) (0.003) = 15.2 ml O_2/dl blood

$C(a\text{-}v)O_2$ = 19.77 − 15.2 = 4.57 ml O_2/dl blood

Cardiac output = minute oxygen consumption/$C(a\text{-}v)O_2$ = 200/4.57 = 43.76 dl/minute = 4.376 liters/minute

27. B. $C(a\text{-}v)O_2 = VO_2$/cardiac output; if VO_2 doubles and cardiac output is held constant then $C(a\text{-}v)O_2$ also must double.

28. A. False

B. True

C. True

D. True

E. False

29. (Diffusing capacity) × $P(A\text{-}a)O_2$ = minute oxygen consumption (21)x = 210

x = 10 mm Hg

30. Bicarbonate pumping into the CSF by the choroid plexus has compensated for his chronic respiratory acidosis.

31. B

32. A
33. B. Carbon dioxide enters the CSF and brain ICF rapidly, while hydrogen ion enters slowly.
34. C
35. Increases, A. Blood pH rises as hyperventilation blows off carbon dioxide. Hyperventilation occurs in order to maintain PaO_2 in the face of the low PO_2 in the atmosphere.
36. B
37. Metabolic acidosis due to low total body bicarbonate (pH may or may not be acidic, but bicarbonate levels are depressed until the kidney can reclaim enough bicarbonate to replace that excreted on the mountain).

BIBLIOGRAPHY

Ganong, W. F. *Review of Medical Physiology* (12th ed.). Los Altos, CA: Lange, 1985.

Guyton, A. C. *Textbook of Medical Physiology* (7th ed.). Philadelphia: Saunders, 1986.

Stanford University School of Medicine Faculty. *Respiratory Physiology Syllabus* (lecture handout), 1984.

West, J. B. *Best and Taylor's: Physiological Basis of Medical Practice* (11th ed.). Baltimore: Williams & Wilkins, 1985.

West, J. B. *Respiratory Physiology: The Essentials* (3rd ed.). Baltimore: Williams & Wilkins, 1985.

7 Acid-Base Physiology

Charles H. Tadlock

GENERAL CONCEPTS

ACIDS AND BASES DEFINED. Brønsted defined acids as proton donors and bases as proton acceptors. For each acid or base there is a conjugate base or acid from which it differs by the gain or loss of a proton.

REGULATION OF pH. Three basic mechanisms constitute the body's ability to regulate acid-base status.

1. Excretion of hydrogen ions and reabsorption of bicarbonate ions by the kidneys
2. Excretion of carbon dioxide by the lungs
3. Buffering of hydrogen ions by the weak acids and bases that comprise the body's buffering systems

These regulatory mechanisms are necessary because free hydrogen ion concentration, $[H^+]$, is important in the regulation of enzyme activity and in the maintenance of fluid and electrolyte balance. Intracellular $[H^+]$ is the true arbiter of the effects of acid-base status on the regulation of body systems. Unfortunately, $[H^+]$ of the intercellular fluid (ICF) is difficult to measure and to interpret at this time; therefore, the assumption is made that $[H^+]$ in the ICF and the $[H^+]$ in the extracellular fluid (ECF) compartments are in balance, and that changes in one are reflected, at least qualitatively, by changes in the other. This assumption is not always valid.

Body $[H^+]$ ranges from 10^{-1} moles per liter in the gastric juice to 10^{-9} moles per liter in the pancreatic fluid. In the plasma, normal $[H^+]$ is approximately 4×10^{-8} moles per liter. The use of a logarithmic scale to simplify discussions of acid-base is appropriate. The negative logarithm to the base ten of the $[H^+]$ in moles per liter is called the pH.

$$ph = -\log [H^+]$$

Normal arterial pH ranges from 7.35 to 7.45 with venous blood being predictably slightly more acidic. Estimates place ICF pH at approximately 7.0.

The body's ECF is normally alkaline. It is the function of the body regulatory mechanisms to maintain this alkalinity in the face of endogenous and exogenous sources of acid and alkali including metabolic production of acids.

HENDERSON-HASSELBALCH EQUATION. If HA represents an acid and A^- its conjugate base, then the acid dissociation reaction is

$$HA \rightleftarrows H^+ + A^- \tag{7-1}$$

A strong acid dissociates completely liberating one mole of H^+ for each mole of acid present in solution. A weak acid dissociates only partially. Each acid has a dissociation constant (K_a) which describes its tendency to dissociate.

$$K_a = \frac{[H^+]\,[A^-]}{[HA]} = \frac{[\text{products}]}{[\text{reactants}]} \tag{7-2}$$

By solving for $[H^+]$

$$[H^+] = \frac{K_a [HA]}{[A^-]} \tag{7-3}$$

One can now solve this equation for pH by taking the negative logarithm of each side of the equation.

$$-\log [H^+] = -\log K_a + \log \frac{[A^-]}{[HA]}$$

$$pH = pK_a + \log \frac{[A^-]}{[HA]} \tag{7-4}$$

This form of the acid-base equilibrium is called the Henderson-Hasselbalch equation. It may also be written as

$$pH = pK_a + \log \frac{[\text{proton acceptor}]}{[\text{proton donor}]}$$

DAILY ENDOGENOUS ACID PRODUCTION. Normal metabolism yields a variety of organic acids. Breakdown of carbohydrate, fat, and protein yields lactic and pyruvic acid. Lactic acid may accumulate in large quantities as a consequence of anaerobic metabolism during heavy exercise or due to some cause of tissue hypoxia. Oxidation of sulfur-containing amino acids yields H^+ and SO_4^{2-} and oxidation of phosphoproteins including nucleoproteins yields H^+ and PO_4^{3-}. These sources of fixed acid contribute approximately 1 meq/kg/day of H^+ ions to the body (an average of 20–70 meq/day). This daily acid production is normally excreted into the urine by the kidneys.

A far larger source of acid production is oxidative metabolism. Oxidative metabolism results in the formation of approximately 12,500 meq H^+ per day as illustrated by the breakdown of glucose in the Kreb's cycle:

$$C_6H_{12}O_6 + 6\ O_2 \rightleftarrows 6\ H_2O + 6\ CO_2$$

$$CO_2 + H_20 \overset{CA}{\rightleftarrows} H_2CO_3 \rightleftarrows H^+\ HCO_3^-$$

The carbon dioxide produced is predominantly excreted by the lungs (i.e., it is a volatile or nonfixed acid). The partial pressure of carbon dioxide in the arterial plasma ($PaCO_2$) is thereby maintained at 40 mm Hg with venous blood having an average partial pressure ($PvCO_2$) of 45 mm Hg. Carbonic anhydrase (CA) is the enzyme which catalyzes the hydration of carbon dioxide to carbonic acid.

Protein metabolism results in production of excess fixed acid which is eliminated by the kidneys via formation of an acidic urine. Consumption of fruits, in contrast, results in excess production of alkali which must be eliminated via the formation of an alkaline urine.

QUESTIONS

1. Calculate the pH in
   A. Gastric juice
   B. Pancreatic fluid
   C. 1 molar HC1
   D. 1 molar acetic acid ($pk_a = 4.7$)

2. Calculate the $[H^+]$ at a pH of 7.1. How does this compare to the $[H^+]$ of blood?
3. At what pH will the $[H^+]$ of blood be halved?
4. A strong base has a ______ conjugate acid, while a strong acid has a correspondingly ______ conjugate base.
5. Identify the following as either an acid or a base, then identify its conjugate base or acid, respectively, at physiological pH.
   A. HCl
   B. CH_3COO^-
   C. HCO_3^-
   D. NH_4^+
   E. H_3O^+
   F. H_2SO_4
   G. PO_4^{3-}
6. Define fixed acid and unfixed (volatile) acid.
7. Carbon dioxide in the plasma is carried primarily as
   A. Dissolved CO_2
   B. HCO_3^-
   C. H_2CO_3
8. Lactic acid is eliminated by
   A. Excretion by the kidneys
   B. Metabolism to HCO_3^- by the kidneys, liver, and muscle
9. How are beta-hydroxybutyric acid and acetoacetic acid eliminated?

ROLE OF BUFFERS

BUFFERS DEFINED. A buffer is a solution of two or more chemicals which minimizes changes of pH in response to addition of acid or base. Many buffers consist of a weak acid and its conjugate base. Such a buffering system is able to bind or release H^+ in solution and thus minimize fluctuations in $[H^+]$ over its effective buffering range (Fig. 7-1).

A buffer is most effective (i.e., the change in pH per unit of acid or base added is least) when the pH of the solution equals the pK_a of the buffer. The pK_a of a

Fig. 7-1. *Titration curves for the bicarbonate and inorganic phosphate buffers in a closed system. Under these conditions, when the concentration of carbon dioxide cannot be kept low through diffusion to the outside, the bicarbonate system is less efficient as a buffer than phosphate. It is the ability of the lungs to eliminate the carbon dioxide that makes the bicarbonate system such an efficient physiological buffer. Reprinted with permission from R. F. Pitts.* Physiology of the Kidney and Body Fluids *[3rd ed.]. copyright © 1974 by Year Book Medical Publishers, Inc., Chicago.)*

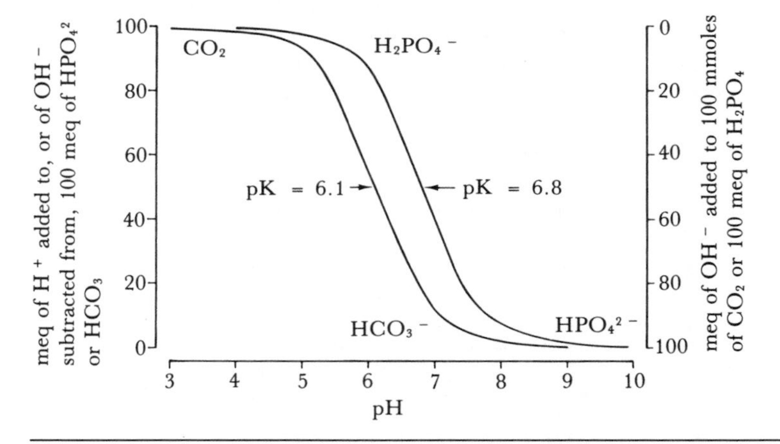

buffer is that pH at which equal amounts of the buffer exist as the acidic and the basic forms. This can be shown using the Henderson-Hasselbalch equation:

$pH = pK_a + \log [base]/[acid]$, if the $pH = pK_a$ then,

$pH - pK_a = O = \log [A^-]/[HA]$ and

$10^0 = 1 = [A^-]/[HA]$ or $[A^-] = [HA]$

Intuitively, such a buffer couple should be most effective for buffering addition of either an acid or a base when half of the buffer exists as the acidic and half as the basic form, i.e., when the $pH = pK_a$. It follows that the ideal extracellular buffer would have a pK_a of 7.4. In fact, buffers are quite effective when the pH of the solution is within 1 pH unit of the pK_a of the buffer and somewhat effective when the pK_a is within 2 pH units. Figure 7-1 illustrates the typical sigmoidal titration curve of a buffer. See Table 7-1 for the pK_as of important physiological buffers.

The physiological role of buffers is to stabilize $[H^+]$ during the time required for the excretion of excess alkali or acid by the kidneys and lungs. The efficacy of a buffered system over that of a nonbuffered system in minimizing pH changes in the physiology range is apparent when comparing the titration curves in Figure 7-2. If a strong acid such as HC1 is added to a nonbuffered system, then for each molecule of a strong acid added to the solution, H^+ also increases by one. Hence relatively small additions of acid lead to marked decreases in pH. In the presence of a buffer, the increase in $[H^+]$ resulting from the addition of the strong acid acts by mass action to protonate the conjugate base of the buffer forming more of the buffer acid. The protons or H^+ ions which bind to the buffer are no longer free in solution and, therefore, do not contribute to the $[H^+]$. The pH does drop slightly, of course, because the ratio $[A^-]$ to $[HA]$ has decreased as A^- is converted to HA by the strong acid, and a small amount of the additional H^+ remains free. Conversely, if a strong base is added, the buffer releases H^+. Thus fluctuations in $[H^+]$ are minimized over the effective range of the buffer.

THE ISOHYDRIC PRINCIPLE. The isohydric principle states that buffers in a common solution must be in equilibrium with one another because they are all

Table 7-1. *pK_as of Representative Buffers*

Name of Buffer	Buffer Pair	pK_a
Acetic acid	CH_3COO^-/CH_3COOH	4.7
Acetoacetic acid	$CH_3COCH_2COO^-/CH_3COCH_2COOH$	3.8
Ammonia	NH_3/NH_4^+	9.2
β-Hydroxybutyric acid	$CH_3CH_2CHOHCOO^-/CH_3CH_2CHOHCOOH$	4.8
Carbonic acid*	HCO_3^-/H_2CO_3	3.57
	HCO_3^-/CO_2	6.1 (pK_a')
	CO_3^{2-}/HCO_3^-	9.8
Hemoglobin		
Deoxyhemoglobin	$Hb^{n-}/HHb^{(n-1)-}$	7.9
Oxyhemoglobin	$HbO_2^{n-}/HHbO_2^{(n-1)-}$	6.7
Lactic acid	$CH_3CHOHCOO^-/CH_3CHOHCOOH$	3.9
Phosphoric acid	$H_2PO_4^-/H_3PO_4$	2.0
	$HPO_4^{-2}/H_2PO_4^-$	6.8
	PO_4^{3-}/HPO_4^{-2}	12.4

* Carbon dioxide does not fit the Brønsted definition of an acid. It is an acid anhydride. Because carbon dioxide is found in such high concentrations in blood, it is potentially a much larger source of H^+ than its intermediary, H_2CO_3. The pK_a of H_2CO_3 above is 3.57. However in the presence of CO_2 the apparent pK_a (pK_a') is 6.1.

Fig. 7-2. *Titration curves resulting from addition of strong acid or base to water* (solid line) *or a weak buffer solution with pK_a of 7.0* (dashed line). *Note that over a wide pH range (including the range compatible with life) small changes in $[H^+]$ lead to large changes in pH in an unbuffered system, hence the almost flat line between pH 3.0 to 11. In contrast, the buffered system is quite effective in minimizing pH changes in response to addition of acid or alkali over its effective buffering range.*

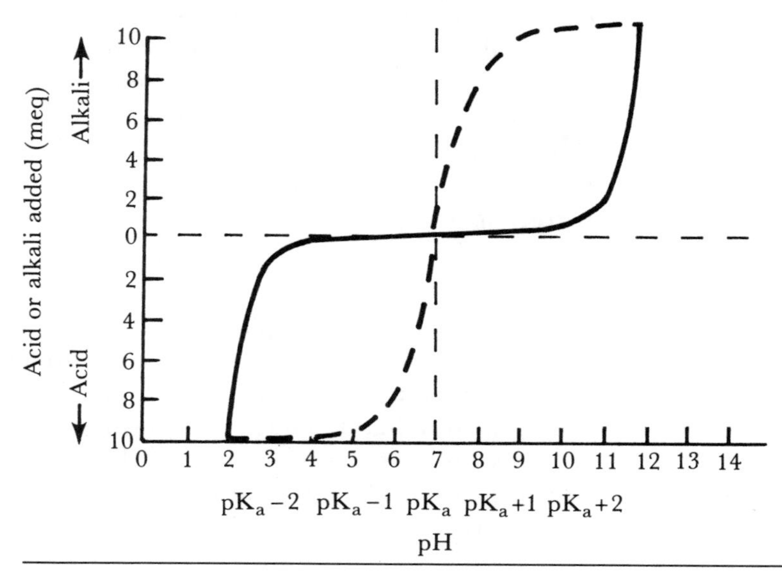

in equilibrium with the same H^+. In chronic states, it is assumed that qualitatively similar changes occur in all body water compartments when an alteration in acid-base status occurs. Thus the ratio of base to acid for each buffer changes proportionately as H^+ is added or subtracted from the body.

QUESTIONS

10. Calculate the ratio of A^-/HA when
    A. $pH = pK_a$
    B. $pH = pK_a + 1$
    C. $pH = pK_a + 2$
    D. $pH = pK_a + 3$
    E. $pH = pK_a - 1$
    F. $pH = pK_a - 2$
    G. $pH = pK_a - 3$
11. The region of the titration curve corresponding to the effective buffering range is characterized by
    A. Steep slope
    B. A width corresponding to plus or minus one pH unit from the pK_a of the buffer
    C. Both of the above
    D. Neither of the above
12. $$K_{a1}\frac{0.03\ PCO_2}{[HCO_3^-]} = K_{a2}\frac{[H_2SO_4]}{[HSO_4^-]} = K_{a3}\frac{[HA]}{[A^-]}$$
    is a statement of the
    A. Law of mass action
    B. Henderson-Hasselbalch equation
    C. Isohydric principle
    D. Ohm's law
    E. Stoichiometry
13. At any given pH, reduced blood (deoxyhemoglobin) contains more or less HCO_3^- than oxygenated blood. Why?

BUFFERING SYSTEMS

CARBONIC AND NONCARBONIC ACID LOADS. When discussing the buffering capacities of various buffering systems, it is important to define what type of acid load is being discussed. By definition, a buffer system cannot buffer an acid-base disturbance for which it itself is responsible. Because the HCO_3^--CO_2 system is not only the major extracellular buffer, but also the major endproduct of oxidative metabolism, acid-base disturbances are generally divided into those in which a disturbance is due to carbon dioxide (respiratory disturbance) and those in which the disturbance is due to some other cause (metabolic disturbance). Because carbon dioxide is responsible for respiratory disturbances, the HCO_3^--CO_2 system cannot contribute toward buffering a respiratory disorder and must be ignored when determining the contribution of various buffering systems to correcting a respiratory disorder.

THE PLASMA BUFFERS: BICARBONATE AND PROTEIN. The principal plasma buffers are bicarbonate and the plasma proteins with a very small contribution by ammonia and phosphate because these are present in quite small concentrations.

The bicarbonate–carbonic acid system is the most important plasma and extracellular buffer despite its relatively inefficient pK_a of 6.1. This system alone accounts for 90 percent or more of the buffering capacity of the plasma for metabolic acid loads. The bicarbonate–carbonic acid system is important for four major reasons.

1. HCO_3^- is present in high concentration, 24 meq per liter.
2. The lungs control the level of carbon dioxide in the blood by controlling the rate of ventilation.
3. Similarly, the kidneys govern HCO_3^- levels by governing the rates of H^+ secretion and bicarbonate reabsorption.
4. The formation of carbon dioxide by oxidative metabolism is the major source of acid production in the healthy state (12,500 meq/day of H^+).

The Henderson-Hasselbalch equation for the HCO_3^+-H_2CO_3 system is also unique. The general equation

$$pH = pK_a + \log[HCO_3^-]/[H_2CO_3] \tag{7-5}$$

while accurate is misleading. As the equation

$$CO_2 + H_2O \overset{CA}{\rightleftarrows} H_2CO_3 \rightleftarrows HCO_3^- + H^+$$

indicates, the total acid available for reaction is not equal to the $[H_2CO_3]$ but to the sum of $[H_2CO_3]$ + CO_2 dissolved. Because carbon dioxide is present at approximately 200 times the concentration of H_2CO_3, the total concentration of acid is approximately equal to the concentration of dissolved carbon dioxide. The total amount of carbon dioxide (TCO_2) in solution is approximately equal to the PCO_2 times solubility of its plasma, 0.03 mmol CO_2/liter/mm Hg.

$$TCO_2 = [H_2CO_3] + [CO_2]\text{in solution} = PCO_2 \times 0.03 \text{ mmol } CO_2\text{/liter/mm Hg} \tag{7-6}$$

Substituting this into the Henderson-Hasselbalch equation:

$$pH = 6.1 + \log \frac{[HCO_3^-]}{PCO_2 \times 0.03} \tag{7-7}$$

Clinically, pH and PCO_2 are measured directly and HCO_3^- is calculated using this equation; however, during rapid changes in acid-base status, the equilibrium conditions required by this equation do not hold true. The $[HCO_3^-]$ should then be calculated by determining the TCO_2 of the blood using the following equation:

$$TCO_2 = [HCO_3^-] + [\text{dissolved } CO_2] + [\text{carbamino } CO_2] + [H_2CO_3]$$

$$[HCO_3^-] = TCO_2 - [\text{dissolved } CO_2] - [\text{carbamino } CO_2] - [H_2CO_3]$$

Carbonic anhydrase catalyzes the hydration of carbon dioxide enabling equilibrium to be established rapidly.

$$CO_2 + H_2O \overset{CA}{\rightleftarrows} H_2CO_3$$

It is present in high concentration in erythrocytes, in renal tubular cells, and on the luminal aspect of proximal tubular membranes.

Proteins are weak acids at physiological pH. They are important buffers in the plasma and red cells, and constitute the major intracellular buffer. Because its imidazole group has a pK_a of 6.0, histidine is the only amino acid with significant buffering capacity in the physiological range, i.e., between pH 6 and 8. The pK_a of cysteine's sulfhydryl group is 8.3, which is closer to normal blood pH than the pK_a of the imidazole group, but this sulfhydryl group is often in disulfide linkages (such as in the dimer cystine) and, therefore, unavailable to act as a buffer.

Phosphate is also present in plasma but due to its low concentration (1–2 meq/liter) is not an effective buffer there. Table 7-2 illustrates the relative contributions of the plasma buffers in response to a metabolic acid load.

RED CELL BUFFERS. The protein hemoglobin is found in extremely high concentrations in red cells (15 g/dl). Although hemoglobin is technically an intracellular buffer, the red cell membrane is so permeable to H^+ that hemoglobin is functionally a plasma buffer.

Every liter of venous blood carries 1.68 mmoles of extra carbon dioxide. This *additional* carbon dioxide is carried as follows (Fig. 7-3):

65 percent is carried as HCO_3^-
27 percent as carbamino hemoglobin
8 percent is dissolved

Table 7-2. *Buffering of a Metabolic Acid Load*

Compartment	Major Buffers (% buffering in compartment)		% of Total Body Buffering
Plasma	Bicarbonate	(92)	13
	Plasma proteins	(8)	
	Inorganic phosphate	(<1)	
Red cell	Hemoglobin	(60)	6
	Bicarbonate	(30)	
	Inorganic phosphate and intracellular proteins	(10)	
Interstitial fluid	Bicarbonate	(>99)	30
	Inorganic phosphate	(<1)	
Intracellular fluid and bone	Proteins and inorganic phosphate		51

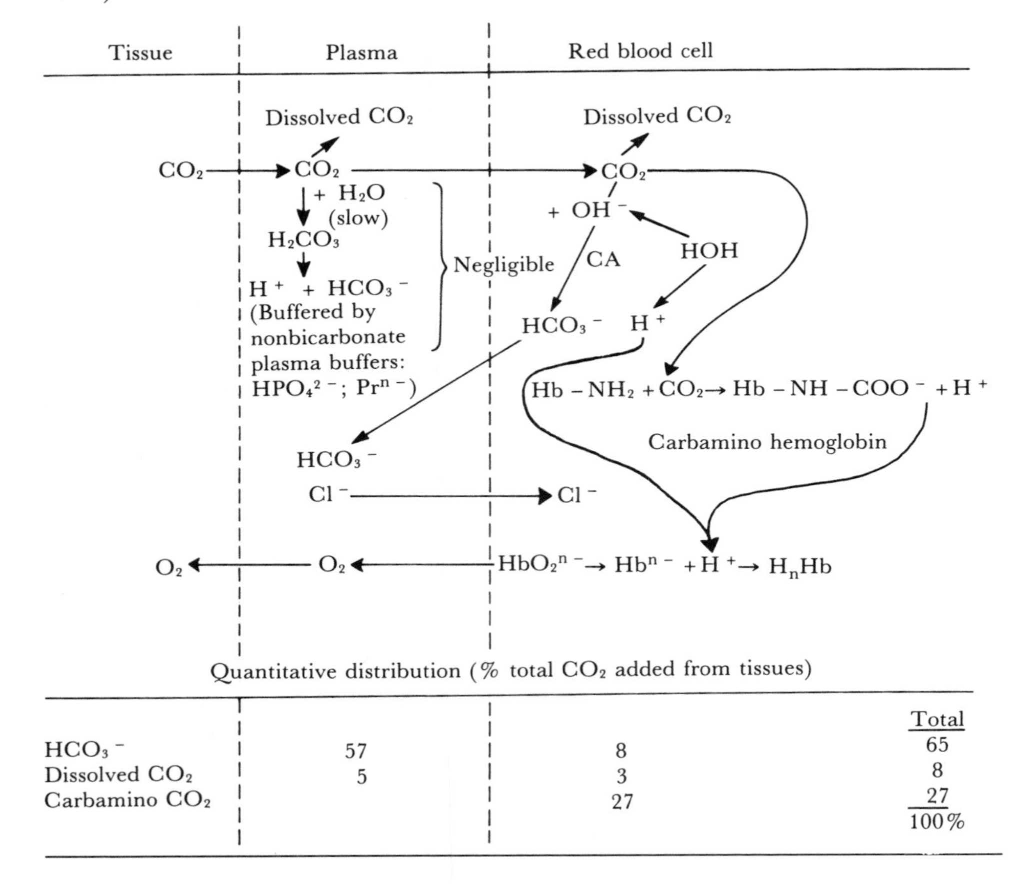

	Plasma	Red blood cell	Total
HCO_3^-	57	8	65
Dissolved CO_2	5	3	8
Carbamino CO_2		27	27
			100%

Fig. 7-3. *Transport of carbon dioxide and buffering of H^+ by the blood. (Adapted from H.W. Davenport.* The ABC of Acid-Base Chemistry *[6th ed.]. Chicago, © 1974 University of Chicago Press, and E. J. Masoro, and P. D. Siegel.* Acid-Base Regulation: Its Physiology, Pathophysiology and the Interpretation of Blood-Gas Analysis *[2nd ed.]. Philadelphia: Saunders, 1977.)*

The H^+ formed by the hydration of carbon dioxide in erythrocytes and by the formation of carbaminohemoglobin (Fig. 7-3) is buffered primarily by hemoglobin.

Oxygenated hemoglobin (HBO_2^{n-}) is a stronger acid than deoxygenated hemoglobin (Hb^{n-}). Both H^+ and carbon dioxide decrease the affinity of hemoglobin for oxygen. The reaction $HbO_2^n \rightarrow Hb^{n-} + O_2^-$ would cause a tremendous rise in pH if carbon dioxide and, therefore, H^+ were not added at the same time because Hb^{n-} is a stronger base than HbO_2. Release of CO_2 by the tissues results in the production of approximately 1.68 mmoles per liter of H^+ as CO_2 dissociates to form H^+ and HCO_3^-. Because pH depends only upon the concentration of hydrogen ion, their increase in H^+ would be expected to cause a precipitous drop in pH. Instead, the added H^+ is buffered by Hb^{n-}, thus minimizing the pH change. Hemoglobin is the most abundant nonbicarbonate extracellular buffer and is more powerful in the short run than the bicarbonate system.

Both hemoglobin and CA are located within erythrocytes. Like the plasma proteins, hemoglobin's buffering action is due primarily to its content of the amino acid histidine. Five percent of hemoglobin is histidine by weight. The concentration of hemoglobin-bound histidine in whole blood is, therefore, 45 mmoles/liter blood given a hemoglobin concentration of 14 g per deciliter. The pK_a of hemoglobin is 6.68, making hemoglobin a better buffer than a simple solution of histidine. The protonation of the imidazole group of hemoglobin is coupled to the deoxygenation of hemoglobin.

$$HbO_2 + H^+ \rightleftharpoons HHb + O_2$$

Both oxy- and deoxyhemoglobin may be protonated; however, because the pK_as are 6.68 and 7.93 respectively, the unloading of oxygen from hemoglobin is associated with increased affinity for H^+. The net reaction, therefore, is as indicated. Deoxyhemoglobin also has increased affinity for carbon dioxide. Carbon dioxide binds to the amino acid terminals of hemoglobin creating carbaminohemoglobin.

$$Hb\text{–}NH_2 + CO_2 \rightleftharpoons Hb\text{–}NH\text{–}COO^- + H^+$$

Ten percent or more of the total carbon dioxide contents of venous blood may be carried as carbaminohemoglobin (Fig. 7-3). Because carbaminohemoglobin carries oxygen less avidly, the increase in the PCO_2 encountered in the tissues causes oxygen to be released by changing the affinity of hemoglobin for oxygen and thus shifting the oxyhemoglobin dissociation curve to the right. Increased $[H^+]$ and 2,3= DPG also cause a similar shift in the oxyhemoglobin dissociation curve causing increased unloading of oxygen.

While the red cells are traversing the pulmonary capillaries, the PO_2 increases to approximately 90 mm Hg. Oxygen reacts with hemoglobin resulting in the formation of HbO_2^{n-} and the release of H^+. Concurrently, carbaminohemoglobin dissociates releasing carbon dioxide. The hydrogen ions liberated inside the red cell immediately react with HCO_3^- to form H_2CO_3, and this in turn is dehydrated in the reaction catalyzed by CA to yield additional carbon dioxide and water. Carbon dioxide then diffuses out of the red cells and is blown off by the lungs.

As the red cells enter the tissue capillaries, the PO_2 drops and hemoglobin begins to release oxygen. As carbon dioxide diffuses from the cells into the plasma and red cells, it causes the pH to drop. The increased H^+ and PCO_2 both contribute to additional unloading of oxygen. Hemoglobin acts both to buffer the H^+ and to transport carbon dioxide.

The red cell membrane is permeable to HCO_3^-; therefore, as HCO_3^- is formed and diffuses out of the cell along its new concentration gradient in the tissue capillaries, it is necessary for some other anion to enter the cells to maintain electroneutrality. Thus chloride shifts from the plasma and ECF fluids into the red cells against its concentration gradient but with the new electric gradient created by the efflux of HCO_3^-. This is called the chloride shift.

EXTRACELLULAR FLUID BUFFERS. Bicarbonate is the only significant extracellular buffer outside of the blood. Because the interstitial fluid volume is much greater than that of the plasma, its capacity to buffer metabolic acid loads is correspondingly greater. The ECF is rapidly cleared of protein and has a very low concentration of phosphates; therefore, the ECF is unable to help buffer a carbonic acid load. Equilibration with the interstitial fluids requires several minutes in contrast to the blood which equilibrates quite rapidly.

INTRACELLULAR FLUID BUFFERS. Proteins and phosphates are the main intracellular buffers with a small contribution made by intracellular bicarbonate. The pK_a of the $H_2PO_4^- = HPO_4^{2-}$ system is 6.8, very close to the average pH inside cells which is believed to be 7.0. This coupled with the high concentration of phosphate in cells accounts for its role in intracellular buffering.

$$H_2PO_4^- \rightleftarrows HPO_4^{2-} + H^+$$

Both phosphorous acid and hypophosphorous acid are lipid insoluble.

Intracellular buffers ultimately account for 50 percent of the buffering of a metabolic acid load and 95 percent of a carbonic acid load (because the bicarbonate system is unable to help in that case). Due primarily to the lipid barrier imposed by the cell membranes, intracellular buffering requires long equilibration times (hours) in contrast to the rapid equilibration associated with extracellular buffers (seconds to minutes) in responding to a metabolic acid-base disturbance. An exception to this is the hemoglobin system, which, due to the leakiness of the red cell membrane, equilibrates in tandem with the extracellular buffers. Figure 7-4 reviews the handling of a metabolic acid load.

RESPIRATORY ACID LOADS. The discussion thus far has centered on the response of body buffers to a metabolic acid load. The major differences in the response of body buffers to a respiratory acid load are as follows.

The bicarbonate–carbonic acid system is the culprit, it must, therefore, be discounted in discussions of respiratory acid buffering. Because the bicarbonate system is the major extracellular buffer, ECF buffering obviously contributes little to alleviating a respiratory acid load.

Furthermore, because carbon dioxide is lipid soluble, intracellular phosphate and protein can act to buffer respiratory disturbances without the delay associated with metabolic disturbances in which a charged molecule must generally equilibrate across the cell membranes.

Acute carbonic acid loads are, therefore, essentially entirely buffered by intracellular buffers with some contribution by nonbicarbonate extracellular buffers, primarily plasma proteins and inorganic phosphates. The role of hemoglobin in buffering carbonic acid loads has already been discussed. Roughly 95 percent of a carbonic acid load is buffered intracellularly by hemoglobin and inorganic phosphates in the red cells and by proteins and inorganic phosphates in the remainder of the ICF.

Table 7-2 illustrates the contribution of each of the various buffers to buffering a metabolic acid load while Table 7-3 briefly reviews the differences between metabolic and respiratory acid load buffering.

TUBULAR FLUID BUFFERS. Ammonia, bicarbonate, phosphate, sulfate, and organic anions all contribute to buffering in the nephron. The three most important tubular buffers and their respective pK_as are: NH_3-NH_4^+ with a pK_a

Table 7-3. *Metabolic Acid Buffering Versus Respiratory Acid Buffering*

Metabolic acidosis	50% of buffering by ECF and red cells, 50% ultimately buffered by ICF Rapid buffering of acid by hemoglobin and bicarbonate in the blood (seconds) with longer equilibration times required for equilibration with the interstitial fluid (approximately ½ hour) and intracellular fluid (hours) Bicarbonate system can and does contribute
Respiratory acidosis	95% of buffering occurs intracellularly, 5% extracellularly Rapid equilibration occurs as lipid soluble CO_2 diffuses into cell compartments to be buffered by hemoglobin, proteins, and phosphates The bicarbonate system cannot contribute

of 9.2, $H_2PO_4^-$-HPO_4^{2-} with a pK_a of 6.8 and HCO_3^-/H_2CO_3 with a pK_a of 6.1. The extent to which each is titrated depends on

1. The amount of H^+ secreted
2. The concentration of each buffer
3. The pK_a of each of the buffers present
4. The pH of the tubular fluid

BUFFERS SUMMARIZED. All body fluids are equally good at buffering a metabolic acid load; thus, at equilibrium, approximately 55 percent of an acid load is buffered by the ICF and 45 percent by the ECF. Equilibration with the ECF requires approximately half an hour while equilibration with the ICF may require many hours. Due to the high concentration of hemoglobin found in red cells, red cells are somewhat more effective than plasma on a volume basis.

In respiratory acidosis, in contrast, 95 percent of the acid load is buffered by intracellular buffers and 5 percent by extracellular buffers.

QUESTIONS

14. Respiration refers to two separate but mutually dependent physiological processes, either of which can lead to an alteration in the amount of carbon dioxide in the blood. Describe them.
15. Clinically, both carbamino carbon dioxide and H_2CO_3 are ignored when evaluating arterial blood gas measurements. How large an error does this entail?
16. Which of the following arterial blood gas parameters is not measured directly:
    A. pH
    B. PCO_2
    C. HCO_3^-
    D. All of the above
17. A patient has the following arterial blood gases: pH, 7.40; PO_2, 90; PCO_2, 40; and HCO_3^-, 24.
    A. Calculate the ratio of bicarbonate molecules to carbon dioxide molecules, then do the same for carbon dioxide to carbonic acid and bicarbonate to carbonic acid.
    B. Calculate the $[H^+]$. How does this compare to the other products and reactants?
    C. Estimate the following ratio using the data compiled above: $HCO_3^-/CO_2/H_2CO_3/H^+$.
18. Plasma proteins consist primarily of albumin. Normal serum albumin levels vary between 3.5 and 5.5 g per deciliter.
    A. Assuming that albumin accounts for all the buffering capacity of the plasma proteins and that albumin is 5 percent histidine by weight, what is the buffering capacity of plasma protein when the albumin concentration is equal to 3.5 g per deciliter?
    B. How does this compare with the amount of bicarbonate available? Why, then, do plasma proteins not contribute more to the buffering of metabolic acid loads?
19. How is carbon dioxide carried in arterial blood?
20. A. Calculate the resulting pH change if 1.68 mmoles of H^+ is released per liter of blood by the tissues and no hemoglobin or other nonbicarbonate buffer is available for buffering (remember, the bicarbonate-carbonate system cannot buffer itself).
    B. Calculate the pH if 14 g hemoglobin per deciliter is available for buffering, ignoring the presence of other plasma proteins and inorganic phos-

Fig. 7-4. *Handling of the fixed inorganic acid, HCl, by intact dogs. For the sake of clarity, polyvalent anions such as proteins and hemoglobin have been drawn with a single negative sign. A slanted arrow next to carbon dioxide indicates that the carbon dioxide is quickly excreted through the lungs. The percentages enclosed in rectangles indicate the approximate proportion of the total acid load that is buffered by each mechanism. Twenty-four hours after infusing acid, about 25 percent of acid load has been excreted in the urine as titratable acid and NH_4^+. Extracellular pH and ionic composition are nearly normal; therefore, 75 percent of the administered acid must be sequestered and buffered in tissue cells and bone. Two to six days after infusing acid, the remaining 75 percent of the administered acid is slowly released from tissue cells and bone and excreted in the urine. (From H. Valtin.* Renal Function: Mechanisms Preserving Fluid and Solute Balance in Health *[2nd ed.]. Boston: Little, Brown, 1983. pp. 210–211.)*

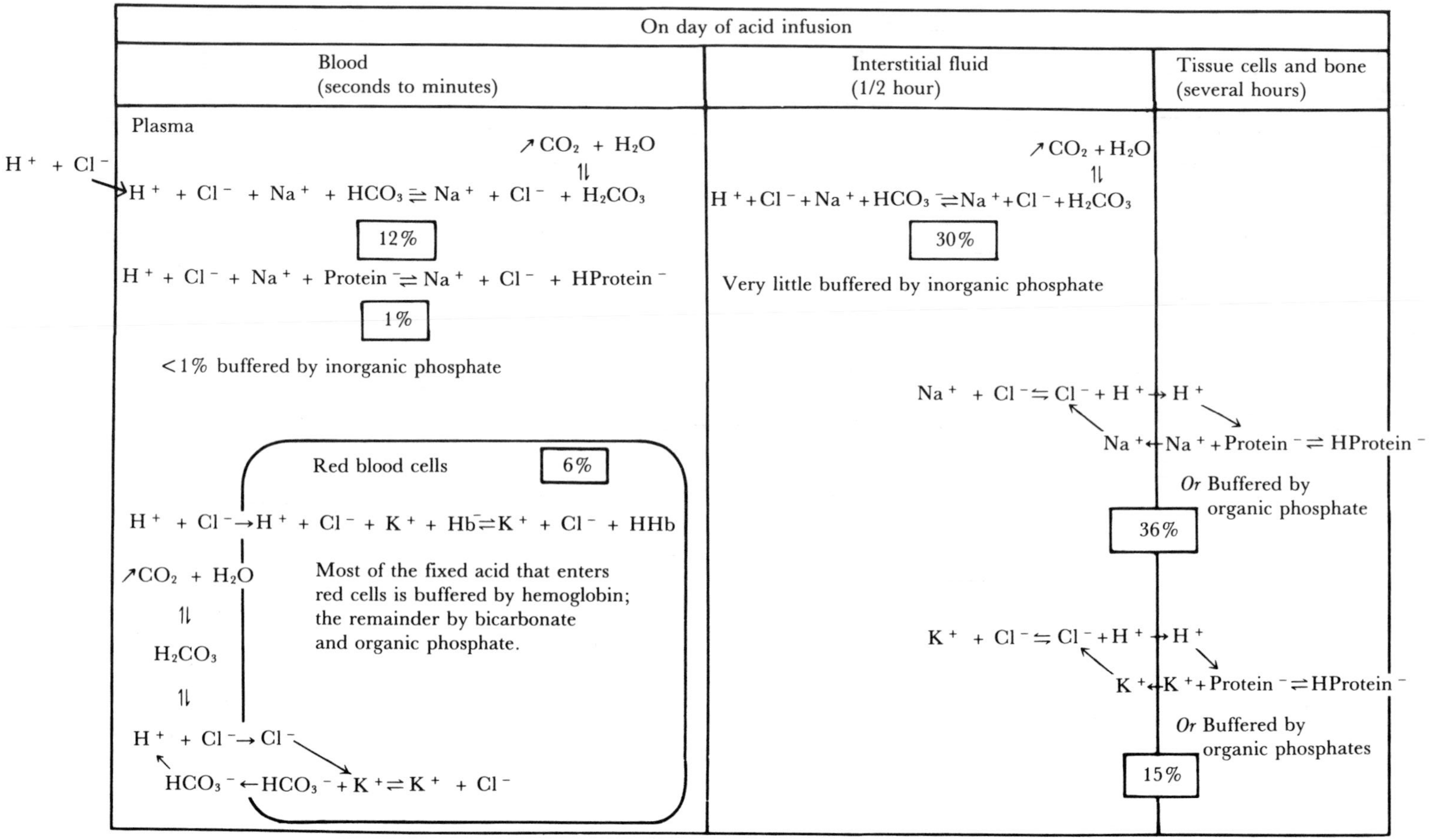

On day of acid infusion

Blood (seconds to minutes)

Plasma

$H^+ + Cl^- \rightarrow H^+ + Cl^- + Na^+ + HCO_3^- \rightleftharpoons Na^+ + Cl^- + H_2CO_3 \rightleftharpoons \nearrow CO_2 + H_2O$

12%

$H^+ + Cl^- + Na^+ + Protein^- \rightleftharpoons Na^+ + Cl^- + HProtein^-$

1%

<1% buffered by inorganic phosphate

Red blood cells

6%

$H^+ + Cl^- \rightarrow H^+ + Cl^- + K^+ + Hb^- \rightleftharpoons K^+ + Cl^- + HHb$

$\nearrow CO_2 + H_2O \rightleftharpoons H_2CO_3 \rightleftharpoons H^+ + Cl^- \rightarrow Cl^-$

$HCO_3^- \leftarrow HCO_3^- + K^+ \rightleftharpoons K^+ + Cl^-$

Most of the fixed acid that enters red cells is buffered by hemoglobin; the remainder by bicarbonate and organic phosphate.

Interstitial fluid (1/2 hour)

$H^+ + Cl^- + Na^+ + HCO_3^- \rightleftharpoons Na^+ + Cl^- + H_2CO_3 \rightleftharpoons \nearrow CO_2 + H_2O$

30%

Very little buffered by inorganic phosphate

$Na^+ + Cl^- \leftrightharpoons Cl^- + H^+ \rightarrow H^+$

Tissue cells and bone (several hours)

$Na^+ \leftarrow Na^+ + Protein^- \rightleftharpoons HProtein^-$

Or Buffered by organic phosphate

36%

$K^+ + Cl^- \leftrightharpoons Cl^- + H^+ \rightarrow H^+$

$K^+ \leftarrow K^+ + Protein^- \rightleftharpoons HProtein^-$

Or Buffered by organic phosphates

15%

phates. Hint: Hemoglobin contains 5 percent histidine by weight and the molecular weight of histidine is 156 g per mole with a pK_a of 6.0.

C. What percentage of the added H^+ is buffered by hemoglobin?

D. The _____ system accounts for _____ percent of plasma buffering capacity and _____ account for only _____ percent.

E. Of HBO_2^{n-} and Hb^{n-} which is the stronger acid and which is the stronger base?

F. What are the two major buffers in blood? What are their relative contributions in buffering carbon dioxide production?

21. Considering only whole blood, what percentage of a carbonic acid load is buffered in

A. Plasma

B. The red cell

22. A. What other body fluid compartment is important in buffering a carbonic acid load?

B. The hydration of carbon dioxide and the subsequent dissociation of H_2CO_3 leads to the formation of equal amounts of H^+ and HCO_3^-. How does this excess H^+ get buffered?

23. Due to the rapid influx and subsequent hydration of carbon dioxide,_____ moves out of red cells along its concentration gradient and _____ moves into the red cells along its electrical gradient. This sequence of events occurs in the _____ capillary system while the _____ series of events occurs in the _____ capillary system.

24. Describe the effect of each of the following maneuvers on arterial, intracellular, and CFS pH and PCO_2:

A. Decreasing the rate and depth of respiration below normal levels

B. Intravenous infusion of HC1

C. Intravenous infusion of $NaHCO_3$

RENAL HANDLING OF H^+ AND HCO_3^-

ROLE OF THE KIDNEY. The role of the kidney in acid-base regulation is to match the rate of acid excretion to the highly variable and essentially unregulated rate of dietary acid intake and metabolic production and to compensate for respiratory disorders. The kidney responds to acid-base imbalances by forming and excreting a fluid with either an excess or a deficit of acid in order to return the $[H^+]$ of the blood to normal. The kidney is able to alter acid-base status in 3 basic ways:

1. Reabsorption of HCO_3^-
2. Titration of urinary buffers
3. Formation and excretion of ammonia (NH_3)as ammonium (NH_4^+)

Methods (2) and (3) result in the net excretion of H^+ while the first method results in the reabsorption of bicarbonate. The gain of HCO_3^- is equivalent to the loss of H^+, and, conversely, the loss of HCO_3^- is equivalent to the gain of H^+ (review the Henderson-Hasselbalch equation, equation 7-4). Furthermore, the initial source of most of the H^+ secreted is via the hydration of carbon dioxide, not secretion of free H^+ by the tubular cells (Fig. 7-5A). Thus H^+ secretion (and, therefore, bicarbonate reabsorption) can be almost entirely abolished by CA inhibitors. Some net excretion of H^+ does occur in the distal tubule where a low capacity but high gradient H^+-Na^+ exchange pump acts to acifidy the urine (Fig. 7-5B).

Fig. 7-5. *Mechanisms for H^+ secretion in the proximal tubule and distal nephron. In both regions, carbon dioxide from metabolic processes, the peritubular capillaries, or the tubular lumen reacts with water to form H_2CO_3, a reaction catalyzed by carbonic anhydrase (CA) in the epithelial cells. The H_2CO_3 dissociates to form H^+ and HCO_3^-. The H^+ is secreted by secondary active transport via the Na^+/H^+ exchange process (antiport) in the proximal tubule (indicated by the circle) and via an active process in the distal nephron. The HCO_3^- diffuses across the basal-lateral surfaces into the peritubular capillaries, so that for every H^+ secreted, a HCO_3^- is returned to the systemic circulation. Filtered bicarbonate = 4500 meq/day, reclaimed by proximal tubule = 4500 meq/day, new bicarbonate generated by the distal tubule = 50 meq/day, and net acid excretion = 50 meq/day.*
A. Reabsorption of filtered HCO_3^- via H^+ secretion. The secreted H^+ reacts with HCO_3^- in the tubular fluid to form H_2CO_3, which then dissociates into carbon dioxide and water, i.e.,

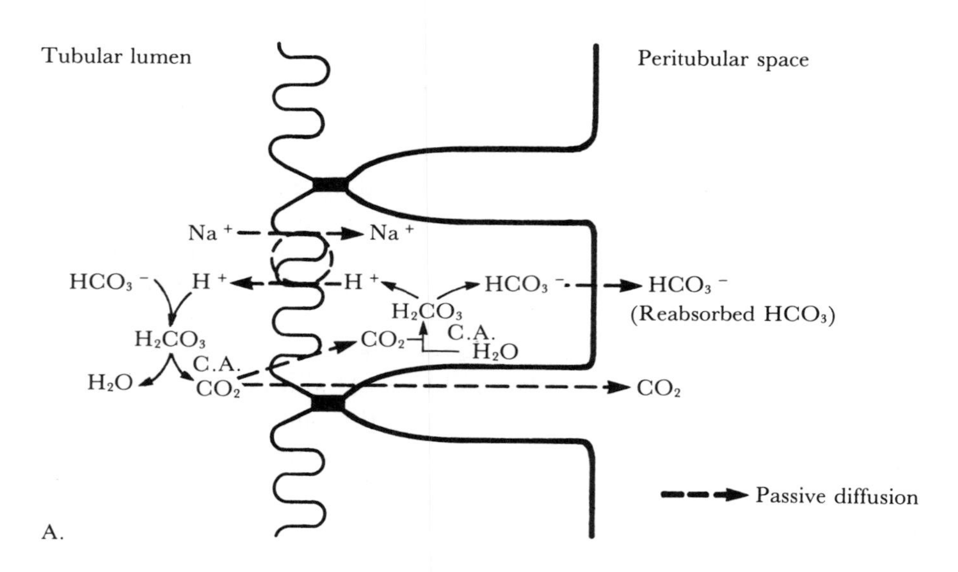

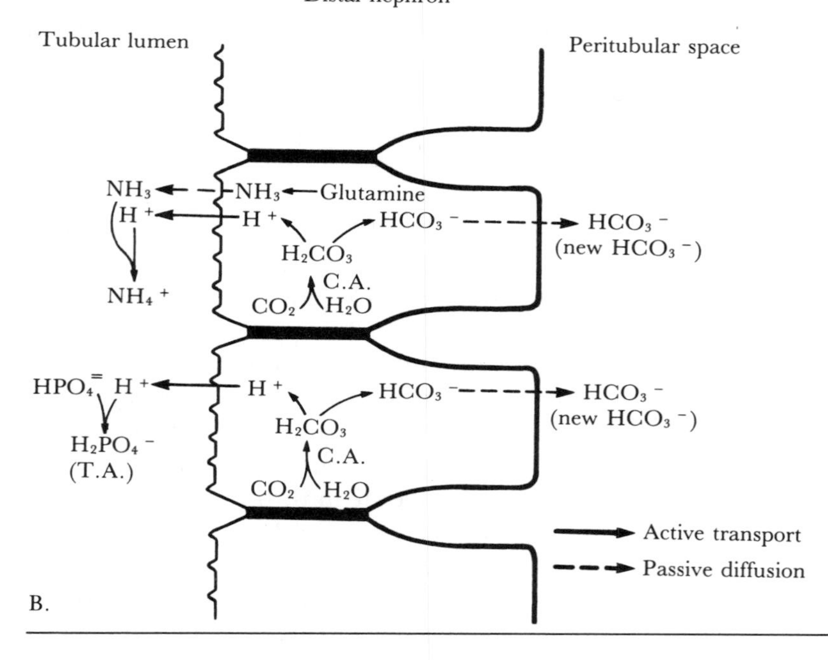

Figure 7-5 (Continued)

a HCO_3^- ion is lost from the tubular fluid. Because the H^+ secretion process adds a HCO_3^- ion to the peritubular fluid and peritubular capillaries, however, the next result is HCO_3^- reabsorption. The proximal tubule is illustrated here, because approximately 90 percent of the filtered HCO_3^- is reabsorbed in the proximal tubule. However, similar reactions occur in more distal portions of the nephron (except for the absence of CA in the luminal surface) and are responsible for the reabsorption of most of the remaining HCO_3^- in the tubular fluid. An active Na^+-K^+ exchange pump in the basolateral membrane (not shown) actively pumps Na^+ from the cells into the peritubular fluid. This supplies the gradient for Na^+ reabsorption from the tubular fluid. (From J. B. West. Best and Taylor's: Physiological Basis of Medical Practice *[11th ed.]. P. 522. © 1985 The Williams & Wilkins Co., Baltimore.)*

B. Generation of new HCO_3^- via H^+ secretion. The HCO_3^- generated in the H^+ secretion process represents a new HCO_3^- if the secreted H^+ ion (a) reacts with NH_3 synthesized by the epithelial cells to form NH_4^+ or (b) reacts with HPO_4^{-2} and other titratable acids in the tubular fluid to form $H_2PO_4^-$ and to acidify the remaining titratable buffers. Although the distal nephron is illustrated here, similar reactions can occur in the proximal tubule (except that proximal tubular H^+ secretion is primarily coupled to Na^+ reabsorption and is gradient limited). Most of the H^+ secreted in the proximal tubule reacts with HCO_3^- in the tubular fluid and, therefore, accomplishes the reabsorption of filtered HCO_3^- rather than the generation of new HCO_3^-. The distal tubular H^+ secretory pump can create a 1000-fold $[H^+]$ gradient enabling it to acidify the urine. (From J. B. West. Best and Taylor's: Physiological Basis of Medical Practice *[11th ed.] P. 524. © 1985 The Williams & Wilkins Co., Baltimore.)*

REABSORPTION OF HCO_3^-. The proximal tubule reabsorbs filtered bicarbonate by secreting H^+ and subsequently reabsorbing it as carbon dioxide (Fig. 7-5A). Greater than 75 percent of filtered bicarbonate is reabsorbed in the proximal tubule; however, the pH decreases only slightly, and, therefore, only a relatively small amount of other buffers is titrated. The H^+ secretory pump is located in the luminal membrane and is a high-capacity, low-gradient pump. Thus, while it is able to pump large numbers of H^+ ions, it is unable to create a pH gradient (i.e., it cannot acidify the urine). Hydrogen ions from the cell are exchanged for Na^+ from the tubule fluid thus maintaining electroneutrality. The radium ion moves down its concentration gradient into the cell and thus secondarily actively transports H^+ against its concentration gradient into the lumen.

The H^+ secreted into the lumen reacts with the filtered HCO_3^- to form carbonic acid. The carbonic acid is then broken down into carbon dioxide and water in the presence of CA on the luminal membrane of the proximal tubule. As a result, carbon dioxide is then in greater concentration in the tubular fluid than in the cell, and it passively diffuses down its concentration gradient into the cells. As carbon dioxide diffuses into the cells, its increased concentration causes it to encounter intracellular CA and, by mass action, to drive the reaction toward the formation of H^+ and HCO_3^-. The H^+ is secreted into the tubular fluid (note that there is no *net* secretion of H^+ in this process). The bicarbonate, now in high concentration inside the cell, is transported into the peritubular space as sodium bicarbonate in a facilitated passive transport process fueled by an active sodium pump in the peritubular membrane. Intracellular Na^+ is replenished as Na^+ from the tubular fluid enters the cells passively along the concentration gradient formed by the active transport of Na^+ out of the cell into the peritubular fluid.

The net result is the reabsorption of sodium bicarbonate from the tubular fluid and a slight fall in the pH of the urine. There is no change in the PCO_2 and only the minimal excretion of H^+ associated with the inadvertent titration of other buffers by hydrogen ions secreted by the proximal tubule.

The extent of H^+ secretion in the proximal tubule (and in the rest of the renal tubule as well) is to a large extent governed by the availability of buffer because only by reacting with buffers can a significant amount of H^+ be excreted before the limiting concentration gradient is reached. Even in the distal tubule the pH gradient limits urinary pH to 4.5.

The total amount of bicarbonate reabsorbed is proportional to the amount filtered, and this in turn is equal to the product of the glomerular filtration rate and the serum concentration of bicarbonate. This relationship is valid until the concentration of bicarbonate in the tubular fluid exceeds the proximal tubule's ability to reabsorb it (approximately 28 meq/liter). Normally more than 75 percent of the bicarbonate filtered is reabsorbed in the proximal tubule. Additional bicarbonate reabsorption occurs in the loop of Henle.

EXCRETION OF H^+: THE DISTAL CONVOLUTED TUBULE AND COLLECTING DUCTS. The distal convoluted tubule and collecting ducts contain a luminal H^+ secretory pump (Fig. 7-5B). It is able to acidify the urine and titrate the remaining urinary buffers. The gradient is limited to approximately a 1000-fold H^+ gradient between the ICF and the tubular fluid. Hydrogen ion is secreted into the tubular fluid and reacts with phosphate and the other titratable acids (sulfate and organic acids). The relatively impermeable epithelium of the distal convoluted tubule and collecting ducts coupled with a high-gradient H^+ secretory pump allows the kidney to form an acidic urine. This pump is low capacity.

TITRATABLE ACIDITY. Titratable acidity (TA) refers to the amount of acid that can be titrated by adding alkali (OH^-) to the urine until a pH of 7.4 is achieved. The amount of alkali needed to return urinary pH to 7.4 is equal to the amount of acid excreted bound to urinary buffers with pK_as such that the buffers bind significant amounts of H^+ at pHs below 7.4. The major titratable acids are phosphate and sulfate (bicarbonate should not normally be present in apprecia-ble amounts). Organic acids such as creatinine, beta-hydroxybutyric acid, and acetoacetic acid may become important in disease states (e.g., in diabetes mellitus). The amount of H^+ free in solution is negligible in comparison with the amount bound to urinary buffers. Because the pK_a of ammonia is so high (9.2), addition of alkali does not titrate ammonia; thus, its contribution to H^+ excretion must be considered separately.

AMMONIA AND ION TRAPPING. If renal H^+ excretion were limited to the formation of titratable acid, insufficient H^+ ion would be excreted before reaching the limiting pH of 4.5. The amount of H^+ excreted would then be limited by the amount of phosphate and other titratable acids. Because the amount of these acids present in the urine is diet dependent and not subject to regulation, total H^+ excretion would not be subject to regulation either.

Due to its high pK_a (9.2), the ammonia-ammonium system allows the kidneys to excrete H^+ as the neutral salt NH_4Cl. Measurements of TA would not reveal the contribution of this system because titration to pH 7.4 would only affect 1 to 2 percent of the ammonia buffer present.

$$NH_3 + HCl \rightleftarrows NH_4Cl \quad pK_a = 9.2$$

Within the renal tubule cells, carbon dioxide reacts with water to form H^+ and HCO_3^-. The H^+ is secreted into the tubular fluid where it reacts with ammonia to form ammonium while the bicarbonate is reabsorbed into the peritubular fluid.

The NH_3 in the tubular fluid is derived from the breakdown of glutamine and other amino acids by renal tubular cells (Fig. 7-5B). Being lipid soluble, the ammonia diffuses out of the cells and into the tubular fluid along its concentra-tion gradient. In the tubular fluid, H^+ readily combines with NH_3 to form

NH_4^+. This occurs at all urinary pHs because the pK_a of ammonia is so high; however, the ratio of NH_3 to NH_4^+ decreases as the tubular fluid becomes more acidic, trapping the lipid insoluble ammonium within the tubular fluid. As the ammonia (as ammonium) is trapped in the tubular fluid, additional ammonia diffuses out of the cells to replace it. The trapped ammonium is subsequently excreted, primarily as ammonium chloride. This process is called ion trapping or nonionic diffusion.

The net function of the NH_3-NH_4^+ system is to excrete H^+ and supplement HCO_3^- stores. This probably occurs in all parts of the nephron. The mechanism by which NH_3 production and excretion is increased in acidic states is unknown; however, as the pH of the urine becomes more acidotic, more of the ammonia becomes trapped in the urine as ionized ammonium. This in turn increases the gradient favoring diffusion of the uncharged ammonia molecule out of the cells into the tubular fluid and may in some unknown fashion trigger increased ammonia production by the cells.

The final determinant of ammonia excretion is the ratio of peritubular blood flow to tubular fluid flow. When urine pH is equal to plasma pH, more ammonia would diffuse into blood than into the tubular fluid because the rapid blood flow relative to urine flow would dilute the ammonia and, therefore, create a greater concentration gradient (a larger sink effect). Normally the urine pH is acidic, however, so most of the NH_3 produced by the renal tubular cells enters into the tubular fluid by nonionic diffusion. About three-quarters of the daily endogenous load of nonvolatile acid is excreted as ammonium, the remainder is excreted as titratable acids.

REGULATION OF RENAL ACID AND HYDROGEN ION EXCRETION. Renal acid excretion is dependent on

1. The concentration of tubular fluid buffers. If the urine is already maximally acidic, then H^+ excretion is directly proportional to the amount of buffer available in the urine.
2. The glomerular filtration rate (GFR). Changes in GFR alter the filtered load of bicarbonate and, therefore, the rate of bicarbonate reabsorption and the filtered load of titratable acids.
3. Carbonic anhydrase activity.
4. The PCO_2 in the blood and the blood pH. Increased PCO_2 and decreased pH favor proton secretion by the renal tubular cells which results in increased bicarbonate reabsorption, increased ammonia production and excretion, and increased TA.
5. ECF volume. Volume contraction favors secretion of H^+ (contraction alkalosis) while volume expansion does not.
6. Potassium balance. Hydrogen ion secretory capacity is inversely proportional to body potassium stores. If either H^+ or K^+ is decreased, then the other is forced to move excessively. Decreased plasma levels of K^+ (hypokalemia) cause K^+ to move out of cells. In order to maintain electroneutrality, H^+ then moves into cells simultaneously causing an intracellular acidosis and an extracellular metabolic alkalosis determined by arterial blood gases. Hypokalemia also causes decreased aldosterone secretion and increased renin secretion causing ECF volume to be decreased.
7. Aldosterone causes K^+ or H^+ to be exchanged for Na^+ resulting in salt water accumulation and the excretion of H^+ and K^+. If either H^+ or K^+ is present in less than normal accounts, the other cation is forced to move excessively.
8. Parathyroid hormone causes increased bicarbonate excretion and decreased titratable acid excretion.

25. What effect does the body's state of hydration have on the ability of the kidneys to excrete H^+? In a water conserving state? In a water excess state?
26. In a dehydrated state, the concentration of _____ is increased while the concentration of _____ ion is decreased, leading to a metabolic _____. This is called _____ alkalosis.
27. Which of the following occurs to net acid excretion when a carbonic anhydrase inhibitor is administered:
    A. Increases
    B. Decreases
    C. No change
    D. The change in net acid excretion then depends on a number of other factors (name them).
28. What changes would be expected in urinary pH and the composition of the urinary buffers.
29. Which of the following is/are true of the amount of bicarbonate reabsorbed?
    A. Amount reabsorbed = amount of H^+ ions buffered by bicarbonate in the proximal tubule.
    B. Amount of bicarbonate reabsorbed = plasma concentration of bicarbonate × glomerular filtration rate (GFR) − amount of bicarbonate excreted.
    C. 75 percent is reabsorbed in the proximal tubule.
    D. Additional reabsorption occurs in the loop of Henle.
    E. Bicarbonate wasting occurs when the serum concentration exceeds 28 meq per liter.
30. The H^+ pump in the distal tubule is a high-gradient, low-capacity pump. The high gradient results in the formation of an acidic urine. What are the consequences of the pump's low capacity?
31. What effect would a tubular fluid pH of 8.0 have on ammonia excretion?
32. In diabetic ketoacidosis without compromised renal function, what would happen to ammonia production and excretion by the kidney?
    A. Decrease 100-fold
    B. Decrease 10-fold
    C. Increase 100-fold
    D. Increase 10-fold
    E. No change would be apparent
33. Define titratable acid.
34. What is the ratio of NH_3 to NH_4^+ at
    A. pH 7.4
    B. pH 7.2
    C. pH 6.2
    D. pH 5.2
35. Net acid excretion is equal to
    A. $TA + NH^+ + HCO_3^-$
    B. $TA + NH_4^+ - HCO_3^-$
    C. $TA + NH_4^+$
    D. $TA + H_2PO_4 + NH_4^+$

    Net H^+ excretion (i.e., ignoring bicarbonate reabsorption) would be equal to which of the above?
36. Nonionic diffusion is important in promoting the urinary excretion of weak acids and weak bases. In which of the following will the rate of excretion be increased by alkalinization of the urine? By acidification of the urine?
    A. NH_3
    B. N-acetylsalicylate overdose (aspirin)
    C. Phenobarbital

D. Creatinine ($pK_a = 5.0$)
E. Beta-hydroxybutyric acid

37. Rate of urinary ammonium excretion is influenced by
   A. pH of the urine
   B. Chronicity of the acidosis
   C. Relative flow rates of peritubular blood versus tubular fluid
38. There are two mechanisms by which body bicarbonate stores are replenished. What are they?
39. The rate at which filtered bicarbonate is reabsorbed is affected by
   A. Filtered load of bicarbonate
   B. Expansion or contraction of the ECF
   C. Arterial PCO_2
   D. $[Cl^-]$ in the plasma
   E. Hormones (ACTH and parathyroid)

REGULATION OF RESPIRATION

The initial responses to a fluctuation in acid-base status include an almost instantaneous response by the ECF buffers followed by a somewhat slower equilibration with intracellular buffers (final equilibration with the ICF may require many hours). The kidneys in turn require many hours to several days to act. If, despite the buffering system response, the $[H^+]$ changes measurably, then the respiratory center is immediately stimulated to alter the rate and depth of breathing. The resulting change in PCO_2 causes the $[H^+]$ to return toward normal (see Chapter 6 for a thorough discussion of central regulation of respiration and the effect of the blood-brain barrier).

An increase in the PCO_2 in the body's fluids causes a decrease in the pH (acidosis). Alternatively, a decrease in the PCO_2 causes an increase in the pH (alkalosis). Because the PCO_2 is directly proportional to the rate and depth of respiration, the respiratory system is able to alter pH by altering respiration. The other major determinant of body fluid PCO_2 is the rate of metabolism in the cells. If the metabolic rate increases, then carbon dioxide production increases proportionally. This leads to both an increase in PCO_2 and an acute decrease in pH (a respiratory acidosis if the lungs fail to compensate).

Metabolic production of carbon dioxide depends primarily upon diet. The amount of carbon dioxide given off relative to the amount of oxygen consumed is called the respiratory exchange ratio (RER) or respiratory quotient (RQ).

$$RQ = \frac{\dot{V}CO_2}{VO_2} = \frac{CO_2 \text{ produced/minute}}{O_2 \text{ consumed/minute}} \quad (7\text{-}8)$$

The minute oxygen consumption ($\dot{V}O_2$) is normally 250 to 350 ml per minute. The normal minute carbon dioxide production ($\dot{V}CO_2$) is 200 to 250 ml per minute. Carbohydrates such as sugar have a RQ of 1.0 because their breakdown in the Kreb's cycle results in the release of one carbon dioxide molecule for each oxygen molecule consumed.

$$C_6H_{12}O_6 \rightleftarrows 6CO_2 + 6H_2O \qquad \dot{V}O_2$$

The RQ for fat is 0.7 and that for protein 0.8. On a mixed diet the RQ is approximately 0.8.

High $PaCO_2$, high $[H^+]$, and low PaO_2 all stimulate the carotid body chemoreceptors which results in a reflex increase in alveolar ventilation and a corresponding increase in carbon dioxide excretion. Similarly, the respiratory center in the medulla is sensitive to $[H^+]$ (and to $PaCO_2$ to the extent that it dissociates to form H^+). A decrease in pH or an increase in $PaCO_2$ may result in an increase in respiration 3 to 7 times normal levels. The response to increasing the pH is less dramatic, generally resulting at most in a reduction in respiration to 50 to 75 percent of normal levels. Note that fluctuations in serum pH may or may not lead to similar changes in the CSF and ICF depending on the cause of the disturbance. A discrepancy may occur due to the high lipid solubility of carbon dioxide and the low lipid solubility of H^+ resulting in extremely different equilibration times.

If the primary acid-base disturbance is metabolic in origin, then the lung's response cannot return the $[H^+]$ to normal. It can, however, partially compensate for the abnormality. A complete response is not possible because as the pH returns to normal, the driving force behind the lung's response decreases. Generally, the respiratory response to a metabolic abnormality results in correction of 50 to 75 percent of the abnormality.

Overall, the ability of the respiratory system to respond to a metabolic acid-base disturbance is approximately 1 to 2 times that of all the chemical buffers combined.

The arterial PO_2, $PaCO_2$ and $[H^+]$ all effect the rate and depth of alveolar ventilation. Changes in any one of these may enhance or antagonize the response to changes in another. Thus, although PaO_2 normally has little effect on the rate and depth of ventilation, when arterial PO_2 decreases below 60 mm Hg (corresponding to a hemoglobin saturation of 89%), hypoxia may drive respiration. This may result in a $PaCO_2$ as low as 20 mm Hg. In chronic obstructive pulmonary disease (COPD), patients frequently retain carbon dioxide and may become narcotized by the high carbon dioxide levels reached if their hypoxic drive is abolished (for example by increasing their F_IO_2 thus eliminating the chronically low arterial PO_2 levels that have been driving their respirations).

In summary, a rise in arterial PCO_2 or $[H^+]$ or a drop in arterial PO_2 increases the level of respiratory center activity, while changes in the opposite direction have a slight inhibitory effect. Chemoreceptors in the medulla and in the carotid and aortic bodies initiate the impulses which stimulate the respiratory center. This in turn results in a compensatory increase or decrease in alveolar ventilation.

QUESTIONS

40. V/Q mismatch may result in an increase in PCO_2. This will cause
    A. Metabolic alkalosis
    B. Respiratory alkalosis
    C. Metabolic acidosis
    D. Respiratory acidosis
41. Which of the following could result in a paradoxical intracellular acidosis?
    A. Decreased respiratory drive resulting in an increase in PCO_2
    B. Infusion of hydrochloric acid
    C. Infusion of sodium bicarbonate

PRIMARY ACID-BASE DISORDERS

FOUR PRIMARY ACID-BASE DISORDERS. There are four primary acid-base disorders:

1. Alveolar hypoventilation resulting in a primary respiratory acidosis
2. Alveolar hyperventilation resulting in a primary respiratory alkalosis
3. Metabolic acidosis
4. Metabolic alkalosis

RESPIRATORY ACIDOSIS. Respiratory acidosis may result from neurological dysfunction (e.g., a decrease in central ventilatory drive or spinal cord damage), airway obstruction (e.g., COPD, reactive airway disease), muscular dysfunction (e.g., myasthenia gravis), decreased surface area for respiratory exchange (e.g., emphysema), V/Q mismatch (e.g., aspiration), decreased lung compliance (e.g., pneumonia, pleural effusion), trauma (e.g., pneumothorax, chylothorax), or physical malformation (e.g., scoliosis). Any factor that interferes with respiratory gas exchange can lead to carbon dioxide retention and respiratory acidosis. By prodigious effort, one can hold one's breath and induce a respiratory acidosis. As the pH falls, however, one begins to lose consciousness and breathing resumes.

RESPIRATORY ALKALOSIS. Respiratory alkalosis may occur as a result of an iatrogenic episode or as a response to a metabolic disturbance. Occasionally, psychoneurosis results in overbreathing and respiratory alkalosis. Acutely, moving to a high altitude with its concomitant low PO_2 results in increased respiration and respiratory alkalosis.

METABOLIC ACIDOSIS. Metabolic acidosis refers to all abnormalities of acid-base imbalance other than respiratory acidosis which results in a decrease in pH below normal. The sine qua non of a metabolic acidosis is an acid pH coupled with an abnormally low bicarbonate concentration (resulting from buffering of the acid responsible for the acidosis by bicarbonate). The PCO_2 may be normal (in the absence of respiratory compensation) or decreased (if either a normal or abnormal respiratory response is present) but should not be elevated unless a mixed disorder is present. Metabolic acidosis may result from

1. Failure of the kidneys to excrete endogenous acids
2. Increased formation of endogenous acids (as in diabetes mellitus)
3. Administration of exogenous acids
4. Loss of alkali from the body (from gastrointestinal or renal losses for example)

METABOLIC ACIDOSIS AND THE ANION GAP. Metabolic acidosis is often subdivided into those with a normal anion gap and those with an increased anion gap. The anion gap is calculated by subtracting the sum of the chloride and bicarbonate concentrations from the sodium concentration. Because sodium is the major extracellular cation, and chloride and bicarbonate are the major extracellular anions (and because the remaining cations and anions fortuitously cancel out), this difference is usually 0 to 15. Many causes of metabolic acidosis involve strong acids which contribute their conjugate bases to the anion pool, thus increasing the anion gap to greater than normal values. In contrast, administration of ammonium chloride does not result in an increased anion gap because chloride is one of the anions measured in determining anion gap.

$$\text{Anion gap} = [Na^+] - [Cl^-] - [HCO_3^-]$$

METABOLIC ALKALOSIS. Metabolic alkalosis is relatively rare in comparison to metabolic acidosis. Administration of diuretics, with the exception of CA inhibitors, and alkaline drugs such as sodium bicarbonate may result in a metabolic alkalosis. Prolonged vomiting of gastric contents can result in a hypochloremic metabolic alkalosis as hydrogen ions and chloride ions are lost in the vomitus. Hydrogen ions are secreted by gastric cells in much the same fashion as renal tubular cells with concomitant reabsorption of sodium bicarbonate. Furthermore, because bicarbonate and chloride are the two major anions in the body, it is a general principle that loss of one of these anions is counterbalanced by avid retention of the other in order to maintain electroneutrality. Increased aldosterone levels may precipitate a metabolic alkalosis.

QUESTIONS

42. A Stanford medical student (a sea-level scutboy) joins a Himalayan expedition and climbs (or is carried) rapidly to 4 miles above sea level. Describe his blood gases on Himalayan room air. How will these change over the next 2 weeks? On returning to Stanford, another blood gas is drawn. What acid-base abnormality is found?
43. What acid-base disturbance results from the following:
    A. Chronic administration of hypertonic sodium chloride by peripheral IV
    B. Massive diarrhea
    C. Vomiting of stomach contents
    D. Vomiting of intestinal contents
    E. Administration of a CA inhibitor
    F. Ingestion of paraldehyde, salicylate, ethylene glycol, or methanol
    G. Diabetes mellitus
    H. Severe exercise or peripheral hypoxia (e.g., postcardiopulmonary bypass)
    I. Hypokalemia
44. For the previous problem, determine whether the anion gap would be normal or increased.

COMPENSATION OF ACID-BASE DISTURBANCES

The principle governing compensation of a disturbance in the acid-base system is that the unaffected system acts to minimize the fluctuation in pH. Thus, if the disturbance is metabolic, the lungs immediately respond by hypo- or hyperventilation as appropriate. If the disturbance is primarily that of the respiratory system, then the kidneys respond (albeit more slowly) by holding on to or by dumping bicarbonate. This is called the slow renal response because, in contrast to chemical buffering and respiratory compensation, it is the slowest, taking hours to days to achieve full effect. Renal response is usually divided into acute and chronic stages. The acute renal or metabolic response for a respiratory disorder is actually not a response at all but simply a reflection of the amount of carbon dioxide present in all forms including HCO_3^-. The chronic response requires that the kidneys have time to retain or discard HCO_3^- as required.

Compensation by one system for an inadequacy in the other is not 100 percent effective. It will result in moving the pH 50 to 75 percent of the way back to normal. This works relatively well for respiratory disturbances; however, the compensatory response for metabolic disturbances is neither predictable nor dependable. Table 7-4 gives the normal compensatory responses (acute and

Table 7-4. *Normal Compensation for Acid-Base Disturbances*

Respiratory	Acute		Chronic	
$PaCO_2$ (mm Hg)	pH	HCO_3^- (meq/liter plasma)	pH	HCO_3 (meq/liter plasma)
↑ 10	↓ 0.08	↑ 1	↓ 0.04	↑ 3
↓ 10	↑ 0.08	↓ 2	↑ 0.04	↓ 5

Metabolic acidosis: For every 1 meq fall in HCO_3^- a compensating 1.0 mm Hg fall in $PaCO_2$ occurs unless ventilation is compromised (down to $PaCO_2$ of 5–10 mm Hg). (From T. Raffin, and J. McGee. Stanford Respiratory Therapy Department Handout, 1980.)

chronic) for each acid-base disorder. Alternatively, an acid-base nomogram such as that pictured in Figure 7-6 can be used to evaluate the arterial blood gas values.

An acid-base disorder is mixed when both the renal and respiratory systems are contributing to the problem. Thus, if the PCO_2 is elevated and the bicarbonate is low, a mixed respiratory and metabolic acidosis is present. Similarly, if a mixed problem is not present then appropriate compensation by the uninvolved system should be apparent. When plotted on an acid-base nomogram, such a mixed disturbance would yield values between those expected for each of the primary disorders present.

The following is a four-step method for determining acid-base status on the basis of arterial blood gases:

1. What is the pH? If the pH is alkaline (pH > 7.42), then the problem is an alkalosis. If the pH is acidotic (pH < 7.38), then the problem is an acidosis.
2. Which is the problem, bicarbonate or PCO_2? If bicarbonate is the problem, then the disorder is metabolic. If carbon dioxide is the problem (i.e., abnormal), then the malfunction is in the lungs.
3. Check the compensatory response, is it appropriate? Has a mixed disorder been overlooked? If the problem is a respiratory one, is the renal response consistent with an acute or a chronic response? Is this consistent with the history?
4. Calculate the anion gap.

Fig. 7-6. *Acid-base nomogram. Shown are the 95 percent confidence limits of the normal respiratory and metabolic compensations for primary acid-base disturbances. (From B. M. Brenner, and F. C. Rector.* The Kidney *(3rd ed.). Philadelphia: Saunders, 1986. P. 473.)*

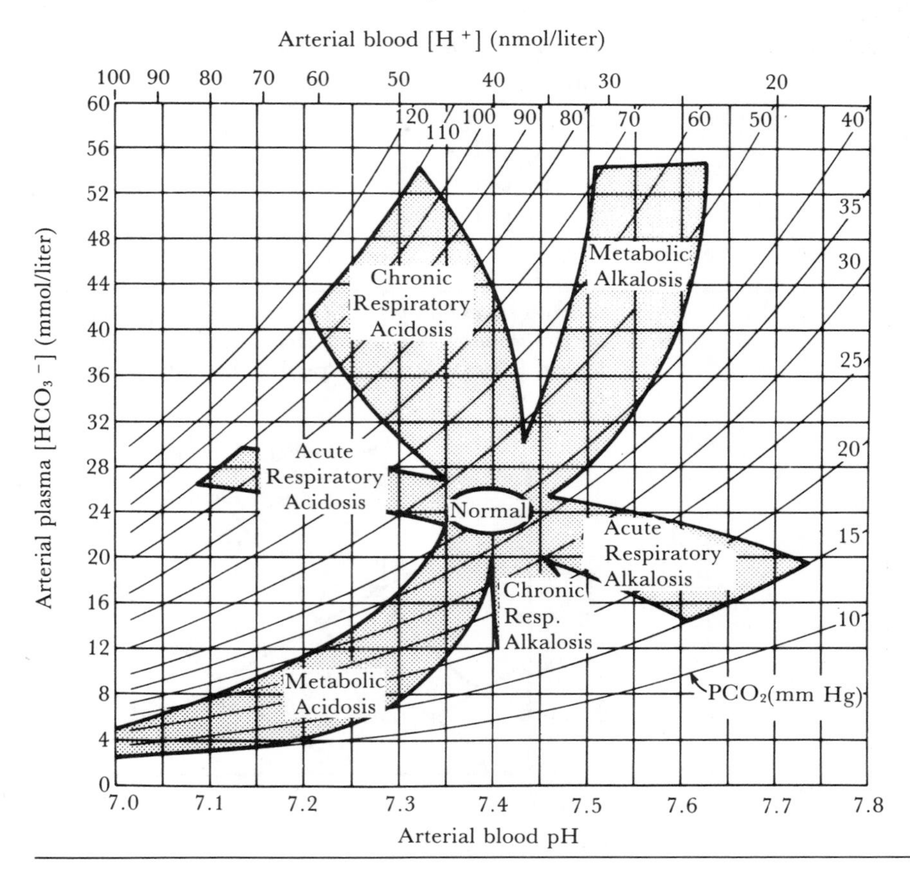

QUESTIONS

45. Fill in the following chart indicating what the primary acid-base problem is.

	pH	*HCO_3^-*	*PCO_2*	*Acid-base Status*
A.		14.6	30	
B.	7.43		39	
C.	7.10		55	
D.	7.51	20		
E.		31	42	
F.	7.30		65	
G.	7.58	31		
H.	7.40	30		
I.	7.48	30		

46. Identify the points on Figure 7-7 which correspond to
    A. Chronic metabolic acidosis
    B. Acute respiratory acidosis
    C. Chronic respiratory acidosis
    D. Chronic metabolic alkalosis
    E. Acute respiratory alkalosis
    F. Chronic respiratory alkalosis
    G. Normal arterial point

CLINICAL CORRELATES

The major effect of acidosis is to depress the CNS which may culminate in coma and death. A metabolic acidosis results in increased rate and depth of respiration while in respiratory acidosis it is the depressed rate and depth of respiration which is in fact causing the problem.

Alkalosis causes overexcitability of the nervous system which may result in death due to tetany and convulsions. Both the central and peripheral nervous systems are affected. The peripheral nervous system is usually more sensitive. Tetany can frequently be demonstrated in the forearms and in the muscles of the face prior to the onset of CNS symptoms or tetany of the respiratory muscles.

ANSWERS

1. A. pH = 1
   B. pH = 9
   C. pH = 0

Fig. 7-7. *Changes in arterial pH, PCO_2, and $[HCO_3^-]$ in the four primary acid-base disturbances and in the compensatory responses to these disturbances. (From J. B. West.* Best and Taylor's: Physiological Basis of Medical Practice *[11th ed.]. P. 529. © The Williams & Wilkins Co., Baltimore.)*

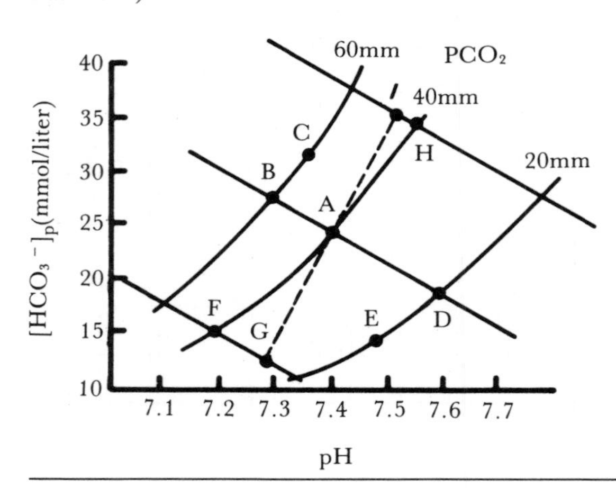

D. $K = 10^{-pK} = 1.74 \times 10^{-5}$ and $K = [H^+][RCOO^-]/[RCOOH] = 1.74 \times 10^{-5}$. If acetic acid is a weak acid, then only a small fraction dissociates. In that case, [RCOOH] does not change significantly and remains approximately 1 mole per liter. Furthermore, $[H^+]$ is approximately equal to $[RCOO^-]$ because acetic acid is the only significant source of H^+ present (the H^+ in water contributes only 10^{-7} moles per liter of H^+). Hence:

$$[H^+]^2 = K[RCOOH] = 1.74 \times 10^{-5}$$

$$[H^+] = 4.16 \times 10^{-3} \quad pH = -\log p[H^+] = 2.3$$

The assumptions made that the $[H^+]$ already in aqueous solution would not contribute significantly to the $[H^+]$ of the acetic acid (1.0×10^{-7} versus 4.16×10^{-3}) and that the amount of RCOOH undergoing reaction would also be insignificant relative to the amount present (4.16×10^{-3} versus 6.023×10^{23}) prove reasonable thus confirming the validity of our assumptions.

2. $[H^+] = 10^{-pH} = 10^{-7.1} = 7.94 \times 10^{-8}$ or approximately 80 nmolar. Blood is normally 40 nmolar so at a pH of 7.1, twice as much H^+ is present as at pH 7.4.
3. At pH = 7.4, $[H^+]$ = 40 nmolar. Half this concentration is 20 nmolar concentration or pH = 7.7.
4. Weak, weak
5. A. Acid (Cl^-)
   B. Base ($CH_3\,COOH$)
   C. Base (H_2CO_3)
   D. Acid (NH_4^+)
   E. Acid (H_2O)
   F. Acid (SO_4^{2-}) and base (H_2SO_4)
   G. Base (HPO_4^{2-})
6. A fixed acid, such as H_2SO_4, is an acid which does not exist in equilibrium with a volatile product or reactant. A nonfixed acid, such as carbonic acid, exists in equilibrium with a volatile substance, for example carbon dioxide in this case. Because the volatile component can exist in more than one phase, its concentration in one phase (e.g., liquid) is subject to fluctuation based on changes in its concentration in the other phase (e.g., gaseous).
7. B
8. B
9. When present in low concentrations, these substances are metabolized by the liver to bicarbonate. However, when their concentrations in the blood exceed the ability of the kidney to reabsorb them, they are eliminated in the urine (a process called ketonuria). In disease states such as diabetes mellitus these substances may accumulate beyond the ability of kidneys to excrete them sufficiently rapidly. This state is called diabetic ketoacidosis.
10. A. 1
    B. 10
    C. 100
    D. 100
    E. 0.1
    F. 0.01
    G. 0.001
11. C
12. C
13. The PCO_2 is higher in venous blood and, because carbon dioxide is in equilibrium with HCO_3^-, more bicarbonate is present as well. More base

is also available in the form of deoxyhemoglobin, and this serves to buffer the H^+ released as the additional carbon dioxide undergoes hydration. Just the addition of base is sufficient to account for some increase in bicarbonate because by the isohydric principle all bases in the same solution are in equilibrium. As the additional base equilibrates, H^+ ion comes off the other buffers increasing the concentrations of their conjugate bases.

14. Respiration refers to the exchange of carbon dioxide for oxygen by the lungs. It also refers to the process of oxygen consumption and carbon dioxide production at the cellular level. Both of these processes interact to determine the level of carbon dioxide in the body.
15. Carbonic acid is present in such small quantities that ignoring it is easy; however, carbamino carbon dioxide is present in significant levels even in arterial blood. A 5 percent error is, therefore, produced.
16. C
17. A.

$$HCO_3^-/CO_2 = \frac{24 \text{ mmoles/liter}}{40 \text{ mm Hg} \times 0.03 \text{ mmoles/liter/mm Hg}}$$

$$= \frac{24 \text{ mmoles/liter}}{1.2 \text{ mmoles/liter}} = 20/1$$

$[H_2CO_3] = [H^+][HCO_3^-]/K_a$ and $K_a = 10^{-pKa} = 1.58 \times 10^{-4}$

$[H_2CO_3] = (40 \times 10^{-9})(24 \times 10^{-3})/1.58 - 10^{-4} = 6.06 \times 10^{-6}$ moles/liter or 6.06×10^{-3} mmoles/liter

$$CO_2/H_2CO_3 = \frac{1.2 \text{ mmoles/liter}}{6.06 \times 10^{-3} \text{ mmoles/liter}} = 198/1$$

$$HCO_3^-/H_2CO_3 = \frac{24 \text{ mmoles/liter}}{6.06 \times 10^{-6} \text{ moles/liter}} = 3960/1$$

    B. Hydrogen ion is present at a concentration of 40 nmoles/liter or 4×10^{-8} moles/liter. It is the least common product or reactant with a concentration less than 1/100th that of carbonic acid.

    C. $HCO_3^-/CO_2/H_2CO_3/H^+ = 4000/200/1/0.01$

18. A. 3.5 g/dl = 35 g albumin/liter

$0.05 \times 35 = 1.75$ g histidine/liter plasma

1.75 g histidine/ (156 g histidine/mole) = 0.011 moles histidine or 11 mmoles histidine. At pH 7.4 almost all of the histidine exists as the unprotonated form. The ratio of the unprotonated to protonated forms of histidine may be calculated using the Henderson-Hasselbalch equation for histidine:

$pH = pK_a + \log [His^-]/[His]$

$\log [His^-]/[His] = pH - pK_a$

$[His^-]/[His] = 10^{pH - pK_a} = 10^{1.4} = 25$

Therefore, 1/26th of the total histidine is protonated and 25/26ths unprotonated.

    B. Approximately 11 mmoles of histidine are available for buffering per liter of plasma compared to 24 mmoles of bicarbonate per liter of plasma. Calculated in this fashion roughly 30 percent of the buffering capacity of plasma for a metabolic acid load would be due to albumin. In fact, only about 8 percent of plasma buffering is attributed to plasma proteins primarily because the excess carbon dioxide produced can be blown off by the lungs.

19. Ninety percent as bicarbonate, 5 percent as carbaminohemoglobin, 5 percent dissolved as carbon dioxide, and a minute amount as carbonic acid.
20. A. At a pH of 7.4, the $[H^+] = 4 \times 10^{-8}$ moles per liter. If 1.68×10^{-3} moles H^+ is added then:

    $$pH = -\log(1.68 \times 10^{-3} + 4 \times 10^{-8}) = 2.77$$

    B. In titrating hemoglobin starting at pH 7.4 one need consider only the imidazole groups of histidine until the pH rises above 7.6 or falls below 5.5 because the pK_a values of other groups prohibit their contributing significantly. Hemoglobin is 5 percent histidine by weight and assuming [Hb] of 14 g per deciliter or 140 g per liter.

    g His/liter = 0.05 × 140 g/liter = 7.0 g/liter

    moles His/liter = 7.0 g/(156 g/mole) = 0.045 moles His/liter = 45 mmoles His/liter

    At pH 7.4 the protonated form of the imidazole group accounts for 1.7 mmoles (1/26th of 45, see Question 18) and the unprotonated form for 43.3 mmoles per liter (25/26ths). If 1.68 mmoles HCl is added then 3.38 mmoles are protonated and 41.2 unprotonated and applying the Henderson-Hasselbalch equation

    $$pH = 6.0 + \log 41.2/3.38 = 7.09$$

    The contribution of other blood buffers accounts for an additional increase in pH to 7.32. Finally, the increased affinity of Hb^{n-} (pK_a 7.93) for H^+ accounts for a further increase in pH to approximately 7.37.

    C. The concentration of H^+ increased from 40 nmolar to approximately 80 nmolar (pH = 7.1) after addition of the acid. Thus, only 40×10^{-9} moles per liter acid was unbuffered. $(1.68 \times 10^{-3})/(1.68 \times 10^{-3} + 4.0 \times 10^{-8}) \times 100 = 99.998\%$ was buffered or only 0.002% was unbuffered by hemoglobin, even ignoring the contribution of other buffers and the pK_a change as hemoglobin desaturates.

    D. Bicarbonate, 70 percent, plasma protein, 30 percent

    E. Deoxyhemoglobin is a stronger base; oxyhemoglobin is a stronger acid.

    F. Hemoglobin and bicarbonate. Only hemoglobin can buffer a carbonic acid load.
21. A. Approximately 11 mmoles of histidine from plasma proteins are present per liter of plasma or approximately 6.0 mmoles per liter whole blood (see Question 18) while 45 mmoles of histidine from hemoglobin are present per liter of blood.

    B. No more than 12 percent of a carbonic acid load will be buffered by the plasma. Actually, at least 90 percent or more of the buffering of a carbonic acid load must occur in the red cell because phosphate is also present in the red cells at higher concentration than in the plasma.
22. A. ICF; however, equilibration requires hours so the ICF contributes little acutely except for the red cells.

    B. Predominantly by hemoglobin.
23. HCO_3^-, Cl^-, tissue, opposite, pulmonary
24. A. The pH will drop equivalently in the arterial blood, the ICF, and the CSF as the increased PCO_2 rapidly equilibrates.

    B. The increased arterial $[H^+]$ stimulates ventilation which causes the arterial PCO_2 to drop. Because the H^+ cannot easily pass the lipid cell membrane, the ICF and the CSF do not acutely equilibrate with the new higher $[H^+]$; however, the decrease in PCO_2 immediately causes an equivalent decrease in the ICF and CSF and a paradoxical intracellular alkalosis occurs as carbon dioxide and therefore hydrogen ions are lost despite the continued arterial acidosis.

C. The excess bicarbonate reacts with H^+ to form carbon dioxide (catalyzed by CA in the red cells) driving $[H^+]$ down drastically. Respiratory drive may decrease as the pH increases causing an additional increase in $PaCO_2$ as a consequence of depressed respiration. The carbon dioxide, however, unlike the bicarbonate, is free to diffuse into the ICF and CSF where it becomes hydrated to form H^+. A paradoxical intracellular acidosis ensues in the presence of a metabolic alkalosis as measured in the blood.

Neither H^+ nor bicarbonate can enter cells or CSF rapidly, but carbon dioxide can and does. Rapid changes in arterial pH due to metabolic causes may cause paradoxical changes in the CSF and ICF (particularly brain cells).

25. Volume contraction favors secretion of H^+ ion while volume expansion disfavors it.
26. Buffers, H^+, alkalosis, contraction
27. C
28. The urine pH would become alkaline with high levels of bicarbonate present and decreased amounts of ammonia. Other buffers would be unchanged in composition and concentration.
29. All are true.
30. The distal tubule pump is unable to acidify the urine if greater than normal amounts of base are present (e.g., bicarbonate in proximal tubular acidosis).
31. Ammonia trapping would preferentially occur in the arterial blood (pH 7.4) rather than the tubular fluid. Furthermore, blood flow is much greater than tubular fluid flow creating a greater sink effect. Finally, ammonia production by the renal tubular cells would decrease by an unknown mechanism. All of these would contribute to a decrease in ammonia excretion.
32. C
33. A titratable acid is one with a pK_a such that titration of the urine to a pH of 7.4 with base would account for its contribution to H^+ ion excretion.
34. A. 0.016
    B. 0.01
    C. 0.001
    D. 0.0001
35. B
    C
36. A. Acidification
    B. Alkalinization
    C. Alkalinization
    D. Neither
    E. Alkalinization
37. A, B, and C
38. Reabsorption of filtered bicarbonate by the proximal tubule, and excretion of H^+ with concomitant formation of bicarbonate by the distal tubule.
39. All of these affect bicarbonate reabsorption. Note that chloride and bicarbonate are the major anions in the plasma. A change in one generally results in the opposite change in the other so as to maintain electroneutrality. Their concentrations are, therefore, generally inversely proportional.

40. D
41. C
42. Initially, our scutboy will hyperventilate as the decreased PO_2 (PO_2 at 20,000+ feet = 73 mm Hg and PaO_2 = 40 mm Hg) causes his respiratory rate to increase acutely to 65 percent greater than normal. Over the next few days, as he acclimatizes, his respiratory rate will increase to 3 to 7 times normal as his body becomes used to the lower $PaCO_2$ and higher pH this entails. Thus, he will experience an acute respiratory alkalosis. Over the next 2 weeks, his kidneys will adjust to the situation by excreting bicarbonate in an attempt to normalize his blood gases. On returning to Stanford-level, he will suddenly experience an acute metabolic acidosis (pH may be decreased or normal) as his respiratory drive returns to normal while he continues to have below normal levels of total body bicarbonate. This situation will normalize over the next few days as his kidneys act to conserve bicarbonate.
43. A. Metabolic acidosis
    B. Metabolic acidosis
    C. Metabolic alkalosis
    D. Metabolic acidosis
    E. Metabolic acidosis
    F. Metabolic acidosis
    G. Metabolic acidosis
    H. Metabolic acidosis
    I. Metabolic alkalosis
44. A. Normal. (Note that a normal anion gap in the presence of a metabolic acidosis implies that the patient is hyperchloremic.)
    B. Normal
    C. Not applicable
    D. Normal
    E. Normal
    F. Normal
    G. Increased
    H. Increased
    I. Normal
45. A. 7.31, metabolic acidosis with appropriate compensation
    B. 25, within normal limits
    C. 16.5, mixed respiratory and metabolic acidosis
    D. 20, acute respiratory alkalosis
    E. 7.49, metabolic alkalosis, appropriate compensation variable
    F. 31.5, chronic respiratory acidosis
    G. 34, mixed respiratory and metabolic alkalosis
    H. 50, mixed disorder: respiratory and metabolic acidosis
    I. 20, chronic respiratory alkalosis
46. A. G
    B. B
    C. C
    D. I
    E. D
    F. E
    G. A

    Metabolic processes generally are chronic. Thus, both metabolic acidosis and metabolic alkalosis would probably develop along the dashed lines connecting A to G and A to I, respectively.

BIBLIOGRAPHY

Brenner, B. M. and Rector, F. C. (eds.). *The Kidney* (3rd ed.). Philadelphia: Saunders, 1986.

Ganong, W. F. *Review of Medical Physiology* (11th ed.). Los Altos, CA: Lange, 1983.

Guyton, A. C. *Textbook of Medical Physiology* (7th ed.). Philadelphia: Saunders, 1986.

Friedman, P. J. *Biochemistry: A Review with Questions and Explanations* (2nd ed.). Boston: Little, Brown, 1982.

Stanford University School of Medicine Faculty, *Renal Physiology Syllabus* (lecture handout), 1984.

Stanford University School of Medicine Faculty, *Respiratory Physiology Syllabus,* 1984.

Valtin, H. *Renal Function: Mechanisms Preserving Fluid and Solute Balance in Health* (2nd ed.). Boston: Little, Brown, 1983.

West, J. B. *Best and Taylor's: Physiological Basis of Medical Practice* (11th ed.). Baltimore: Williams & Wilkins, 1985.

West, J. B. *Respiratory Physiology: The Essentials* (3rd ed.). Baltimore: Williams & Wilkins, 1985.

8 Neurophysiology

Jack M. Percelay

The neurology questions on the national boards cover neuroanatomy, neurophysiology, neuropathology, and biochemical aspects of neurochemical function. This chapter reviews the essential neurophysiology with minimal references to neuroanatomy; therefore, a concurrent review of neuroanatomy is strongly recommended. In particular, this text does not emphasize histology, specific neurological pathways, or the localization of lesions.

THE SOMATIC SENSORY SYSTEM

Sensory receptors are transducers that convert energy in the environment into action potentials in neurons. Anatomically, a sensory receptor may be either part of a modified sensory nerve ending, such as a pacinian corpuscle, or a specialized cell, such as a cochlear hair cell. Receptors respond to a given stimulus by altering ion permeability and producing a receptor potential, a nonpropagating local potential. The high sensitivity of a receptor to a particular form of energy accounts for the extraordinary specificity of receptors. As is true throughout the nervous system, when the local receptor potential reaches threshold, an action potential is produced. This action potential travels along a neuronal pathway specific for both sensory modality and, to varying degrees, location. Stimulation of a nerve fiber anywhere along this pathway, electrically or chemically, centrally or peripherally, produces a conscious sensation for that modality and that location.

The pacinian corpuscle is a modified sensory nerve ending responding to touch. It consists of the unmyelinated tip of a sensory nerve surrounded by concentric lamellations of connective tissue resembling an onion (Figure 8-1). When pressure is applied to the corpuscle, an electrical potential is produced in the nerve. Small amounts of pressure produce only a receptor or generator potential. As the applied pressure increases, the magnitude of the receptor potential increases. When the receptor potential reaches a threshold of approximately +10mv, an action potential is generated in the sensory nerve. Further increase in pressure results in an increased frequency of action potentials. As the strength of a stimulus increases, the stimulus tends to spread over a larger area and activates the surrounding receptors as well as increasing the frequency of action potentials. The relationship between stimulus intensity, receptor potential, and rate of firing is shown in Figure 8-2.

Mechanistically, compression of the external capsule of the pacinian corpuscle tends to distort the central core. This increases the permeability of the nerve membrane to Na^+ in proportion to the intensity of the stimulus, perhaps by

Fig. 8-1. *The pacinian corpuscle. Note how the outer lamellations serve to buffer the central core from mechanical distortions. (From W. R. Lowenstein. Biological Transducers.* Sci. Am. *203:98, 1960.*

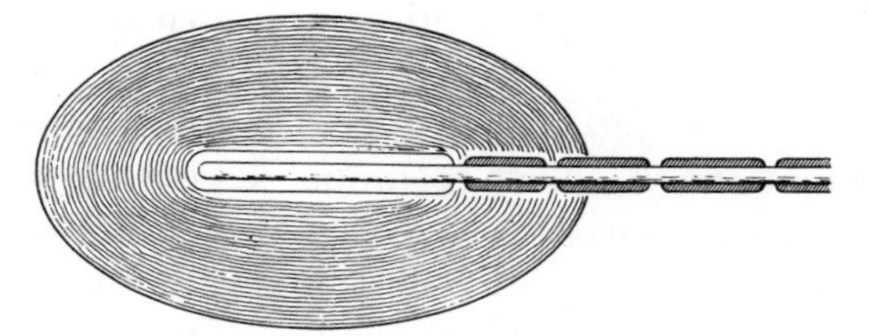

Fig. 8-2. *Relationship between stimulus intensity, receptor potential, and action potentials. Receptor potential is proportional to stimulus intensity for both amplitude and duration. Once the receptor potential exceeds the threshold, action potentials result with frequency reflecting stimulus intensity. (Reprinted by permission of the publisher from* Principles of Neural Science *[2nd ed.], by E. K. Kandel and G. H. Schwartz, P. 24, 1985, by Elsevier Science Publishing Company, Inc.)*

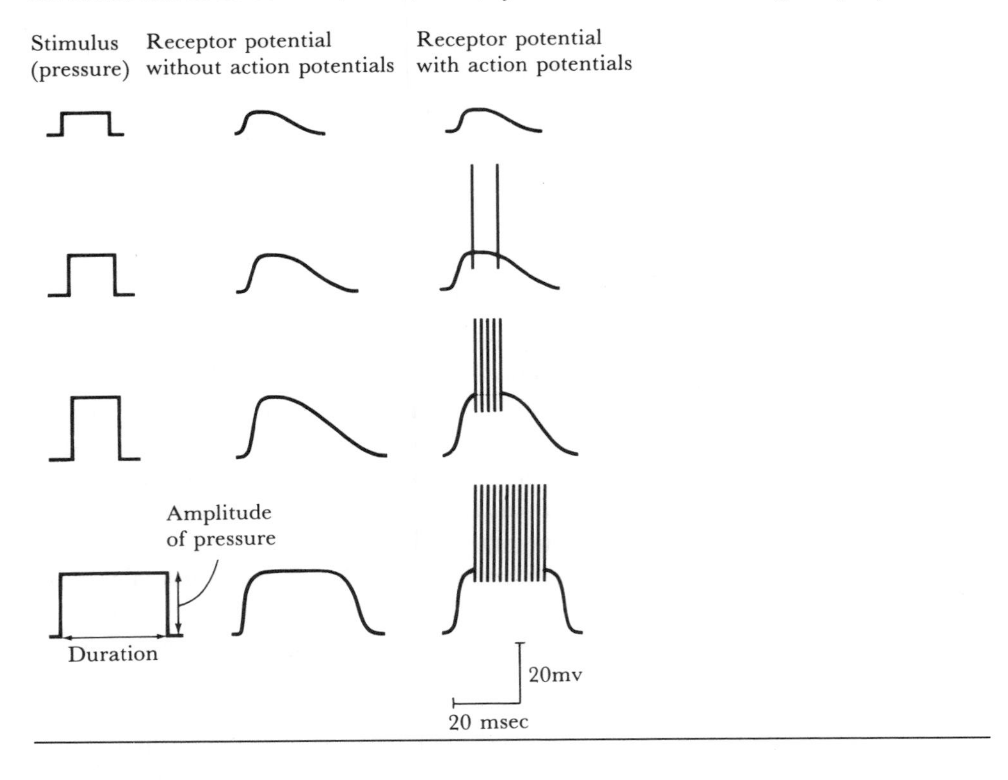

mechanically opening ion channels. The resultant Na^+ influx produces the generator potential. When the generator potential is large enough to depolarize the closest node of Ranvier beyond threshold, an action potential results.

All sensory receptors adapt either partially or completely to their stimuli over time. Thus, a constant stimulus produces a progressively decreasing response. Two processes contribute to adaptation: (1) adjustments in the structure of the receptor itself, and (2) modifications in the ion channels of the nerve fibers, a process termed accommodation. Depending upon their rate of adaptation, receptors can be classified as either phasic or tonic. Phasic receptors adapt rapidly and thus transmit valuable information about a change in stimulus strength. They react strongly while a change is taking place, and impulse frequency is proportional to the *rate* of change. When combined with information about the body's current status, information about rate change allows the nervous system to predict the future status of the body. The pacinian corpuscle is a phasic receptor. It responds to touch, a change in pressure, but is useless for transmitting information about constant pressure. Tonic receptors adapt slowly and incompletely and thus continuously apprise the brain of a given stimulus. Examples of tonic receptors include joint capsule receptors and carotid baroreceptors, which provide continuous information about joint position and arterial blood pressure respectively.

The major pathways for transmitting sensation from cutaneous sensory receptors are the dorsal columns (lemniscal system) and the anterolateral or spinothalamic tracts. These systems are contrasted in Table 8-1. A key feature of the dorsal columns is the faithfulness of transmission, in terms of both stimulus intensity and geographic location. Large myelinated fibers transmit this infor-

Table 8-1. *Comparison of the Dorsal Column/Medial Lemniscal and Anterolateral/Spinothalmic Systems*

	Dorsal Column	Anterolateral
Submodalities	Touch-pressure Position sense Kinesthesia Vibration	Pain Temperature Crude touch
Location in spinal cord	Dorsal columns	Anterolateral columns
Type of fibers	Large myelinated	Small myelinated Small unmyelinated
Somatotopic organization	Strictly maintained throughout pathway	Present, but not strictly maintained
Level of decussation	Medulla	Spinal cord
Brain stem terminations	Ventral posterior lateral nucleus and posterior nuclear group of thalamus	Brainstem reticular formation Ventral posterior lateral nucleus and posterior nuclear group of thalamus
Cerebral terminations	Primary and secondary somatic sensory cortices and somatic sensory association area	Primary and secondary sensory cortices and somatic sensory association area

mation rapidly. The information from the spinothalamic tract is more crude. It generally requires a less rapid response from the organism, and is transmitted by smaller, more slowly conducting fibers.

Higher levels of the nervous system also participate in sensory processing. For example, two-point discrimination depends on cells in the dorsal column nuclei that have small excitatory receptive fields with inhibitory surrounds. Such receptive fields can be explained by lateral inhibition. The effect is to sharpen contrast between adjacent stimuli and is depicted graphically in Figure 8-3. Stereognosis, the ability to identify objects by handling them, depends on information of touch and pressure but also requires a large cortical component.

Fig. 8-3. *Lateral inhibition. In the absence of lateral inhibition, two closely applied stimuli* (dotted lines) *are perceived as the sum of their pressures, and it is difficult to determine if one large blunt stimulus is applied or two sharp stimuli. With lateral inhibition, the border between the two stimuli is sharpened. (Reprinted by permission of the publisher from* Principles of Neural Science *[2nd ed.], by E. K. Kandel and G. H. Schwartz, P. 325, 1985, by Elsevier Science Publishing Company, Inc.)*

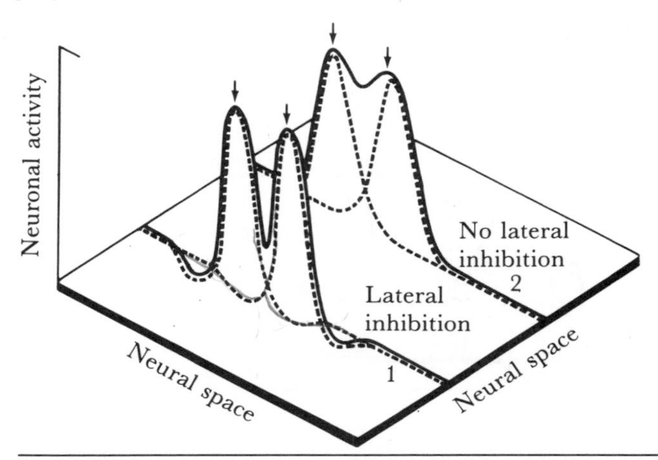

The receptors for pain, nociceptors, are free nerve endings. These receptors may be activated by either mechanical, chemical, or thermal stimuli, and adapt slowly if at all. The intensity of a pain response is related to the extent of tissue damage. There are two pathways for pain transmission. Acute, discretely localized pain (e.g., pin prick) is transmitted by small myelinated A-delta fibers in the lateral spinothalamic tract. More diffuse, chronic pain (e.g., burning) is transmitted by small unmyelinated C fibers in the medial spinothalamic and spinoreticular pathways. It is believed that the cortex plays an important role in interpreting the quality of pain, even though pain perception may be a function of subcortical centers. Pain receptors in the viscera are sparsely distributed and visceral pain is thus poorly localized. Because visceral pain fibers travel along fibers of the autonomic nervous system, many autonomic changes (e.g., nausea, vomiting) are associated with visceral pain. Viscera are much more sensitive to distention and ischemia than to incision or temperature.

Referred pain from a viscus to a somatic structure that embryologically shares the same dermatome results from visceral and somatic afferents converging on the same spinothalamic neurons in the dorsal horns. Because dorsal horn cells are usually activated by cutaneous input, the brain misinterprets the activity of these dorsal horn neurons as somatic rather than visceral pain. The ability of tactile stimulation to reduce pain — shaking a burnt hand — can be explained by higher order processing of pain signals. Tactile stimuli are believed to activate interneurons that inhibit transmission of impulses from the dorsal root pain fibers to spinothalamic neurons. This is the basis of the gate control theory. The brain also has an endogenous pain suppression system which uses enkephalins, endogenous peptides with opioid activity. This illustrates the phenomenon of higher centers modifying afferent input.

QUESTIONS

1. Explain phantom limb pain, pain perceived to arise from an amputated body part, in neurophysiologic terms.
2. What are the two means by which the nervous system records stimulus intensity?
3. What is the effect of a lesion of the dorsal columns?
4. What is the effect of a lesion of the spinothalamic tract?
5. Which would induce a greater change in perceived stimulus intensity, increasing applied pressure from 5 to 10 "units" or from 50 to 60?

Fig. 8-4. *The monosynaptic stretch reflex. Also illustrated is inhibition of the antagonist muscles through an inhibitory* (shaded) *interneuron. This is an example of a polysynaptic reflex. (Reprinted by permission of the publisher from* Principles of Neural Science *[2nd ed.], by E. K. Kandel and G. H. Schwartz, P. 459, 1985, by Elsevier Science Publishing Company, Inc.)*

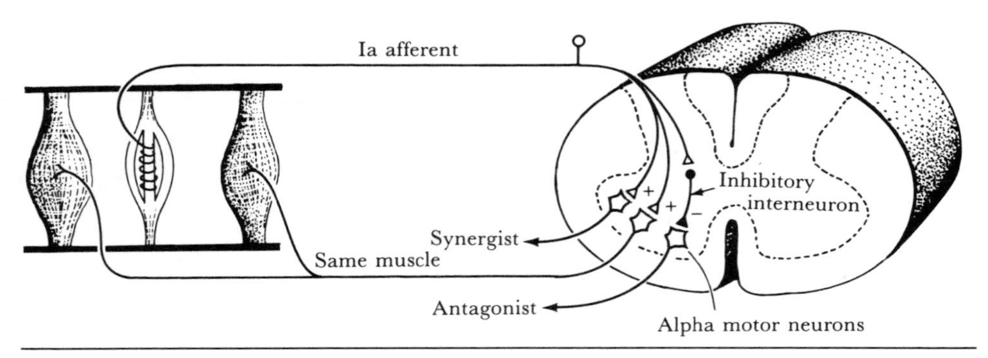

SOMATIC MOTOR FUNCTION

REFLEX ARCS. The reflex arc is the basic unit of integrated neural activity. Reflex arcs consist of afferent sensory input, central integration and processing, and efferent output. The monosynaptic stretch reflex (i.e., the knee jerk) is the simplest reflex arc (Fig. 8-4). Stretch is detected by the muscle spindle. It lies parallel to the extrafusal muscle fibers and consists of small intrafusal muscle fibers with a noncontractile central portion that receives sensory nerve endings (Fig. 8-5). The intrafusal fibers are innervated by small-diameter gamma motorneurons, while the extrafusal fibers of the main muscle are innervated by large diameter A-alpha motorneurons. When the muscle is stretched, the spindle organ is also stretched because it is aligned parallel to the extrafusal fibers. Stretching the spindle excites the sensory nerves, leading to reflex stimulation of the A-alpha motorneurons and contraction of the extrafusal fibers. If the muscle contracts without contraction of the intrafusal fibers, however, the spindle becomes slack because the muscle has shortened while the spindle itself has not. In this case afferent impulses decrease and the contractile state of the extrafusal fibers decreases (Fig. 8-6). Thus, the spindle and its reflex connections constitute a feedback device that operates to maintain muscle length.

Another important feature of this reflex system is the ability of gamma efferent stimulation to indirectly induce contraction of the extrafusal fibers. The gamma motorneurons directly cause contraction of the intrafusal fibers. This stretches the midportion of the muscle spindle, even if the length of the entire muscle does not change. The subsequent activation of Ia fibers leads to a reflex contraction of the extrafusal fibers. In most motor behaviors, alpha and gamma motorneurons are activated together, a process termed alpha-gamma coactivation. Thus, the spindle shortens with the muscle throughout the contraction and by responding to stretch helps to reflexively adjust motorneuron discharge throughout the contraction. The gamma efferent system is also important in the maintenance of muscle tone.

Polysynaptic reflexes make use of centrally located interneurons, excitatory or

Fig. 8-5. *The muscle spindle. The muscle spindle is seen to lie in parallel to the extrafusal fibers. Sensory fibers arise from the spindle to provide information about stretch. The group Ia (annulospiral) primary nerve endings are sensitive to both absolute stretch and to changes in the rate of stretch of the spindle. The group II secondary nerve endings are sensitive to absolute stretch only. Gamma motor neurons innervate the intrafusal fibers located at the end of the spindles. (From A. C. Guyton.* Textbook of Medical Physiology *[7th ed.]. Philadelphia: Saunders, 1985. P. 608.)*

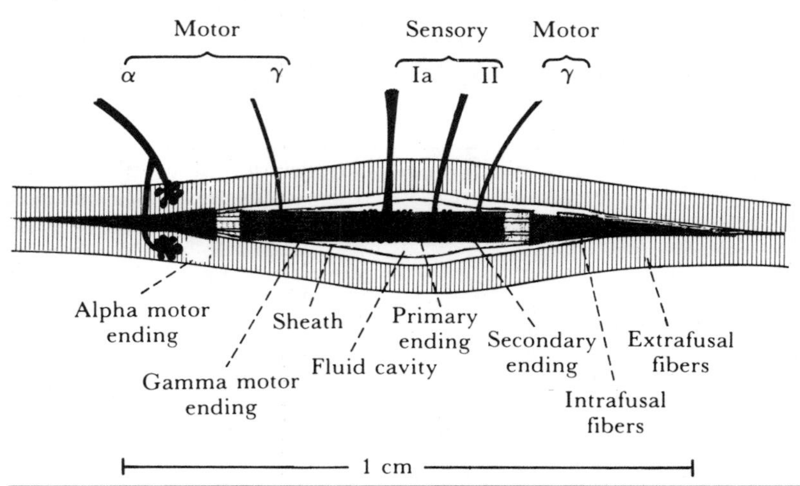

Fig. 8-6. *Effect of extrafusal muscle stretch and contraction on the muscle spindle. See text for explanation. (From W. F. Ganong.* Review of Medical Physiology *[12th ed.]. P. 94. © 1985 by Lange Medical Publications, Los Altos, CA.)*

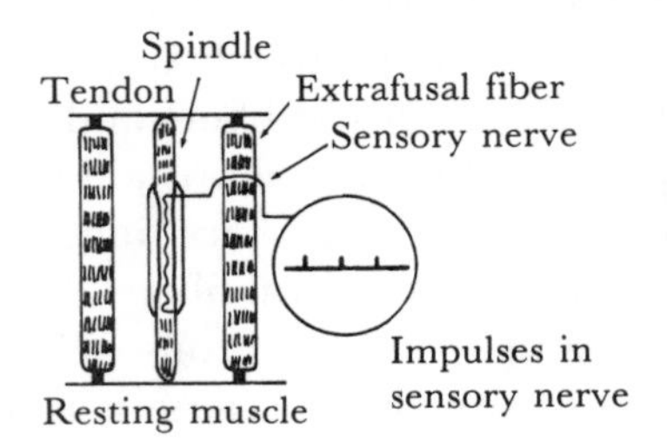

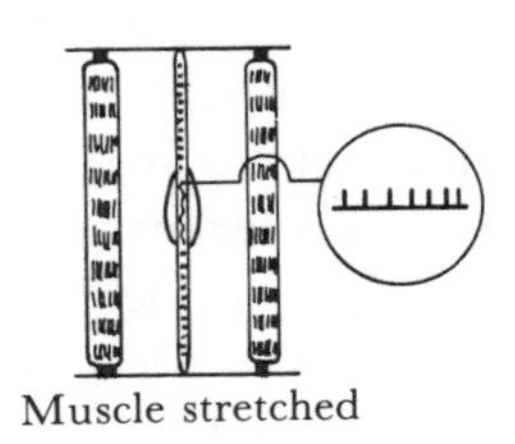

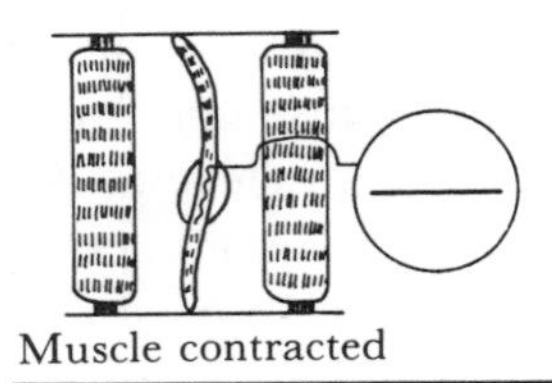

inhibitory, converging or diverging, to provide for more complex reflex behavior. Examples include reciprocal innervation of antagonist muscles in the stretch reflex and the crossed extensor response of the withdrawal reflex.

CONTROL OF POSTURE AND MOVEMENT. Somatic motor activity ultimately depends on the discharge of the spinal motor neurons and the corresponding motor nuclei of the cranial nerves. These motor neurons receive input from a variety of higher centers to control the initiation, coordination, and maintenance of motor activity. These pathways are depicted schematically in Figure 8-7. The physiology of these systems is less well understood than that of the sensory systems, and much of the available information comes from lesion studies and disease states. The three main sources of higher input are the pyramidal system, the extrapyramidal systems, and the cerebellum.

The pyramidal system consists of the corticospinal (CST) and corticobulbar tracts. (Additionally, the rubrospinal tract is closely related in function to the

Fig. 8-7. *Higher centers controlling movement. (From W. F. Ganong.* Review of Medical Physiology *[12th ed.] P. 163. © 1985 by Lange Medical Publications, Los Altos, CA.)*

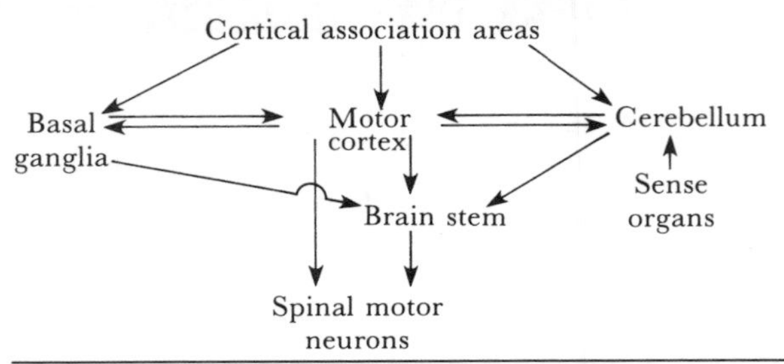

corticospinal tract.) Fibers arise primarily from the precentral motor cortex, descend through the internal capsule, decussate in the pyramids, and travel down the spinal cord to synapse in the cord with either interneurons or lower motor neurons directly. Most CST fibers end on interneurons which allow for the spread of the signal to a variety of motorneurons. This helps provide integrated muscular activity. Some fibers synapse directly on the individual motorneurons. This allows fine muscular control, such as is necessary for delicate hand movements. The human motor homunculus has greatly enlarged hands and mouth, representing the motor cortex necessary for the fine control of manual manipulations and speech. In monkey studies, if a lesion is made in the CST alone, the only deficit noted is in fine dextrous finger movements. Gross movement and postural control are preserved. If, however, the rubrospinal tract is lesioned along with the CST, the typical signs of an upper motor lesion result (Table 8-2). The rubrospinal tract is an additional pathway providing for relatively discrete signals from the cortex to the spinal cord via the red nucleus. In the cord it lies just anterior to the CST. Together with the CST it forms the lateral motor system of the cord which is primarily concerned with movement of the distal extremities. The pyramidal tract also includes descending fibers from the postcentral sensory cortex. These fibers allow the cortex to modify ascending sensory input and serve as a feed forward system for motor control.

The term extrapyramidal is a clinical term that includes those systems regulating movement aside from the pyramids and cerebellum, namely the basal ganglia and the brainstem. Within the brainstem, the ventromedial motor system controls the proximal and axial musculature to stabilize the pelvic and shoulder girdles. A variety of reflexes and inputs are essential to this control. Lesion studies have been useful in demonstrating the integration and hierarchy of postural control from brainstem centers. Results of these studies are displayed in Table 8-3.

The basal ganglia (the caudate, putamen, globus pallidus, substantia nigra, and subthalamic nucleus) are believed to play a major role in the initiation and direction of voluntary movement, particularly of large muscle masses. Their precise function remains unknown, but prior to the initiation of muscle activity, action potentials appear in the basal ganglia before appearing in the motor cortex. As demonstrated in Figure 8-7, the basal ganglia do not project to the spinal cord. They form a closed feedback loop with the motor cortex, receiving input from the motor cortex and discharging their output to the motor cortex via the thalamus. Disorders of the basal ganglia result in increased tone and dyskinesias. The increased tone, as in brainstem animals (Table 8-3, medullary pontine lesion) is a result of decreased gamma motor neuron inhibition. Various dyskinesias and their associated lesion sites are shown in Table 8-4.

The cerebellum functions to modulate and coordinate motor activity. The medial portions of the cerebellum help the medial motor system maintain equilibrium and tone. The lateral cerebellum is more concerned with the lateral motor system and the distal extremities. The cerebellum has extensive connec-

Table 8-2. *Motor Neuron Lesions*

Feature	Upper motor neuron	Lower motor neuron
Tone	Increased	Decreased
Reflexes	Hyperreflexic	Areflexic
Muscle mass	No or slight decrease	Atrophy
Spontaneous activity	Fasciculation	None

Table 8-3. *Motor Deficits from Lesion Studies*

Level of Lesion	Motor Deficits or Capabilities
Spinal Cord	Paralysis Initially reflexes are depressed (spinal shock) but then become hyperactive No righting reflexes
Junction of medulla and pons	Decerebrate rigidity (extension in all extremities) due to decreased inhibition from higher centers causing increased gamma efferent discharge and increased excitability of the motor neuron pool No righting reflexes
Midbrain	Can rise to standing position, walk, right self
Neocortex	Decorticate rigidity (extensor in legs, moderate flexion in arms) from decreased inhibition of gamma efferents Great deal of movement is possible, but animal cannot learn

Table 8-4. *Disorders of the Basal Ganglia*

Disease	Abnormal Movements	Muscle Tone	Primary Anatomic Locus
Parkinsonism	Tremor at rest	Rigidity	Substantia nigra
Huntington's disease	Chorea	Hypotonicity	?
Athetosis	Athetosis	Spasticity Paresis	Lenticular nucleus
Hemiballismus	Ballismus	Marked hypotonia	Subthalamic nucleus

tions with the spinal cord, brainstem reticular formation, vestibular nuclei, and cerebral cortex which enable it to perform its role.

The circuitry of the cerebellum (Fig. 8-8) helps explain its function in modulating motor activity. Input to the cerebellum is excitatory from mossy and climbing fibers. Climbing fibers arise from the inferior olivary nuclei and synapse directly with Purkinje cells, providing a strong excitatory stimulus. All other afferent projections to the cerebellum end as mossy fibers which synapse on granule cells. These granule cells then stimulate Purkinje cells and other inhibitory interneurons (e.g., basket cells and Golgi cells). Both climbing and mossy fibers also send out collateral fibers which excite the deep cerebellar nuclei. The output of the deep cerebellar nuclei to either the cord or thalamus is always excitatory. The final output of the cerebellum is thus a balance of the excitatory effects of the afferent collaterals of the mossy and climbing fibers and the inhibitory effects of the Purkinje cells. Note that there is a built-in time delay between the initial excitement mediated by the collaterals and the subsequent inhibition from the Purkinje cells. This balanced circuitry and extensive input enables the cerebellum to modulate and coordinate motor activity, providing smooth and properly timed motions.

Lesions of the cerebellum result in hypotonia (from decreased gamma efferent discharge), marked ataxia, the breakdown of movements into their component parts, dysdiadochokinesis, dysmetria, and intention tremors.

QUESTIONS

6. What happens to spindle length, Ia discharge rates, alpha motor neuron discharge, and muscle length when the muscle is stretched and when the muscle relaxes?

Fig. 8-8. *Cerebellar circuitry. See text for details. BC = basket cell; GC = Golgi cell; GR = granule cell; NC = nuclear cell; PC = Purkinje cell. (Modified from J. C. Eccles, M. Itah, and J. Szentágothai.* The Cerebellum as a Neuronal Machine. *New York: Springer, 1967.)*

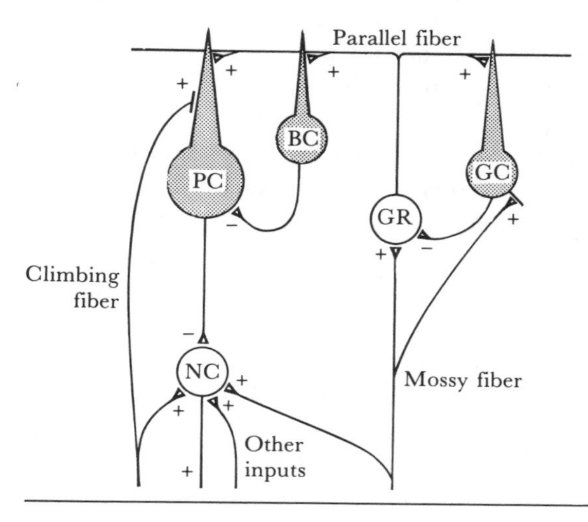

7. What is the effect on muscle tone of decreasing gamma efferent discharge?
8. What type of signals does the cerebellum send to other parts of the brain, excitatory, inhibitory or both?
9. Study Tables 8-2, 8-3, and 8-4. Be able to match motor deficits with lesion sites.

THE AUTONOMIC NERVOUS SYSTEM

The autonomic nervous system (ANS) is the part of the nervous system which innervates smooth muscle, cardiac muscle, and glands. Together with the endocrine system it regulates the visceral functions of the body at a level independent of conscious control. The ANS operates via reflex arcs which are modified by higher centers in the brainstem, hypothalamus, limbic system, and cortex. Conscious input can also influence autonomic output.

Functionally the ANS is exclusively a motor system. Although sensory fibers (visceral afferents) running with the motor fibers (visceral efferents) are crucial for autonomic reflex activity, they are not considered part of the ANS. Unlike the somatic motor system, in which skeletal muscle is innervated directly from a centrally located motor neuron, the ANS is a two motor neuron system. Centrally located preganglionic neurons send out myelinated axons that synapse with peripherally located postganglionic neurons. These postganglionic neurons then send out small unmyelinated axons to the effector cells. Anatomically the ANS is partitioned into parasympathetic (craniosacral) and sympathetic (thoracolumbar) divisions.

THE PARASYMPATHETIC NERVOUS SYSTEM .In the parasympathetic nervous system (PNS), preganglionic fibers exit the third, seventh, ninth, and tenth cranial nerves to supply the head, thorax, and the bulk of the abdominal viscera as well as from the second through fourth sacral spinal roots to supply the pelvic viscera and the remaining abdominal viscera. The vagus nerve (cranial nerve X) alone carries approximately 75 percent of preganglionic parasympathetic fibers. These preganglionic axons take a long but direct path to synapse on to short postganglionic neurons in or near the target organ. Preganglionic parasympathetic fibers release acetylcholine on to nicotinic receptors of parasympathetic postganglionic neurons. The postganglionic neurons in turn release acetylcholine on to muscarinic receptors of the effector organ (Fig. 8-9). Cholinergic stimulation is terminated with the splitting of acetylcholine into acetate and choline by acetylcholinesterases in the synaptic cleft. The small amount of acetylcholine that diffuses away is quickly and completely degraded by serum and erythrocyte cholinesterases. This assures that cholinergic stimulation is short and discrete.

The actions of the PNS are discussed in more detail throughout various chapters of this text. As shown in Table 8-5, parasympathetic stimulation may be either excitatory or inhibitory depending upon the specific organ system. In general, the PNS may be considered a "vegetative" system, conserving energy (e.g., decreasing heart rate) and mediating various "housekeeping" functions (e.g., digestion). Also in contrast to the sympathetic nervous system (see next section), the various components of the PNS act relatively independently of each other, and there is no mass PNS discharge.

THE SYMPATHETIC NERVOUS SYSTEM. In the sympathetic nervous system (SNS), the cell bodies of the preganglionic sympathetic neurons are located in the interomediolateral horns of spinal cord segments T1 to L2. Preganglionic axons reach the paravertebral sympathetic ganglia via the spinal nerves and white rami communicantes. Unlike parasympathetic preganglionic axons, sympathetic ganglion connections diverge and converge extensively. A sympathetic preganglionic axon may (1) synapse with postganglionic neurons in the ganglia it first enters, (2) travel up or down the sympathetic trunk to synapse in other ganglia, and/or (3) pass through the paravertebral ganglia to synapse in one of the outlying sympathetic ganglia. Postganglionic axons exit the paravertebral ganglia via the grey rami communicantes to rejoin the spinal nerve and travel towards their effector organs.

Preganglionic sympathetic fibers release acetylcholine on to nicotinic receptors of postganglionic sympathetic nerve fibers. Postganglionic sympathetic effector compounds include acetylcholine and the catecholamines epinephrine and

Fig. 8-9. *Transmitters and receptors in somatic and autonomic motor neuron transmission. Note that the autonomic nervous system is a two motor neuron system. Ach = cholinergic neuron; NE = noradrenergic neuron; N = nicotinic receptor; M = muscarinic receptor.*

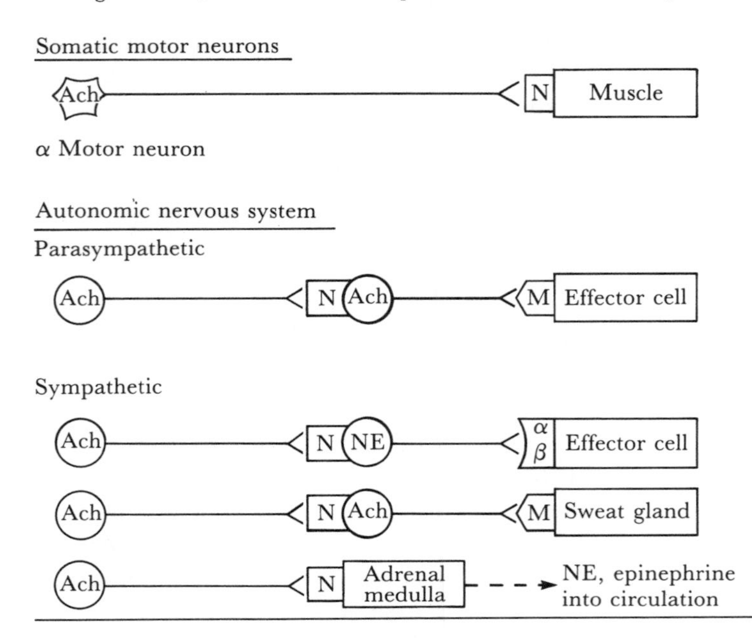

Table 8-5. *Actions of the Autonomic Nervous System*

Organ	Effect of Sympathetic Action[1]	Receptor[2]	Effect of Parasympathetic Action	Receptor[2]
Eye				
Iris				
Radial muscle	Contracts	α		
Circular muscle			Contracts	M
Ciliary muscle	(Relaxes)	β	Contracts	M
Heart				
Sinoatrial node	Accelerates	β_1	Decelerates	M
Ectopic pacemakers	Accelerates	β_1		
Contractility	Increases	β_1	Decreases (atria)	M
Vascular smooth muscle				
Skin, splanchnic vessels	Contracts	α		M[3]
Skeletal muscle vessels	Relaxes	β_2		
	(Contracts)	α		
	Relaxes	M[4]		
Bronchiolar smooth muscle	Relaxes	β_2	Contracts	M
Gastrointestinal tract				
Smooth muscle				
Walls	Relaxes	β_2	Contracts	M
Sphincters	Contracts	α	Relaxes	M
Secretion			Increases	M
Myenteric plexus	Inhibits	α		
Genitourinary smooth muscle				
Bladder wall	Relaxes	β_2	Contracts	M
Sphincter	Contracts	α	Relaxes	M
Uterus, pregnant	Relaxes	β_2		
	Contracts	α		
Penis, seminal vesicles	Ejaculation	α	Erection	M
Skin				
Pilomotor smooth muscle	Contracts	α		
Sweat glands				
Thermoregulatory	Increases	M		
Apocrine (stress)	Increases	α		
Metabolic functions				
Liver	Gluconeogensis	α/β_2[5]		
Liver	Glycogenolysis	α/β_2		
Fat cells	Lipolysis	α_2, β_1[6]		

[1]Less important actions are in parentheses.
[2]Specific receptor type: α = alpha; β = beta; M = muscarinic.
[3]Most blood vessels have uninnervated muscarinic receptors.
[4]Vascular smooth muscle in skeletal muscle has sympathetic cholinergic dilator fibers.
[5]Depends on species.
[6]α_2, inhibits; β_1 stimulates.
From B. G. Katzung, et al. [eds.]. *Basic and Clinical Pharmacology.* (2nd ed.). © 1984 by Lange Medical Publications, Los Altos, CA.

norepinephrine. Most preganglionic sympathetic fibers are adrenergic and release norepinephrine as a neurotransmitter. Sympathetic postganglionic cholinergic fibers innervate thermoregulatory eccrine sweat glands and some vasodilatory skeletal muscle vessels. Preganglionic sympathetic neurons also innervate chromaffin cells of the adrenal medulla. These cells may be considered modified postganglionic neurons which secrete epinephrine (80%) and norepinephrine (20%) into the bloodstream. These circulating catecholamines can also activate adrenergic receptors. Figure 8-9 summarizes the organization of the SNS.

Adrenergic receptors are pharmacologically classified into alpha-1, alpha-2, beta-1, and beta-2 receptors on the basis of the receptor's affinity for various adrenergic agonists and antagonists. Epinephrine and norepinephrine both act strongly at alpha receptors. Epinephrine significantly stimulates both beta-1 and beta-2 receptors while norepinephrine has weak beta-1 and essentially no beta-2 activity.

After norepinephrine is released into the synaptic cleft, it is rapidly removed by (1) active reuptake into the noradrenergic nerve ending, (2) diffusion, and (3) less importantly, degradation by the enzymes monoamine oxidase (MAO) and catechol O-methyl transferase (COMT), both of which are found principally in the liver. Catecholamines released into the circulation from the adrenal medulla are inactivated primarily by MAO and COMT metabolism; their halflives are approximately 10 times greater than that of norepinephrine released by the noradrenergic nerve terminal. Despite the long halflives of the adrenal catecholamines, norepinephrine released from the nerve terminals is the prime effector of sympathetic tone, providing 75 percent of the circulating catecholamines in the basal state. Adrenal epinephrine (20%) and norepinephrine (5%) make up the remaining circulating catecholamines.

As was true for the parasympathetic actions, sympathetic end-organ effects are a function of the effector cell and not of the receptor type (Table 8-5.) For any sympathomimetic agent, the response depends on the agent's affinity for the various receptors, the amount released or dose administered, and reflex homeostatic responses. For example, norepinephrine-mediated vasoconstriction (alpha-1) may increase total peripheral resistance to the point that the baroreceptor reflex induces a vagally mediated bradycardia which overcomes the noradrenergic induced tachycardia (beta-1).

Most sympathetic discharge (for example, that concerned with the maintenance of blood pressure) is discrete; however, the SNS is also capable of a mass discharge, the so-called flight or fight response. This mass activation of the SNS is facilitated by both the adrenal release of catecholamines and the extensive divergence, convergence, and overlapping of preganglionic sympathetic neurons. The net effect is to prepare the organism for emergency activity: heart rate and contractility are increased (beta-1); circulation to skeletal muscle is increased (muscarinic and beta-2 vasodilation) while that to the viscera and skin is decreased (alpha vasoconstriction), thereby preserving the blood flow for the most vital organs and limiting surface bleeding; glycogenolysis in muscle and in liver is accelerated (beta-2); the basal metabolic rate is increased (mediated primarily by epinephrine); and the overall state of alertness is increased (mediated by noradrenergic lowering of the threshold of the reticular activating system).

SPECIAL FEATURES OF AUTONOMIC NEURAL TRANSMISSION. Within the autonomic ganglia extensive neuronal interaction modifies the simple action potential and thereby modulates future responses to neuronal stimulation. The best studied example of this occurs in the sympathetic ganglia. Here the usual fast postsynaptic potential (fast EPSP), the action potential, combines with both a prolonged inhibitory (slow IPSP) and excitatory (slow EPSP) potential to produce a compound or complex potential (Fig. 8-10).

The ANS generally functions with less precise temporal control than does the somatic motor system. Skeletal muscle movements may be completed in fractions of a second while smooth muscle contraction or glandular secretion may act over minutes. This longer autonomic response time is due in part to the contrasting properties of the end effector organs, skeletal muscle and smooth muscle differences (see Chapter 1), and the process of glandular secretion. Additionally, the structure of the postganglionic neuron contributes to the slower response time. First, postganglionic neurons are unmyelinated. Second, unlike the neuromuscular junction in which there is tight apposition of the presynaptic cell to the postsynaptic cell, postganglionic autonomic nerve terminals end as varicosities or swellings that are relatively distant from the postsynaptic effector cell. Thus, more time is required for the neurotransmitters to diffuse this greater distance, slowing transmission. Such varicosities are more characteristic of adrenergic than cholinergic terminals. This greater distance between cells in the adrenergic junction also helps explain the increased reliance on diffusion for the termination of noradrenergic stimulation.

INTERACTIONS OF THE PNS AND SNS. The parasympathetic and sympathetic nervous systems are tonically and simultaneously active and generally have reciprocal effects on organs innervated by both divisions of the ANS. Where present, dual innervation allows a mutual antagonism that provides for fine control of body function. There are three basic mechanisms by which SNS and PNS actions oppose one another. First, the PNS and SNS may operate at the same site to produce opposite results. This is illustrated at the SA node where beta-1 (SNS) stimulation increases the heart rate while vagal (PNS) stimulation slows the heart rate. These effects are mediated by increasing or decreasing the slope of phase 4 depolarization in pacemaker cells (see Chapter 4). Second, the PNS and SNS may act at different sites to produce physiologically opposing actions. For example, alpha stimulation (SNS) of the radial muscle of the iris causes pupillary dilation while muscarinic (PNS) stimulation of the sphincter or circular muscle of the iris causes pupillary constriction. Here, each muscle is innervated by only one division of the ANS, but fine control of pupil diameter is achieved through the two reciprocal systems. Third, as a general principle for dually innervated organs, synapses between parasympathetic and sympathetic neurons act reciprocally both centrally and peripherally to decrease (increase) parasympathetic release when sympathetic neurons are stimulated (inhibited).

Basal rates of parasympathetic and sympathetic activity (e.g., tone) allow the body to reach a given state through activation or inhibition of one system without altering the other system. Returning to the example of the SA node, heart rate may be decreased by reducing sympathetic tone and/or increasing vagal activity.

Fig. 8-10. *Synaptic potentials in postganglionic sympathetic neurons. The fast EPSP, slow IPSP, and slow EPSP combine to produce a compound or complex potential. (From W. F. Ganong.* Review of Medical Physiology *(11th ed.) P. 177. © 1985 by Lange Medical Publications, Los Altos, CA.)*

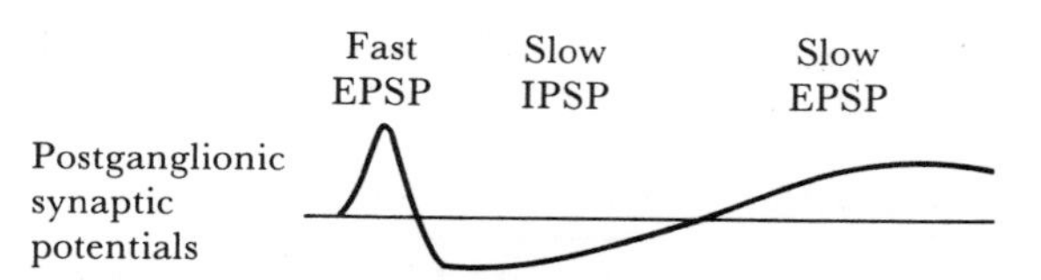

If there were no basal sympathetic discharge, changes in sympathetic release could only result in tachycardia, never bradycardia. This resting tone is crucial to the regulation of organs innervated by only one division of the ANS such as the thermoregulatory sweat glands.

The features of the PNS and the ANS are compared in Table 8-6.

QUESTIONS

10. How are the actions of acetylcholine terminated and norepinephrine terminated?
11. Give two reasons why there is no mass parasympathetic discharge?
12. Which of the following systems would be activated by infusion of a nicotinic agonist: somatic motor, SNS, PNS?
13. In dually innervated organs, which system would be most affected by denervation?
14. Name three methods by which the SNS and PNS oppose one another.
15. Study Table 8-5.

THE SPECIAL SENSES

VISION. The eye contains the same three elements as a camera for recording visual images: a lens system, a means for controlling aperture (the pupil), and film (the retina). The cornea and lens focus the visual image on the retina by bending or refracting incoming light rays. The change in refractive index at the air-cornea interface contributes most of this focusing power. The lens contributes only 25 percent of the eye's refractive power in the unaccommodated state. The importance of the lens is its ability to double its refractive power by increasing its curvature and thus provide for near focus. The lens has an inherent tendency to assume a spherical shape because of the elasticity of the lens capsule. In the unaccommodated state, tension of the lens ligaments

Table 8-6. *Comparison of the Parasympathetic and Sympathetic Nervous Systems*

Feature	Parasympathetic	Sympathetic
Location of central nuclei	Cranial, sacral	Thoracolumbar
Preganglionic fiber	Long, travels directly to innervated organ	Short, makes multiple synapses along the sympathetic trunk
Neurotransmitter of preganglionic neuron	Acetylcholine (nicotinic receptor)	Acetylcholine (nicotinic receptor)
Postganglionic neuron	Short, often found in target organ	Long, cell body in sympathetic trunk (some cell bodies are in outlying ganglia)
Postganglionic neurotransmitter(s)	Acetylcholine (muscarinic receptor)	Norepinephrine Acetylcholine (muscarinic receptor)
Termination of action	Acetylcholinesterase	Acetycholine: acetylcholinesterase Catecholamines: reuptake, diffusion, MAO and COMT
Circulating agonists	None	Adrenal catecholamines and diffused norepinephrine
Mass discharge	None	Flight or fight response

MAO = monoamine oxidase; COMT = catechol O-methyl transferase.

counteracts this tendency and keeps the lens relatively flattened. In accommodation, contraction of the ciliary muscle reduces the tension of the lens ligaments and the lens assumes a more spherical shape, increasing the refractive power. The image that is formed on the retina is inverted and reversed.

The photoreceptors of the retina contain photosensitive compounds which decompose on exposure to light to produce a change in ion permeability and a corresponding change in potential. The outer segment of the photoreceptor membrane is normally highly permeable to Na^+ and has a relatively high resting potential of -30mv. Decreasing Na^+ conductance hyperpolarizes the cell. No action potential is produced at the receptor cell. Instead, this information is transmitted as a local potential to the synaptic end of the receptor cell and leads to a decrease in the rate of transmitter release. The change in transmitter release at the synaptic end of the receptor cells hyperpolarizes horizontal cells and either depolarizes or hyperpolarizes bipolar cells. These responses are also local graded potentials. Action potentials are first produced at the ganglion cells, the axons of which make up the optic nerve. The neural arrangement of the retina is shown in Figure 8-11.

The rod system is very sensitive to light, primarily because of the high degree of convergence of rod receptor cells on to ganglion cells. Rods also adapt more completely to dark than do cones, albeit more slowly. The high degree of convergence in the rod system explains the low visual acuity of night vision. There are three different types of cones, each containing a different photopigment. These three different pigments, blue, green, and red sensitive, mediate color vision. Any given color is encoded by varying combinations of these three component colors. Visual acuity is high in the cone system because of the relatively low number of receptors converging on each ganglion cell. Visual acuity is greatest at the fovea where convergence and the number of structures overlying the photoreceptors are both minimal.

The ganglion cells have circular receptive fields functionally divided into a center and a surround. The center is termed *on* if light shining on the center increases the rate of action potentials and *off* if light decreases the frequency of action potentials. The surround responds oppositely to light from the center, so

Fig. 8-11. *Neural arrangement of the retina. (From A. C. Guyton.* Textbook of Medical Physiology *[7th ed.]. Philadelphia: Saunders, 1985. P. 720.*

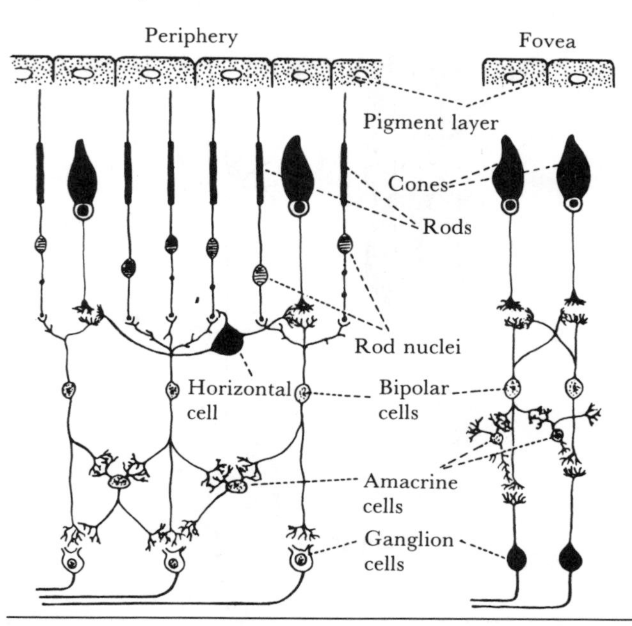

that on centers have off surrounds and vice versa. The inhibition of the center response by the surround is probably due to inhibitory feedback from one receptor to another mediated by the horizontal cells. This is another example of lateral inhibition enhancing the organism's ability to detect contrast. Ganglion and bipolar cells respond best to a difference in illumination between the center and the surround. Higher order color cells also have center-surround receptive fields, but instead of being on-off, they are blue-yellow or red-green. The basic principle is the same.

Axons from the ganglion cells project a detailed spatial representation of the retina to the lateral geniculate nucleus (LGN) which in turn projects a similar point for point representation to the visual cortex where most processing occurs. Pyramidal cells in layer 4 of the cortex receive input from the LGN neurons. The receptive fields of both LGN neurons and these pyramidal cells resemble the center-surround receptive fields of ganglion cells. In the cortex, higher order cells respond to line segments rather than single points, are orientation specific, and receive binocular input. These higher order cells consist of a hierarchy of simple, complex, and hypercomplex cells. The receptive field of simple cells is an orientation- and position-specific line segment. Presumably, each simple cell receives its input from a group of LGN-like cells whose receptive fields are arranged in a row as illustrated in Figure 8-12. Complex cells are orientation specific but are less dependent upon position and respond best to a moving linear stimulus. Their receptive fields are believed to be produced by input from simple cells sharing the same orientation specificity but differing in position specificity. Hypercomplex cells resemble complex cells but are specific for certain lengths of stimuli.

Fig. 8-12. *Receptive fields of simple cells. By combining the input from a set of "aligned" LGN-like cells, each with circular-center surround patterns, the receptive field of the simple cell is produced. The simple cell is most responsive to orientation-specific line segments. (Reprinted by permission of the publisher from* Principles for Neural Science *[2nd ed.], by E. K. Kandel and G. H. Schwartz, P. 374, 1985, by Elsevier Science Publishing Company, Inc.)*

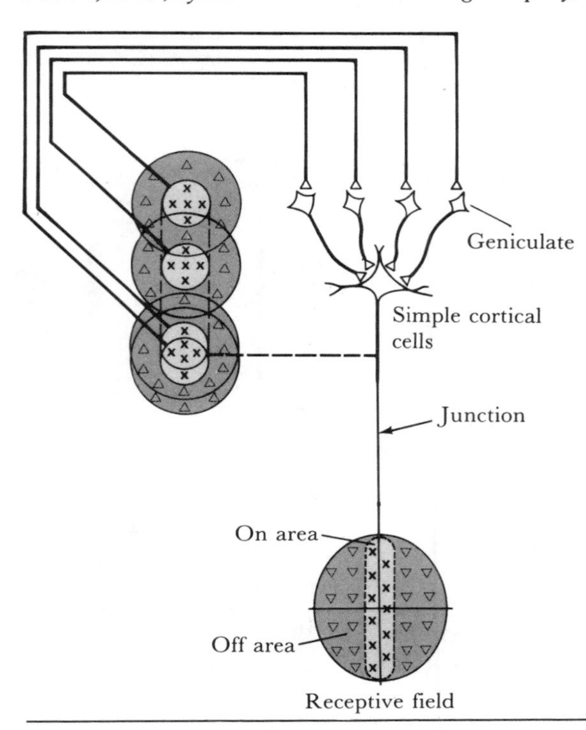

The columnar organization of the cortex facilitates this structure. The cortex is arranged in orientation-specific columns such that cells in any given column respond maximally to stimuli of a specific orientation. These columns can be further subdivided into ocular dominance columns. Binocular input is first present at the level of complex cells and is important for stereoscopic vision and depth perception. Higher order processing of color vision is less well understood, but appears to take place in a specific part of the cortex.

THE VESTIBULAR SYSTEM. The vestibular apparatus consists of the three semicircular canals and the utricle and saccule of the labyrinth. The semicircular canals detect angular acceleration while the utricle and saccule detect linear acceleration and the static position of the head with respect to gravity. The hair cell is the receptor cell common to these systems. Bipolar afferent neurons synapse at the base and sides of each hair cell. Mechanical forces induce bending of the microvilli. Bending of the hairs towards the kinocilia depolarizes the hair cell and increases the rate of nerve impulses in the vestibular nerves. Bending away from the kinocilia has the opposite effect. The hair cells are bathed in endolymph, a K^+-rich extracellular fluid. Movement of microvilli and kinocilia alters K^+ permeability and leads to a receptor potential. The patterns of these nerve impulses are analyzed by the vestibular nucleus and cerebellum to provide information about the position of the head in space essential for the coordination of motor responses, eye movement, posture, and equilibrium.

The semicircular canals are arranged perpendicularly to one another and thus can detect motion in all three planes. The functional anatomy of the semicircular canals is shown in Figure 8-13. When the head begins to rotate, the semicircular canals turn, but the endolymph remains stationary because of its inertia. This creates a relative fluid flow in the canals, which causes movement of the cupula, and thereby bends the hair cells. With continued angular acceleration (e.g., rotation), the inertia of the endolymph is overcome and the endolymph moves with the semicircular canal. There is no relative motion, the cupula returns to its resting position, and impulse rates return to the basal level. When rotation stops, the opposite situation arises: the semicircular canals are stationary but the endolymph is still moving. This relative motion bends the cupula and, therefore, the hairs in the opposite direction, and the impulse rate is appropriately altered.

The hair cells of the utricle and saccule are located in the macula. Here the hair cells are again regularly ordered with respect to the kinocilia, but the organization is planar not linear. The functional equivalent of the cupulae are the otoliths. They respond to linear acceleration, including gravity, in a manner essentially equivalent to the cupula. Because of their mass, they bend and, thereby, stimulate the hair cells. Because the hair cells are arranged in a variety of directions, a single structure can respond to tilt or to linear acceleration in any of several directions.

THE AUDITORY SYSTEM. The human ear is able to detect sound frequencies ranging from 20 to 20,000 Hz over a 140 db range of sound intensity. The longitudinal compressions and rarefactions of sound waves are transmitted from the air to the tympanic membrane, through the auditory ossicles, to the fluid-filled cochlea. The lever action of the ossicles and the large size of the tympanic membrane compared to the oval window provide the 22-fold increase in force of movement at the oval window necessary to transfer the vibrations across the air-fluid interface.

The cochlea is the sense organ for hearing. The motion of the endplate of the stapes on the perilymph of the scala vestibuli establishes a wavelike motion in the

Fig. 8-13. *Functional anatomy of the semicircular canals (Modified from A. C. Guyton.* Textbook of Medical Physiology *(7th ed.). Philadelphia: Saunders, 1985. P. 621; Modified from C. M. Goss [ed.].* Gray's Anatomy of the Human Body *[29th ed.]. Philadelphia: Lea & Febiger, 1973; Modified from Kolmer by Buchanan,* Functional Neuroanatomy. *Philadelphia: Lea & Febiger.)*

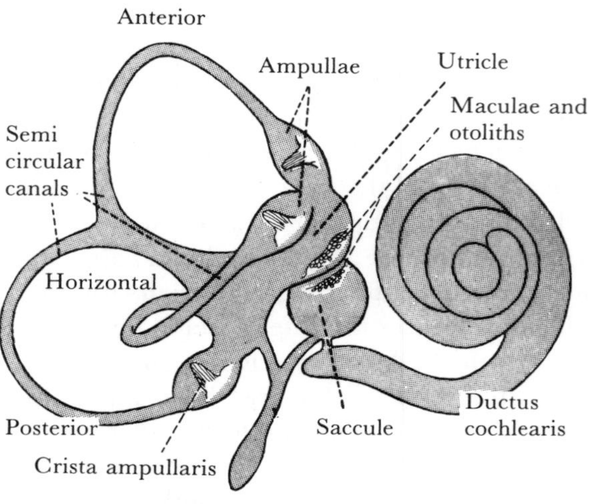

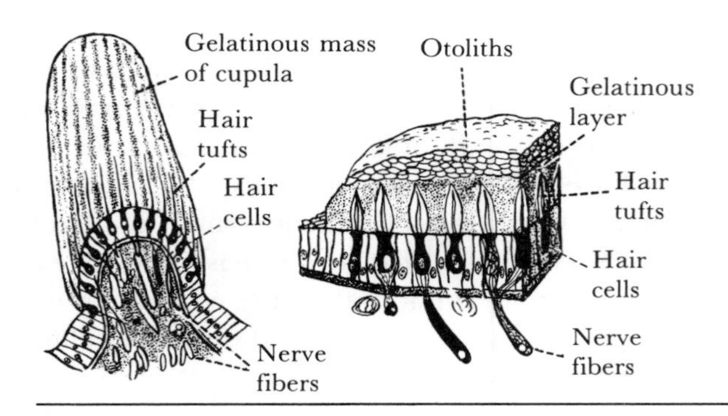

fluid of the cochlea which sets up a traveling wave along the basilar membrane. For any given sound frequency, there is a point along the membrane where displacement (e.g., amplitude of the traveling wave) is maximal. This tonotopic organization of the basilar membrane is the principal means of pitch determination. The organization is quite regular with high-pitched sounds generating maximum vibration at the taut narrow base of the membrane and low-frequency sounds generating maximum vibration at the loose, broad apex.

The organ of Corti rests on the basilar membrane and is thus affected by movements of the membrane. A system of stiff arches, supporting cells, and a stiff anchoring apical membrane (the reticular laminal) ensures that the hair cells move as a rigid unit with the basilar membrane. The microvilli (stereocilia) of the hair cells project out from the apex and are embedded in the tectorial membrane. The tectorial membrane moves independently of the basilar membrane because of its different point of attachment. The relative motion of the basilar membrane with respect to the tectorial membrane produces a shearing action which bends the microvilli (Fig. 8-14). This causes a change in ion permeability which generates a potential change in the hair cell.

The potential change in the hair cell leads to activity in the fibers of the

Fig. 8-14. *Effect of movement of the basilar membrane on the organ of Corti. A. The organ of Corti at rest, illustrating the different points of attachment of the basilar membrane and the tectorial membrane. B. Deflection of the basilar membrane produces a shearing force across the microvilli because of the relative movement of the basilar membrane with respect to the tectorial membrane. This results in bending of the microvilli. Note, however, that the body of the hair cells remains in a fixed position relative to the basilar membrane. (Reprinted by permission of the publisher from* Principles of Neural Science *[2nd ed.], by E. K. Kandel and G. H. Schwartz, P. 374, 1985, by Elsevier Science Publishing Company, Inc.)*

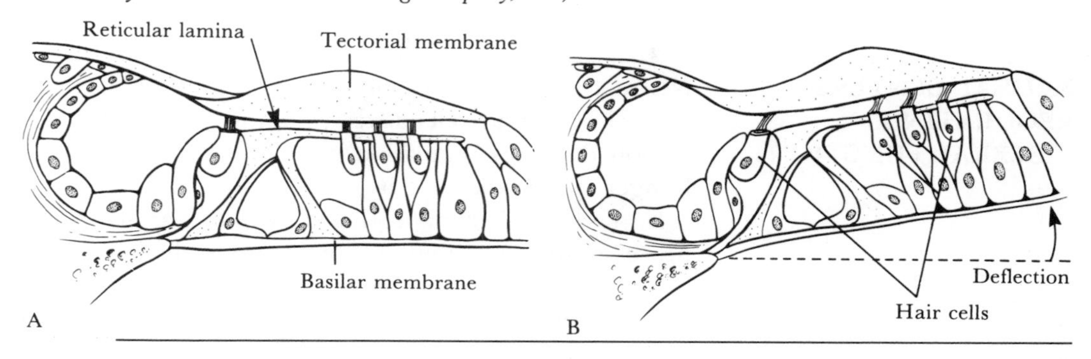

acoustic nerve (cranial nerve VIII) which synapse at the base of the hair cells. Ninety-five percent of the afferent fibers go to the single row of inner hair cells with most fibers innervating only one cell. Thus, each nerve fiber is most sensitive to the characteristic frequency of the hair cell, which it innervates, and nerve fibers are also tonotopically organized. This tonotopic arrangement continues through the higher centers of the auditory pathway; the cochlear nucleus, the superior olivary complex, the inferior colliculus, the medial geniculate, and the auditory cortex. As with other systems, stimulus intensity (loudness) is encoded by frequency of action potentials and total number of receptors stimulated. Because all centers at levels higher than the olivary complex receive bilateral input, lesions of the central auditory pathways do not give rise to mononeural deficits. More importantly, binaural input enables the auditory centers to determine the direction from which a sound emanates. Two mechanisms are used: (1) difference in the intensity of the sound between the two ears, and (2) a lag in the time of arrival of the sound at the two ears (a phase delay).

THE CHEMICAL SENSES. Both taste and smell receptors are chemoreceptors that are stimulated by molecules in solutions in the fluids of the mouth and nose to produce receptor potentials which then lead to action potentials. The nature of the interaction between the molecules and the receptors is not known. Taste buds are the sense organs for taste. The four primary tastes are sour, salt, sweet, and bitter. The pathway of taste sensation resembles that of somatic sensation for the tongue.

Olfactory receptors are neurons located in the olfactory mucous membrane. The basis of olfactory discrimination is unknown. A unique feature of the olfactory pathway is that there is no thalamic relay in transmission of olfactory input to the cortex. Intensity discrimination for both smell and taste is relatively crude and adaptation relatively rapid.

QUESTIONS

16. Give two reasons why bright light improves visual acuity.
17. When is photoreceptor "discharge" maximum?
18. Draw a circuit for complex cell inputs.
19. What structure detects angular acceleration?
20. What structure detects linear acceleration?

21. What structure sorts tone?
22. What structure deflects the hair cells?
23. A loud tone is perceived as less pure than the same tone at a softer volume. Explain.
24. How does the pathway for olfactory perception differ from other pathways?

NEURAL CENTERS REGULATING VISCERAL FUNCTION

Neural regulation of the circulatory, respiratory, renal, endocrine, and gastrointestinal systems are all addressed in the corresponding chapters of the text. This section discusses regulation of other visceral functions, namely temperature, hunger, and sexual activity.

In the body, heat is produced by muscular activity, assimilation of food, and processes that contribute to the basal metabolic rate. Heat is lost from the body by radiation, conduction, convection, and evaporation of water in the respiratory passages and on the skin. Temperature regulation is controlled by the hypothalamus. Temperature-sensitive cells in the anterior hypothalamus are the primary input of body temperature to the system. The information is compared to the hypothalamic set point (normally 37°C), and the appropriate response instituted. If the hypothalamic temperature is greater than 37°C, sweating and cutaneous vasodilation are activated. If the temperature is less than the set point, cutaneous vasoconstriction, shivering, and increased secretion of catecholamines are instituted.

Fever is produced when the hypothalamic set point is raised above 37°C. Tissue breakdown products and bacterial toxins can raise the set point. Polymorphonuclear leukocytes and macrophages are believed to act on bacteria and bacterial breakdown products to produce a substance termed endogenous pyrogen. Endogenous pyrogen is believed to cause fever by inducing formation of prostaglandin E_1 which then acts on the hypothalamus to raise the set point. The temperature receptors then signal that the actual temperature is less than the new set point, and the temperature raising mechanisms are initiated. Despite the fever, the person subjectively feels cold because the body temperature (T_{body}) is less than the set point (T_{set}). When the set point is returned to normal, the situation is reversed and the fever declines.

Regulation of food intake depends primarily upon the interaction of two hypothalamic centers, a feeding center in the lateral nuclei and a satiety center in the ventromedial nuclei. Stimulation of the feeding center causes an animal to eat while stimulation of the satiety center causes cessation of eating. Lesions of the feeding center cause anorexia, and lesions of the satiety center cause hyperphagia. The satiety center inhibits the feeding center when food intake is sufficient for the organism's nutritional status. Nutritional status is probably determined by the level of glucose utilization of cells within the satiety center rather than the level of circulating blood glucose. The amount of adipose tissue in the body is probably important for long-term regulation of feeding.

Sexual activity, particularly human sexual activity, is a very complex phenomenon. Reflexes integrated in spinal and lower brainstem centers make up the act of copulation. Parasympathetic fibers mediate lubrication and erection. Sympathetic fibers mediate ejaculation and emission. The limbic system and hypothalamus help regulate more complex behavioral components as does learned behavior. The sex hormones also influence sexual behavior in complex ways.

QUESTIONS

25. What would be the effect of local heating of the hypothalamus?
26. If the $T_{body} < T_{set}$, how does the person feel subjectively? What responses are activated?
27. What is the role of the ANS in sexual activity?

HIGHER FUNCTIONS

LANGUAGE AND CORTICAL FUNCTION. In humans, language functions depend more on one cerebral hemisphere (the dominant hemisphere) than the other (the nondominant hemisphere). In right-handed individuals the left hemisphere is dominant while in left-handed individuals either hemisphere may be dominant. Neuroanatomic studies reveal that areas specialized for language function are larger in the dominant hemisphere. The aphasias resulting from lesions in the dominant hemisphere (see Table 8-7) have helped researchers discover the way in which language is processed. Wernicke's area receives input from the auditory cortex and interprets this sensory input as meaningful language. This information is then projected to Broca's area via the arcuate fasciculus. Broca's area processes the information and, through its output, coordinates the activity of the motor cortex for the production of meaningful sounds, e.g., speech. The circuitry for processing written language appears to be closely related and in part superimposed upon this above system. Wernicke's and Broca's areas illustrate the importance of association cortex in integrating and transferring information to generate complex processes. The development of multimodal association areas in man is probably the best neuroanatomical correlate of intelligence.

The nondominant hemisphere is more concerned with visuospatial relationships such as music or drawing. Lesions here produce agnosias, the inability to recognize an object by a specific sensory modality even though the sensory modality itself is intact.

The corpus callosum provides communication between corresponding areas of the two hemispheres. Lesions of the corpus callosum illustrate the specialization of the cerebral hemispheres both in terms of right versus left and language versus visuospatial function.

Table 8-7. *Selected Lesions Affecting Higher Function*

Lesion Site	Deficit
Wernicke's area	Fluent aphasia (also called receptive or sensory aphasia) Comprehension impaired Fluent but meaningless speech
Broca's area	Nonfluent aphasia (also called expressive or motor aphasia) Comprehension intact Very limited speech, marked difficulty forming words
Hippocampi	Anterograde amnesia (unable to form new memories)
Amygdala (temporal lobes)	Klüver-Bucy syndrome (hyperoral, hypersexuality, passivity, psychic blindness)
Frontal lobe	Increased distractability Decreased moral and social inhibitions, increase in inappropriate behavior
Parietal lobe	Unilateral inattention and neglect
RAS	Coma

LEARNING AND MEMORY. More is known about the psychology of learning and memory than is known about the corresponding neurophysiology. Lesion studies indicate that learning can occur at subcortical and spinal levels though advanced learning is primarily a cortical function. Learning involves structural changes in the cortex through the development of new synapses. Habituation and sensitization phenomena are also involved. Here strength of response is modified. For example, in the snail *Aplysia,* biochemical modification of ion channels results in a "learned" withdrawal response. Memory is a complex process consisting of encoding, storage in short-term and then long-term memory, and retrieval. The hippocampus appears to be involved in encoding and the temporal lobes in retrieval. Memory does not appear to be stored discretely in any single part of the brain. Current knowledge indicates protein synthesis is somehow involved in the storage of memory.

THE RETICULAR ACTIVATING SYSTEM, WAKEFULNESS, AND SLEEP. The reticular activating system (RAS) is a complex, diffuse network of neurons located in the brainstem reticular formation. The RAS is responsible for producing the alert, awake state. Stimulation of this area in a sleeping animal causes the animal to awaken suddenly and results in generalized activation of the cerebral cortex. The RAS receives inputs from all sensory modalities directly or through collaterals as well as retrograde from the cerebrum. Its outputs are to all portions of the cortex either directly or through thalamic relays. Because of the complex and diffuse connections within the RAS, most RAS neurons are equally activated by various afferent stimuli; they are not modality specific. In turn, activation of the RAS leads to generalized activation of the cortex. However, there is evidence for some selective cortical activation via the thalamic relays. Presumably this plays a role in attention. Permanent damage to the RAS produces coma while temporary inhibition produces sleep.

There are two types of sleep: slow-wave or non-rapid-eye-movement (non-REM, NREM) sleep and paradoxical or REM sleep. Slow-wave sleep is deep restful sleep characterized physiologically by a decreased heart rate, blood pressure, respiratory rate, and basal metabolic rate. It is divided into four stages based on electroencephalogram (EEG) patterns. (The EEG is a recording of the surface potentials or cortical cells analogous to the recording of the surface potentials of cardiac cells recorded by an electrocardiogram.) Brain waves in the early stages are high-frequency and low-amplitude while those in the last stages are low-frequency, high-amplitude, and synchronized. In contrast, REM sleep is characterized by irregular high-frequency beta waves similar to those recorded during alert wakefulness. People awakened from REM sleep often report they were dreaming. Although some muscular twitches persist in REM sleep (e.g., rapid eye movements), muscle tone in REM sleep is decreased compared to NREM sleep. This pronounced hypotonia is the result of active inhibition of muscle tone by centers located in the mid pons. Additionally, the threshold for sensory arousal is increased in REM sleep compared to NREM sleep. Thus, REM sleep appears to be a time of cortical activity disengaged from sensory input and motor output. Typically, REM sleep comprises 25 percent of the total sleep time with REM periods occurring approximately every 90 minutes. In the initial cycles, slow-wave sleep is more prominent, but as the night progresses, there is less NREM stages 3 and 4 and more REM sleep.

THE LIMBIC SYSTEM AND HYPOTHALAMUS. The limbic system and hypothalamus are closely involved with the production of emotions, drives, and affective aspects of sensation, e.g., pleasantness or unpleasantness. Punishment and reward centers located here are believed to play a role in learning and in providing the motivation for the performance or avoidance of certain behaviors. Stimulation of the punishment center produces a rage response while stimulation of the reward centers produces placidity and tameness. The limbic cortex appears to function as an association area for the control of behavior. Phylogenetically, it is among the oldest parts of the cortex and has only five layers compared to the six-layered neocortex. Lesions of the amygdala produce the Klüver-Bucy syndrome (Table 8-7). Additionally, the hypothalamus plays a major role in the control of vegetative functions as discussed in Neural Centers Regulating Visceral Function of this chapter.

QUESTIONS

28. In a right-handed person, what areas of the left cerebral hemisphere are enlarged or more developed compared to the corresponding areas of the right cerebral hemispheres?
29. Compare REM and NREM sleep in terms of muscle tone, muscle activity, and EEG pattern.
30. Study Table 8-7.

ANSWERS

1. Stimulation of nerve fibers that innervated the now absent body part will, when stimulated by pressure, cause sensations which the body interprets as originating from that portion of the body from which the neurons originally received their sensory input. The phenomenon is analogous to stimulating a sensory neuron along its axon. The brain perceives the stimulus as originating from the receptor.
2. The number of receptors activated and the frequency of action potentials.
3. Decreased light touch, vibration and joint position sense. Crude touch and pain and temperature would remain essentially intact.
4. Decreased pain and temperature.
5. The increase from 5 to 10 "units" would produce a greater change in perceived stimulus intensity even though the change from 50 to 60 units is a greater absolute change. Receptor potential and frequency of firing are logarithmically related to stimulus intensity; thus, receptors respond to relative change.
6. If the muscle were stretched, the spindle would be stretched, Ia discharge would increase leading to a reflex increase in alpha motor neuron discharge, and muscle length would shorten with the resulting contraction. The effects are exactly opposite when the muscle relaxes. Note that in either case, the reflex acts to maintain muscle length.
7. Decreasing gamma efferent discharge will result in decreased muscle tone. The gamma efferent system serves to maintain tone.
8. All cerebellar signals are excitatory. The cerebellum is able to modulate motor activity by varying the magnitude, timing, and distribution of its output.
9. No answer
10. Enzymatic cleavage by acetylcholinesterase. Diffusion and active reuptake.
11. There is no acetylcholine in the circulation. PNS innervation is relatively direct and discrete.
12. All. Nicotinic agonists would directly stimulate the receptor at the NMJ. PNS and SNS effectors would be activated as a result of ganglionic stimulation.
13. The PNS would be most affected. Sympathetic stimulation could be maintained by circulating catecholamines.
14. Dual innervation of the same organ with opposing actions. Innervation of separate organs which have physiologically antagonizing effects. Synaptic connections to inhibit neurotransmitter release by the opposite system.
15. No answer

16. Cones are used in bright light; their output is less convergent than that of rods. Additionally, the pupil is constricted in bright light, assuring that incoming light rays pass through the center of the lens where diffraction is minimal.
17. Photoreceptor potential is maximum (most positive, least negative) in the dark. Stimulation leads to hyperpolarization. Recall, photoreceptors produce only a local potential, there is no true "discharge" (i.e., action potential).
18. The figure would resemble Figure 8-12, except simple cells would form the input for complex cells.
19. The semicircular canals, more specifically, the cupula.
20. The utricle and saccule, more specifically, the otoliths.
21. The basilar membrane sorts tone.
22. The tectorial membrane is the structure which actually deflects the hair cells; however, the relative movement is generated by the basilar membrane.
23. While each point on the basilar membrane is most sensitive to a given frequency (ν_0), higher intensity stimuli will also stimulate other areas of the basilar membrane. Thus, at low intensities, only the receptors for ν_0 are stimulated, but at higher intensities, a group of receptors centered about ν_0 is stimulated.
24. There is no thalamic relay. Additionally, olfactory cortex has only four layers as opposed to the usual six.
25. The hypothalamus would sense $T_{body} > T_{set}$, and heat-releasing mechanisms would be activated.
26. The person would subjectively feel cold, and heat-generating and heat-conserving mechanisms (cutaneous, vasoconstrictive, shivering, and increase catecholamine secretion) would be activated.
27. PNS fibers mediate lubrication and erection. SNS fibers mediate ejaculation and emission. Additionally, the autonomic nervous system mediates other aspects of the sexual response such as change in heart rate and cutaneous vasodilation.
28. The language centers (Wernicke's area and Broca's area) are enlarged. Additionally, the left motor cortex is more developed, thereby allowing more precise motor control of the right side.
29. Muscle tone is decreased in REM sleep relative to NREM sleep. Twitches are present in REM sleep and absent in NREM sleep. The EEG of REM sleep is full of desynchronized, small-amplitude, high-frequency beta waves with occasional PGO spikes. The EEG of NREM sleep is made up of slow, synchronized, high-amplitude waves.
30. No answer

BIBLIOGRAPHY

Barr, M., and Kiernan, J. *The Human Nervous System* (4th ed.). Philadelphia: Harper and Row, 1983.

Ganong, W. F. *Review of Medical Physiology* (12th ed.). Los Altos, CA: Lange, 1985.

Gilman, S., and Winana, S.: *Manter and Gatz's Essentials of Clinical Neuroanatomy and Neurophysiology* (6th ed.). Philadelphia: Davis, 1982.

Guyton, A. C. *Textbook of Medical Physiology* (7th ed.). Philadelphia: Saunders, 1985.

Kandel, E. K., and Schwartz, G. H. *Principles of Neural Science* (2nd ed.). New York: Elsevier, 1985.

Nervous System Form and Function (Course syllabus). San Francisco: University of California, 1984.

Index